AF837489

Springer Series in Electronics and Photonics

Editors: D.H. Auston W. Engl T. Sugano

Managing Editor: H.K.V. Lotsch

Volume 21 **Picosecond Electronics and Optoelectronics**
Editors: G.A. Mourou, D.M. Bloom, C.-H. Lee

Volume 22 **High-Speed Electronics**
Basic Physical Phenomena and Device Principles
Editors: B. Källbäck and H. Beneking

Volume 23 **Guided-Wave Acousto-Optics**
Interactions, Devices, and Applications
Editor: C.S. Tsai

Volume 24 **Picosecond Electronics and Optoelectronics II**
Editors: F.J. Leonberger, C.H. Lee, F. Capasso, and H. Morkoc

Volume 25 **Photonic Switching**
Editors: T.K. Gustafson and P.W. Smith

Volume 26 **Guided-Wave Optoelectronics** 2nd Edition
Editor: T. Tamir

Volume 27 **Ultra-Fast Silicon Bipolar Technology**
Editors: L. Treitinger and M. Miura-Mattausch

Volume 28 **Physics of Quantum Electron Devices**
Editor: F. Capasso

Volume 29 **Photonic Switching II**
Editors: K. Tada and H.S. Hinton

Volume 30 **Nonlinear Photonics**
Editors: H.M. Gibbs, G. Khitrova, and N. Peyghambarian

This series was originally published under the title
Springer Series in Electrophysics
and has been renamed starting with Volume 22.

Volumes 1–20 are listed on the back inside cover

K. Tada H.S. Hinton (Eds.)

Photonic Switching II

Proceedings of the International Topical Meeting,
Kobe, Japan, April 12–14, 1990

With 343 Figures

Springer-Verlag

Berlin Heidelberg New York London
Paris Tokyo Hong Kong Barcelona

Professor Kunio Tada, Ph. D.

Department of Electronic Engineering, University of Tokyo,
7-3-1 Hongo, Bunkyo-ku, Tokyo 113, Japan

H. Scott Hinton

AT&T Bell Laboratories, Room 2F-235, 200 Park Plaza,
Naperville, IL 60566, USA

Series Editors:

Dr. David H. Auston

Columbia University, Dept. of Electrical Engineering, New York, NY 10027, USA

Professor Dr. Walter Engl

Institut für Theoretische Elektrotechnik, Rhein.-Westf. Technische Hochschule,
Templergraben 55, D-5100 Aachen, Fed. Rep. of Germany

Professor Takuo Sugano, Ph. D.

Department of Electronic Engineering, University of Tokyo,
7-3-1 Hongo, Bunkyo-ku, Tokyo 113, Japan

Managing Editor: Dr. Helmut K. V. Lotsch

Springer-Verlag, Tiergartenstrasse 17, D-6900 Heidelberg, Fed. Rep. of Germany

TA
1505
.P49
1990

[redacted]

JUN 28 1991

ISBN 3-540-53067-3 Springer-Verlag Berlin Heidelberg New York
ISBN 0-387-53067-3 Springer-Verlag New York Berlin Heidelberg

This work is subject to copyright. All rights are reserved, whether the whole or part of the material is concerned, specifically the rights of translation, reprinting, reuse of illustrations, recitation, broadcasting, reproduction on microfilms or in other ways, and storage in data banks. Duplication of this publication or parts thereof is only permitted under the provisions of the German Copyright Law of September 9, 1965, in its current version, and a copyright fee must always be paid. Violations fall under the prosecution act of the German Copyright Law.

© Springer-Verlag Berlin Heidelberg 1990
Printed in Germany

The use of registered names, trademarks etc. in this publication does not imply, even in the absence of a specific statement, that such names are exempt from the relevant protective laws and regulations and therefore free for general use.

2154/3140-543210 – Printed on acid-free paper

Preface

The 1990 International Topical Meeting on Photonic Switching was held April 12–14, 1990, in Kobe, Japan, in conjunction with the 1990 International Meeting on Optical Computing. It was sponsored by the Institute of Electronics, Information and Communication Engineers in cooperation with the IEEE Lasers and Electro-Optics Society, the Optical Society of America, and the Japan Society of Applied Physics. The attendance was in excess of 340 persons, with 18 countries represented – a testimony to the current international interest in photonic switching. This book contains expanded and more detailed versions of most of the papers presented at the topical meeting.

With the success of optical fibers there is an increased demand for a switching system that can operate directly on the light present in the fibers. The goal is to reduce the total number of optical-to-electrical and electrical-to-optical transformations as far as possible, in addition to exploiting the large spectral and temporal bandwidth offered by such an optically transparent system. This objective has led to an enormous amount of research in guided-wave devices and systems. From another perspective, as the electronics industry has evolved, it has reached a point where integrated circuits (ICs) and printed circuit boards (PCBs) have become pin-out limited. This communications constraint has restricted the architectures possible for both computing and telecommunications systems. Through the use of the spatial bandwidth available in the optical domain, pin-outs on ICs and PCBs have the potential to increase by an order of magnitude. This has led to proposed switching fabrics based on two-dimensional optoelectronic circuits with optical free-space interconnects. These three quantities, spectral bandwidth, temporal bandwidth, and spatial bandwidth, represent the current research directions in photonic switching and are well represented in this book.

The book has been organized into the following ten parts: (I) Introduction and Overview, (II) Electrically Controlled Switches, (III) Optically Controlled Switches, (IV) Free-Space Devices, (V) Free-Space Switching, (VI) Wavelength-Based Devices, (VII) Wavelength-Based Switching, (VIII) Self-Routing Switching, (IX) Space-Division Switching, (X) Systems Considerations. As compared with *Photonic Switching* (Proceedings of the First Topical Meeting, Incline Village, Nevada, March 18–20, 1987), published as Vol. 25 of this series, the number of papers and topics has almost doubled.

In addition, the intervening three years have seen a considerable shift in emphasis, as can be seen from the contents of this book. One apparent change is that the number of system-oriented papers has increased and approached that of device-oriented papers. The number of papers that fall into either category has also increased. This seems to be a good sign for the future of the photonic switching technology. Systems interest is steadily growing, and cooperation and interaction between systems people and device people are being continuously enhanced. This book represents the most detailed and comprehensive account currently available in the field of photonic switching. The reader will thus gain a good overview of the present state of the art in this field.

We would like to thank Mr. Katsuhiko Kurumada, Mr. Tadahiko Yasui, and all the program committee members for their efforts in putting together the technical program. We would also like to thank Prof. Minoru Akiyama, Dr. Hiromasa Ikeda, and all the members of the organizing and steering committees for their efforts in arranging this meeting. Finally, we would like to acknowledge the Institute of Electronics, Information and Communication Engineers for granting us permission to reproduce a number of figures and tables which originally appeared in the technical digest of the topical meeting.

Hongo, Tokyo — *Kunio Tada*
Naperville, IL — *H. Scott Hinton*
July 1990

Contents

Part I Introduction and Overview

Part II Electrically Controlled Switches

Part VI Wavelength-Based Devices

Part VII Wavelength-Based Switching

Part X Systems Considerations

Part I

Introduction and Overview

Trends in Electronic Communication Switching Technologies

T. Aoki

NTT Communication Switching Laboratories,
3-9-11 Midori-cho, Musashino-shi, Tokyo 180, Japan

Abstract. History and trends of communication switching technologies are described from the switching systems developers' viewpoint. On this basis, what we expect from photonic switching technology is discussed.

1. Introduction

The first manual exchange for telephone calls was introduced in the US in 1878. Since then, telecommunication networks have grown gradually in both volume and service menus. Many structural changes, or evolutions, of switching systems have contributed to this progress. B-ISDN, intelligent networks and personal communication are the major issues for the 21st century.

2. History of Communication Switching

Automatic communication switching technology started with electromagnetic systems as shown in Fig. 1. The first switching system introduced was the step-by-step switching system which was characterized by rotary switches directly driven by electrical impulses generated by the subscriber's dialing action. As the set of dialed digits uniquely determined the path between the calling and called parties, the

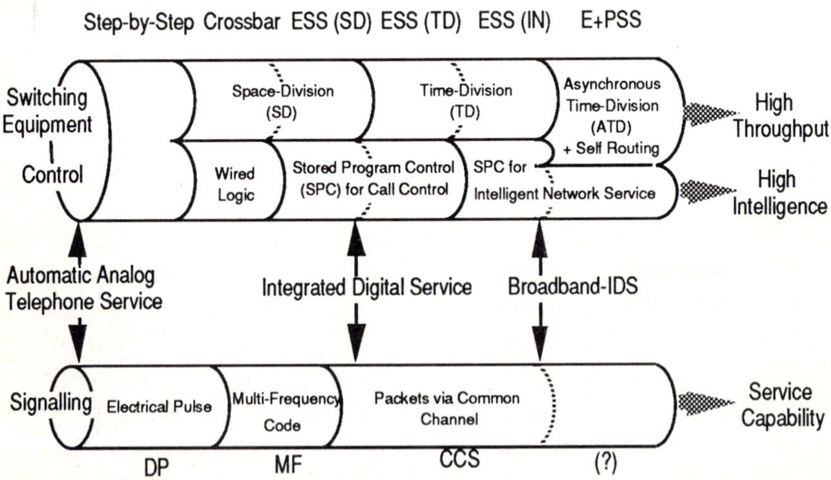

Figure 1 History of switching techniques

Springer Series in Electronics and Photonics, Vol. 29 Photonic Switching II
Editors: K. Tada · H.S. Hinton © Springer-Verlag Berlin, Heidelberg 1990

network had no flexibility in routing. This meant a large blocking probability or a very low usage of trunk lines when the network became large. To provide an alternate routing capability, control functions were separated from speech signal switching equipment. This type of system was called the crossbar switching system because the switches were crossbar switches. Control was executed by wired-logic relay circuits. The dialed number was relayed from one office to another. All the received digits were temporarily stored in a register and then, the controller searched for an optimum path using it. This control scheme was called common control. Multi-frequency code began to be used for signalling in trunk networks.

Electronic technology became available in the early '60s and it showed great performance due to the progress in semiconductor techniques. This led to the first electronic switching systems (ESSs): 1ESS of the US and the D10 system of Japan. In this system, the controller used electronics and a new concept, "SPC (stored program control)", was introduced, which rapidly became a standard for electronic switching systems all over the world. Ever since then, software has played an important role in the network. The common channel signalling (CCS) system was established by using SPC's capability. Switching equipment, however, remained mechanical until time division multiplexing was introduced to switching systems after introduction to transmission systems. The progress of digital techniques based on integrated circuits during the '70s enabled a fully electronic switching system, which resulted in good matching between digital transmission and switching. This system is a TD (time-division) switching system while earlier systems were SD (space-division) systems. The TD systems are now being introduced to establish ISDN in cooperation with the CCS.

As needed for the 800 services in the US or the Free Dial Services in Japan, various information about customers and networks, which is restricted now, should be released widely for flexible services. In ISDN, a variety of services will be produced by customers themselves as well as by network vendors because both terminals at customer sites and networks are becoming more intelligent and flexible for new services. To meet these future demands, the concept of "IN (intelligent network)" was formed and it is now under development. Although ISDN will initially carry 64 kbps traffic, demands for higher speed communication such as NTSC or HD Video and large volume file transfer between remote databases and mainframes are expected to increase in the future. Research toward broadband-ISDN (B-ISDN) is being conducted aiming at commercial introduction early in the 21st century.

3. Progress of the Switching Node

The history above suggests that hardware and software technologies have alternately played important roles in the evolution of switching systems. The first evolution was in hardware architecture from step-by-step control to common control. Next came the software for SPC in the electronic switching system. Then, TD switching resulted from the hardware innovation of LSI. The evolution to INs has its roots in the introduction of SPC, 20 years ago. The key issue of today's development is software such as software-defined services and database management. The most recent evolution to date is hardware-controlled fast switching in B-ISDN. The multiplexing method believed to be most promising in this era is ATM (asynchronous transfer mode), which associates calls with cells by the label in the header. This topic is being intensively researched around the world.

One evolution opened an era of switching systems, during which the communication networks and our lifestyle based on it made enormous progress. This progress stimulated another technology and, at the same time, generated new demands which could not be satisfied by the existing system. Then, the next evolution came to change the switching system. The length of each era supported by one switching system is becoming shorter. We are now developing INs and simultaneously researching ATM-based switching whereas the crossbar switching systems and SD-ESS each lasted more than 10 years.

4. Progress of the Signalling System

Signalling is needed to control connections in networks. A signalling system provides a method of sending and receiving control signals between a terminal and a local switching office, or between two switching offices. The performance of the signalling system is quite important in the network because it carries customers' requests and so greatly affects the service capability provided by the network.

The first signalling system strongly depended on the switching system's routing mechanism. Since the routing in the step-by-step system was done directly by electrical impulses composing the digits of the signal, signals had to be sent in the form of electrical pulses. However, when signals were read from registers which had received and stored them, the signalling system no longer depended much on the switching system. The multi-frequency code system has shortened the time for signalling, includes error checking and enables the signal to propagate in the voice frequency bandwidth. In the common channel signalling system, signals are transported in the form of packets separately from the communication channels. So, sending any signal at any time has become possible.

The progress in signalling systems has been made to satisfy demands of customers in the existing network. These demands will vary with network and social circumstances. Although the newest system, CCS, is now becoming the backbone of ISDN, a new signalling system might have to be studied in the future to support the coming INs and B-ISDN.

5. Prospects of Photonic Switching Techniques

Figure 2 shows a model of a future communication network. It logically has two layers, a transport layer for transmission and switching traffic and an intelligent layer for the control and management of the network and its services. Between these layers, there is a signalling system. The three major areas that are evolving are: hardware for the transport layer, software for the intelligent layer, and the signalling system.

Since photonic switching is a hardware technology, it is reasonable to consider that it should be introduced into the transport layer first, where ATM-based switching will be a standard. The first commercial version of a high-speed ATM switch will probably be mainly based on electronics because no photonic switching technique seems to be available in the near future. However, its performance will be limited if it insists on electronics, and it will not be able to meet the demands of

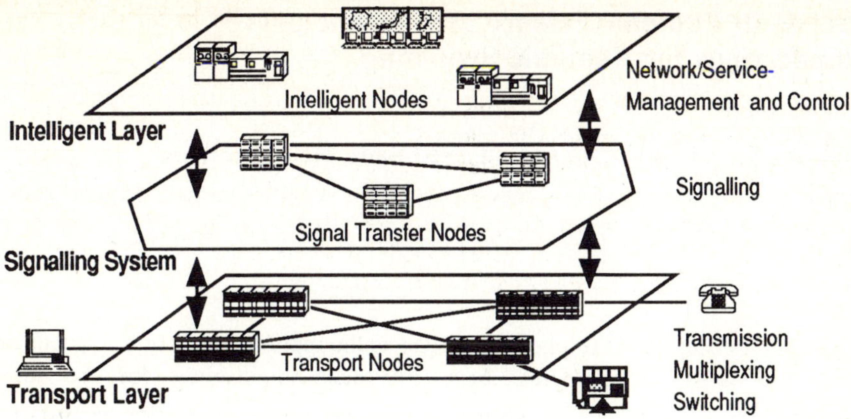

Figure 2 Hierarchical network model

customers when broadband traffic becomes as popular as today's telephone traffic. Practical photonic switching techniques should be studied to improve switching system performance in the ATM networks.

From this viewpoint, photonic switching techniques should be divided into two categories: short-term techniques and long-term techniques. Practical techniques utilizing the high bandwidths of photonics such as optical interconnection, optical SD switching and high-speed optical switching of block-multiplexed data are short-term techniques. Here, a good combination of the high bandwidths of photonics and the high intelligence of electronics is essential. "How to use photonic switching in ATM networks" is the main issue of "system people" and "how to fabricate commercial photonic switching devices" is the issue of "device people". Short-term photonic switching techniques developed should be used widely in commercial systems not only to improve the system performance but also to stimulate photonic switching technology itself. On the other hand, techniques whose operating principles are still under basic study or which would be difficult to introduce into ATM-based networks are classified as long-term techniques. These include frequency switching, two-dimensional parallel switching, coherent signal processing and optical logic. They will need a lot of explorative study on device operation before a practical system is achieved. Some of these techniques should be given priority but it is difficult to choose which.

6. Conclusion

Alternate hardware and software evolution and developments of new signalling systems have supported continuous progress of the telecommunication network and will continue to play important roles for its further progress. Much R & D effort is now devoted to developing systems for post-ISDN; INs and ATM-based switching. Short-term photonic switching techniques hold the key to the improvement of ATM-based switching performance in B-ISDN .

Trends in Photonic Science and Technology: Implications for Photonic Switching

P.W.E. Smith

Bellcore, 331 Newman Springs Rd., Red Bank, NJ 07701, USA

Abstract. Recent trends in photonics, such as the rapid progress in semiconductor laser and OEIC technologies, have important implications for the future development of photonic switching.

1. Introduction

The development of optical fiber transmission has led to the vision of a future high bandwidth telecommunications network capable of providing a wide range of "information age" services. In order to make such a dream a reality, it will be necessary to develop switching, processing, and control technology that will be capable of handling a wide variety of broadband optical signals of different formats in a flexible and economical way. What role can we expect optics to play in providing the switching, control, and "network intelligence"?

The often-quoted strengths of optical devices are: high bandwidth, transparency, and speed. The weaknesses are: large size, high power requirements, and small nonlinearity (difficulty of sensing an optical signal)[1]. In order to compete with existing technology, optical devices and circuits must exhibit at least some of the following advantages: better performance; higher functionality; lower costs / power dissipation; ease of fabrication / packaging, etc; new service capabilities; higher levels of integration.

Although for most optical devices these advantages remain to be proven, there have been in recent years tremendous strides in the development of a photonics technology for the generation, control, and detection of optical signals. There has also been remarkable progress in the integration of photonic and electronic circuits. In this paper we will explore the implications of these technology trends for the future development of photonic switching.

2. Photonics Technology

2.1 Devices

Semiconductor diode lasers are evolving rapidly in many ways. Low cost, reliable, and long life lasers have been demonstrated. Power levels and efficiencies are such that many applications previously requiring other types of lasers are now being

Springer Series in Electronics and Photonics, Vol. 29 **Photonic Switching II**
Editors: K. Tada · H.S. Hinton © Springer-Verlag Berlin, Heidelberg 1990

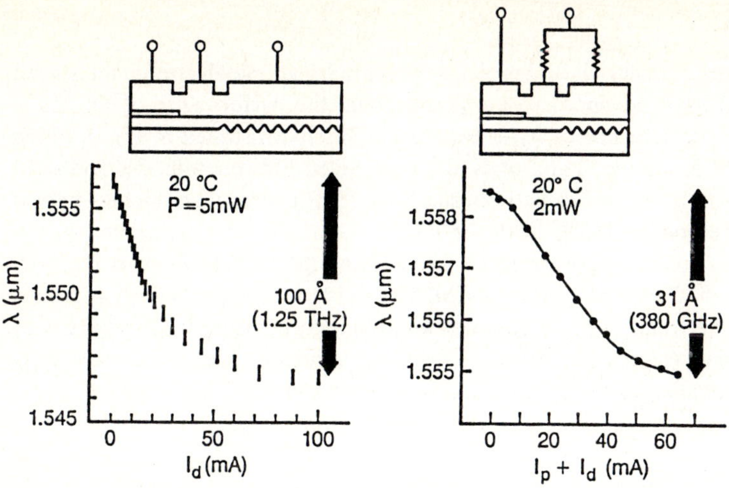

Fig 1. Tunable, three-section distributed Bragg reflector (DBR) semiconductor lasers have demonstrated quasi-continuous tuning ranges of 1.25 THz, and continuous tuning ranges of 380 GHz[2].

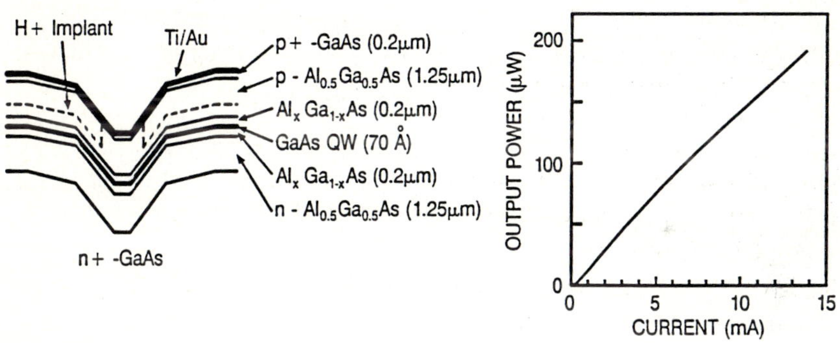

Fig 2. Using a patterned substrate growth technique, Kapon[3] at Bellcore has been able to fabricate single quantum well lasers with threshold currents as low as 0.35 mA.

considered. Tunable, narrow linewidth lasers (Fig 1) are making possible new applications for coherent and wavelength division systems. The recent development of low threshold (< 1 mA) lasers (Fig 2), and arrays of surface-emitting lasers, has opened up new opportunities for optical interconnects and opto-electronic integration.

Similar progress has been made in other photonic device areas. Broad band, high gain amplifiers have been made and tested in a variety of systems experiments. Tunable filters have made possible dense wavelength division multiplexing networks. Detectors have been developed that can be efficiently integrated with optical waveguides.

2.2 Integration

Aided by advances in materials and processing technology, optoelectronic integrated circuits (OEICs) have advanced to the point where the performance of OEICs is approaching the performance of hybrid circuits. This is illustrated in Fig 3 which shows that in 1985, simple OEIC receivers (integrated PIN photodiodes and field effect transistors (FETs)) had sensitivities at least 25dB poorer than the best hybrid receivers. By the end of 1989, however, OEIC technology had improved to the point where OEIC receiver performance was within 4 or 5 dB of the best hybrids. High speed (> 5 Gbit) integrated optical transmitters and receivers have been demonstrated. The trend to more complex circuits incorporating large numbers of individual elements (lasers, transistors, etc) is very evident. One recent example is the three laser wavelength-division multiplexing light source shown in Fig 4. These higher levels of integration offer major advantages in many projected applications.

One technique for optoelectronic integration is a lift-off technique recently developed at Bellcore[5]. This technique allows an optical element (such as a detector or laser) to be removed from the substrate on which it was grown and processed, and attached to another substrate. In this way lasers have been mounted on silicon and glass substrates, and detectors have been integrated with glass and $LiNbO_3$ optical waveguides. With this lift-off technique, hybrid integrated optical circuits can be made using components that are fabricated using their optimum material technology.

Self electro-optic devices (SEEDs) are optical elements whose transmission can be controlled by optical or electrical signals. Recently, large arrays of these elements have been fabricated, and have been used in a demonstration of optical computation[6]. The ability to fabricate large arrays of optoelectronic elements opens up the possibility of information processing systems that exploit the

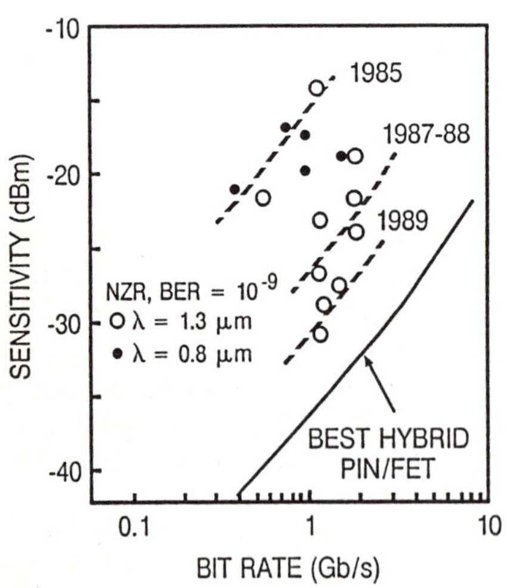

Fig 3. Progress in integrated PIN/FET receiver performance from 1985 to the present.

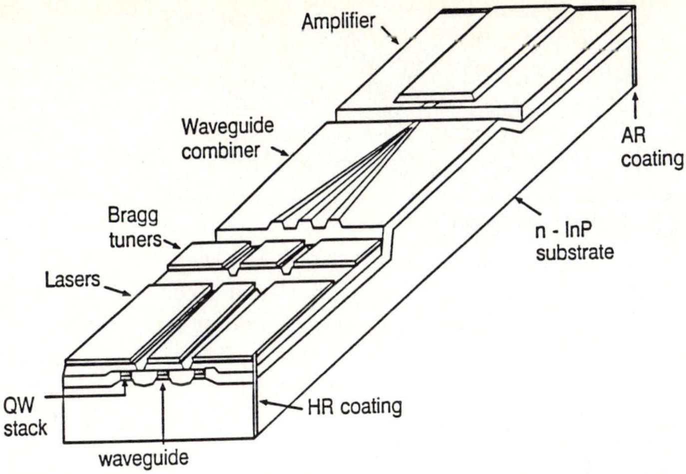

Fig 4. Three laser wavelength division multiplexing light source (after Koren et al.[4]).

parallelism of optics to make large numbers of simultaneous high-bandwidth interconnects.

2.3 Towards Applications

Active programs are underway in many institutions to study the ways in which optical interconnects can be used to overcome problems such as pin-out limitations and clock skew in conventional integrated electronic circuits. Crow at IBM has recently described an optoelectronic circuit designed for optical interconnection of multiprocessor networks[7]. It contains 4 lasers, 4 photodiodes, and 8000 FETs, and operates at a 1 Gb/s rate.

The dense interconnection requirements of neural networks make optics a natural choice for the interconnection of the neuron elements. Optoelectronic implementations of neural networks have been demonstrated to have unique capabilities. Neural networks using holographic elements for optical interconnections have recently been used to demonstrate learning behavior for complex pattern recognition functions.

Large arrays of surface emitting lasers have recently been developed at a number of research laboratories. At Bellcore, we have fabricated a 1000 element array of surface emitting lasers that can be individually electrically addressed, and have used this array to make a rapidly-addressable, high throughput holographic memory read-out (Fig 5). The ability to make large arrays of low threshold, independently-addressable light sources opens up exciting possibilities for many information processing applications.

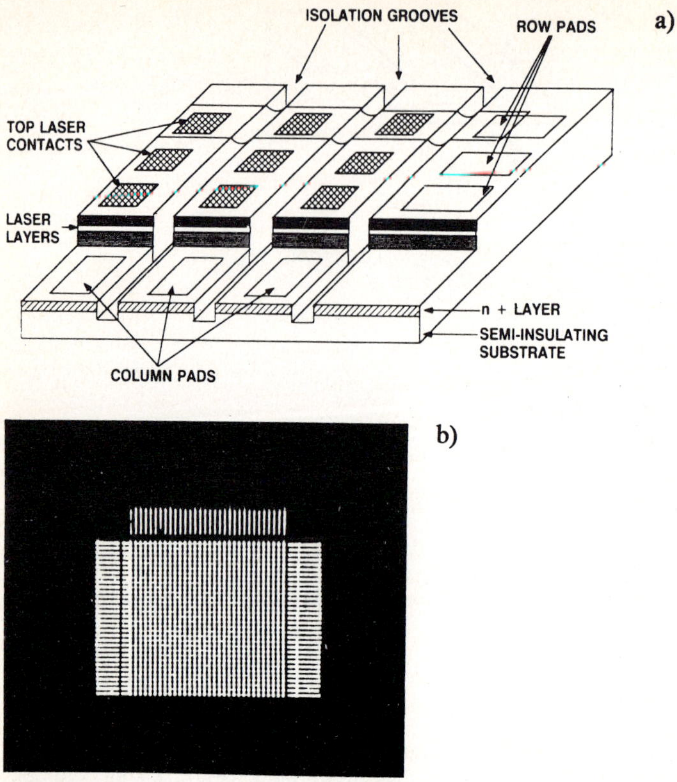

Fig 5. Integrated, 1000 element matrix-addressed surface emitting laser array. a) schematic view; b) photograph of array.

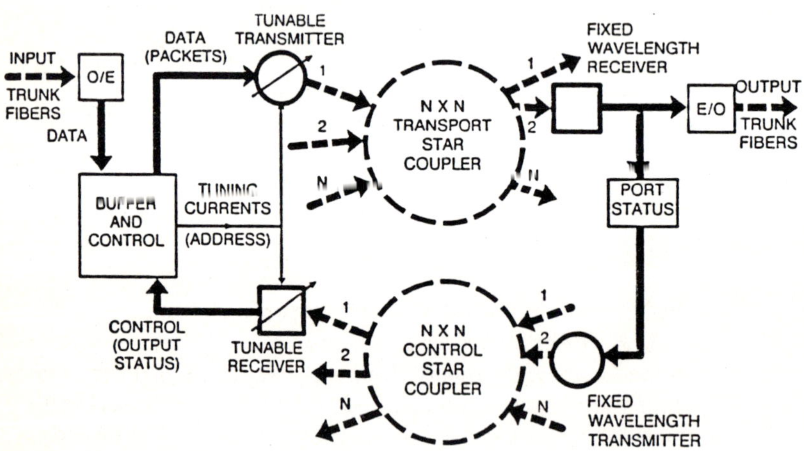

Fig 6. Schematic of hybrid packet switching system architecture (after Arthurs et al.[8]).

A number of optical networks have been proposed and demonstrated that take advantage of the large bandwidth of optical fiber transmission by using time and/or wavelength division multiplexing to increase the throughput of the network. These networks offer new possibilities for switching optical signals. Figure 6 shows a schematic of a hybrid packet switching architecture that relies on wavelength diversity for signal distribution and control.

In the last few years, there have been many proposals published for photonic switching systems, ranging from simple arrays of electrically-controlled optical routing elements to complex packet switching networks with logic and memory circuits[9]. Figures 7 and 8 show two representative examples of some of the wide variety of switching systems that have recently been demonstrated.

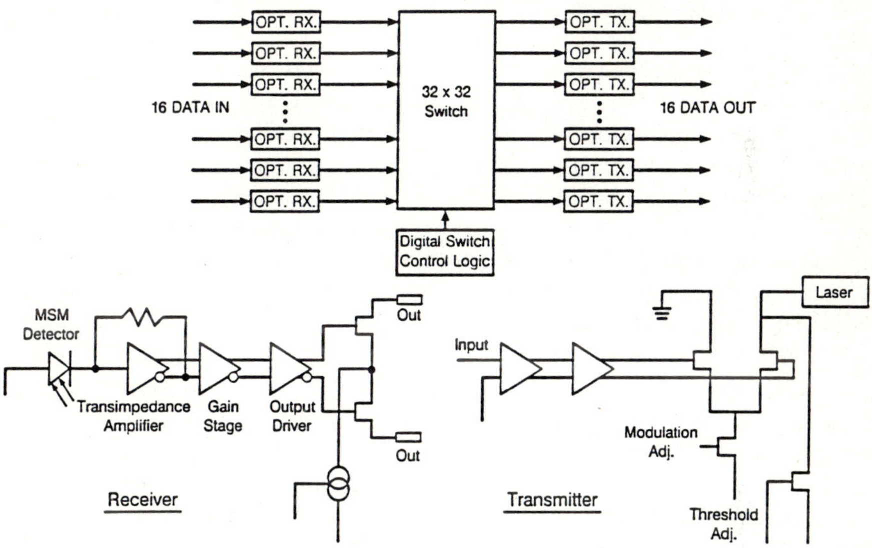

Fig 7. 16 x 16 optical fiber crossbar switch (after K. Pedrotti et al.[10]).

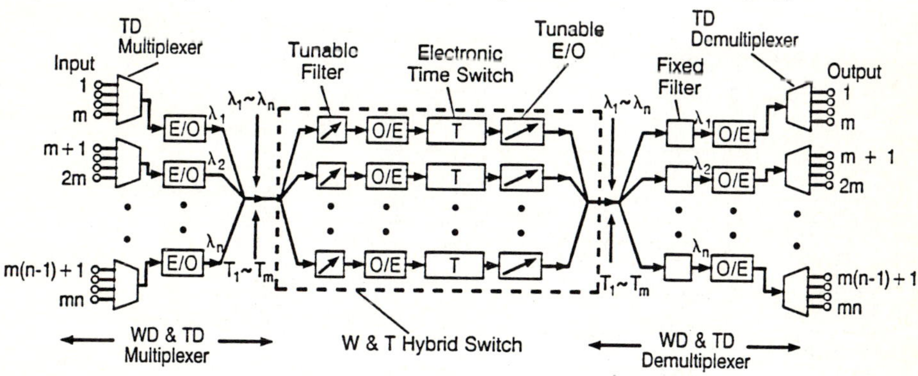

Fig 8. Photonic wavelength-division and time-division hybrid switch (after M. Nishio et al.[11]).

3. Implications for Photonic Switching

What does this mean for photonic switching? If we look at the papers presented at the first two photonic switching conferences (1987 and 1989), and compare them with the papers presented at this conference, we can see that the ideas and demonstrations have evolved, even over this relatively short time, from simple, single switching elements and arrays, to far more complex circuits and systems, involving sophisticated signal-processing operations. The trend is clearly towards more control, more functionality, and more complex processing. This is in part because of continuing progress in electronics technology, and in part because of the increasing complexity of requirements for "information age" networks. It appears likely that the high data rates involved in many projected optical transmission applications will make conventional electronic switching systems inadequate, and new switching technologies must be developed.

Because of the limitations imposed by nonlinear optical materials, and the high optical powers required, digital all-optical processing i.e. "optical logic" is unlikely to find significant applications in photonic switching -- at least not for some time to come. The most attractive prospect would appear to be to implement switching and control functions in a hybrid optical - electronic technology. In this way, each technology can be used for what it does most efficiently: electronics for "logic" and optics for "communications". As the capabilities for integration grow, we will see ever more complex operations performed on a single chip, with electronics providing the logic and control, and optics being used not only for transmission, but also for high-bandwidth interconnections, multiplexing, and perhaps also high speed, high density memory (Fig 9).

The key to the practical implementation of many of the current ideas for photonic switching will be the development of cheap, reliable, and simple optics-to-electronics and electronics-to-optics interfaces -- to the point where optical interconnects become "easier" to implement than wires. At this point, optoelectronic

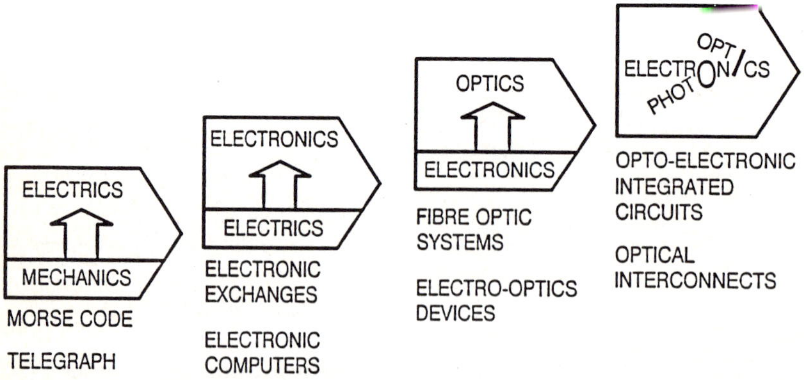

Fig 9. The evolution of communications and computing.

circuits will afford greatly increased functionality with the potential advantages (low cost, reliability, etc) of todays electronic chips. The evolution of the optical fiber network towards the customer (fiber to the home) brings the possibility of high volume requirements for OEIC chips. This "market pull" coupled with the "technology push" should aid in the development of a low cost OEIC technology.

The future "intelligent network" will require new technologies for the operation, control, and switching of data. The powerful data handling capabilities made possible with new generations of optoelectronic circuits incorporating an intimate mix of optical and electronic components should permit the implementation of sophisticated photonic switching operations.

References

[1] see, for example, P. W. Smith, in "Optical Computing and Nonlinear Materials", SPIE, 881, 30 (1988).

[2] see, for example, I. Mito and K. Kobayashi, IEEE J Lightwave Tech 6, 1623, (1988).

[3] E. Kapon, S. Shimony, J. P. Harbison, and L. T. Florez, Appl Phys Lett, 56, (1990)

[4] U. Koren, T. L. Koch, B. I. Miller, G. Eisenstein, and R. H. Bosworth, Appl Phys Lett, 54, 2056, (1989).

[5] E. Yablonovich, T. Gmitter, J. P. Harbison, and R. Bhat, Appl Phys Lett 51, 2222 (1987).

[6] M. E. Prise, N. C. Croft, R. E. LaMarche, M. M. Downes, S. J. Walker, L. D'Asaro, and L. M. F. Chirovsky, Appl Opt 26, (1990).

[7] J. Crow, C. J. Anderson, S. Bermon, A. Callegari, J. F. Ewen, J. D. Feder, J. H. Grenier, E. P. Harris, P. D. Hoh, J. H. Hovel, J. H. Magerlein, T. E. McKoy, A. T. F. Pomerene, D. L. Rogers, G. J. Scott, M. Thomas, G. W. Mulvey, B. K. Ko, T. Ohashi, M. Scontras, and D. Widiger, IEEE Trans Elect Devices, 36, 263 (1989).

[8] E. Arthurs, M. S. Goodman, H. Kobrinski, and M. P. Vecchi, J. Selected Areas in Commun, 6, 1500, (1988).

[9] see, for example, the special issue on photonic switching of the IEEE J Selected Areas in Commun, 6, (1988).

[10] K. Pedrotti, S. Beccue, W. Haber, B. Brar, G. Robinson, and M. K. Kilcoyne, Proceedings of OFC'90, paper WB4, OSA (1990).

[11] M. Nishio, S. Suzuki, N. Shimosaka, T. Numai, T. Miyakawa, M. Fujiwara, and M. Itoh, Proceedings of the II Photonic Switching Conference, 250, OSA (1989).

Role of Optical Amplifiers in Realizing All-Optical Communication Networks

K. Nakagawa

NTT Transmission Systems Laboratories, 1-2356 Take, Yokasuka-shi, Kanagawa-ken 238-03, Japan

Recent optical amplifier studies are outlined featuring Erbium-doped fiber amplifiers. The impact of this new technoklogy is discussed and the future possibilities for all optical communication networks are studied, including trunk lines, subscriber loops and transport node systems.

1. Introduction

Optical fiber cable transmission systems have advanced rapidly over the past ten years due to advances in fibers, lasers and LSI's. A series of trunk transmission systems have been developed and introduced as telecommunication networks in many countries. At present the system transmission capacity has reached the gigabit/s range.

The potential of optical fiber transmission systems is incredibly rich. The implementation of advanced optical technologies using optical amplifiers would enable the creation of more economical and higher quality systems for both transmission and switching, through larger capacities and greater signal-multiplexing capability.

This paper briefly outlines recent studies on optical amplifiers , and describes optical amplifiers' features. The impact of the new amplifier technology is then discussed and the future possibilities for communication systems using optical amplifiers are studied.

2. Communication Technology Evolution

In today's communications technologies, switching and signal processing are done electrically. The related semiconductor electronic technology has made and is continuing to make remarkable progress. Although the development of optical-fiber transmission now permits signals in the form of light to be transmitted cheaply and reliably, the light is converted to electrical signals at each end of the fiber. If the optical signal can be directly amplified with low noise, the repeater will be smaller and cheaper than the existing repeaters. Branching loss can be easily compensated for with an optical amplifier. Thus, an extremely flexible communications network can

Springer Series in Electronics and Photonics, Vol. 29 **Photonic Switching II**
Editors: K. Tada · H.S. Hinton © Springer-Verlag Berlin, Heidelberg 1990

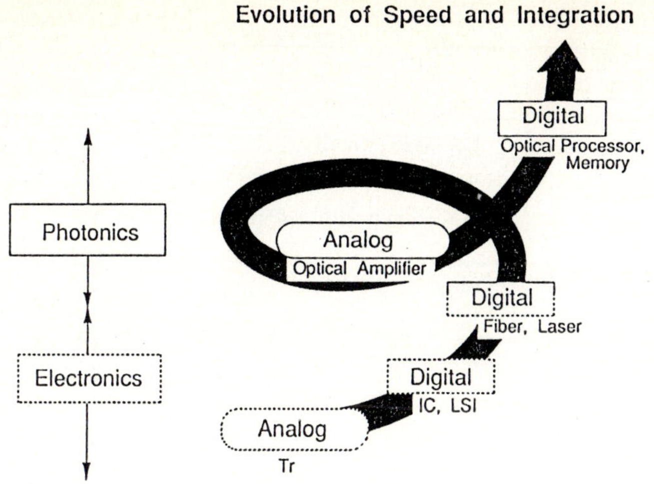

Fig.1 Transmission Technology Progress and Applications

be created because the signals can be easily handled in the optical format without changing it to electrical. The technology trends are summarized in Fig.1.

3. Optical Amplifiers

The types of optical amplifiers available include: rare-earth doped fiber, Raman using a conventional fiber and semiconductor laser, as shown in Table 1. Each of these has its own unique features, and is used for the appropriate applications.

Recent experiments have shown great success in applying Erbium-doped single-mode fibers (EDF) to traveling-wave type optical amplifiers. Nonrepeatered 250 km transmission was demonstrated for these amplifiers at 1.8 Gb/s[1]. Cable length was doubled by using optical amplification together with conventional intensity modulation and direct detection. Optical amplification offers such great advantages that more and more uses will be found for it in the future.

4. Applications of Optical Amplifiers

4.1 Trunk Transmission Systems

The increase in repeater spacing caused by using EDF optical amplifiers will reduce transmission equipment cost, improve line reliability, and simplify maintenance as well as operational functions, because most major telephone offices would be directly connected without outside repeaters. Moreover, ultra-high speed transmission systems, for example, 10 Gb/s exceeds the present F-1.6 G system, are possible if an optical amplifier is

Table.1 Optical Amplifiers and Applications

Optical Amplifiers		Features
Semiconductor Laser Diode		· Compact (<mm) · Various waveband (0.8~1.6μm) · Wide-bandwidth (10THz)
Erbium-Doped Fiber		· High gain at 1.5μm · Large relaxation time (~10ms) · Match to fiber
Nonlinear Fiber	Raman Amp.	· Distributed amplifier · High pump power required (>100mW)
	Brillouin Amp.	· Narrow bandwidth (~100MHz)

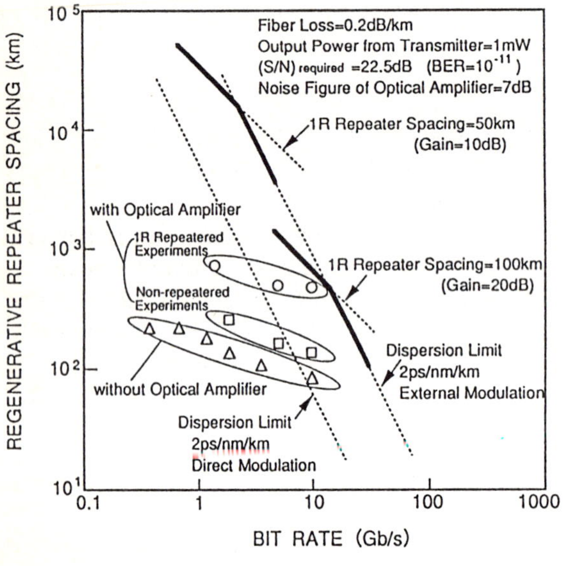

Fig.2 Regenerative Repeater Spacing in IM/Direct-Detection Scheme

used as a 1R repeater or a wideband amplifier. A combination of EDF optical amplifiers and PIN-PDs may eventually replace avalanche photodiodes.

This trend towards longer spans is shown in Fig.2 where the system is designed according to an IM/direct-detection scheme. The great improvement in optical output power and receiver sensitivity made it possible to double the nonrepeatered spacing by using EDF amplifiers. It

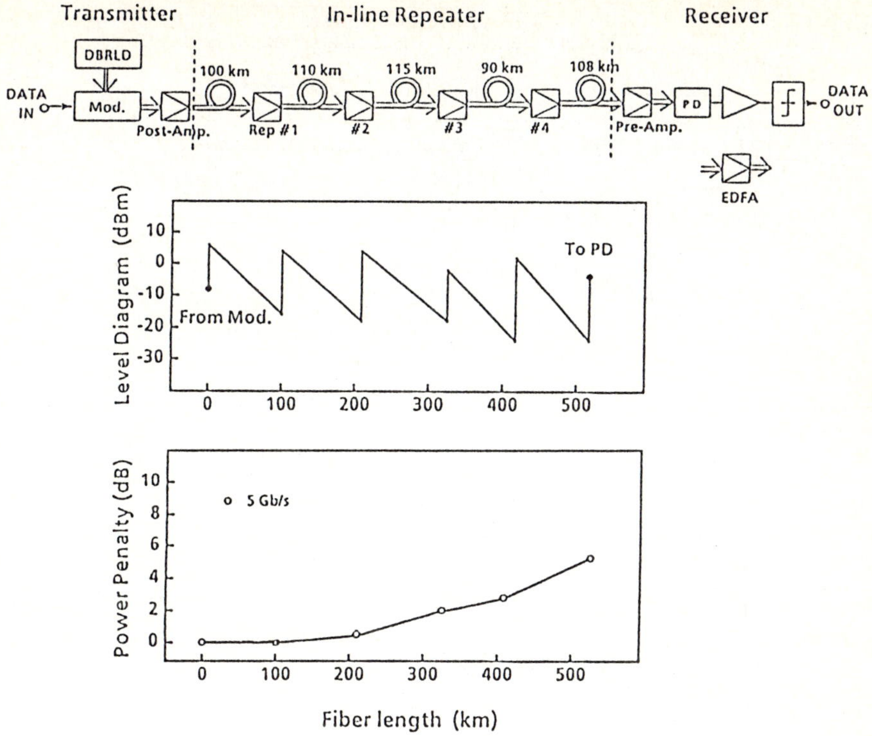

Fig.3 5 Gb/s 523 km transmission experiment and the level diagrams.

also demonstrates that a simple multi-repeatered configuration is feasible with optical amplifiers as shown in Fig.3. Such a system may be designed with a coherent modulation/ demodulation scheme in the near future. The merit of this new configuration is that a bit-rate flexible transmission system will be possible because the repeater does not include retiming and regeneration circuits. Optical processing technologies will lead to the separation of signal transmission and energy supply, thus achieving a higher quality of telecommunication networks.

4.2 Subscriber Loop Systems

Another example of EDF amplifier characteristics is shown in Fig.4. The on/off switching ratio is large when amplification corresponds to the on-state and attenuation to off-state. Amplification and switching performance is applicable in a subscriber loop system, as shown in Fig.5. When a problem such as fiber breakage occurs, an optical fiber line protection switch can easily reconfigure the network by utilizing a fiber active switch. Optical switching by in-line EDF amplifiers is preferable from both the environmental and economical viewpoints.

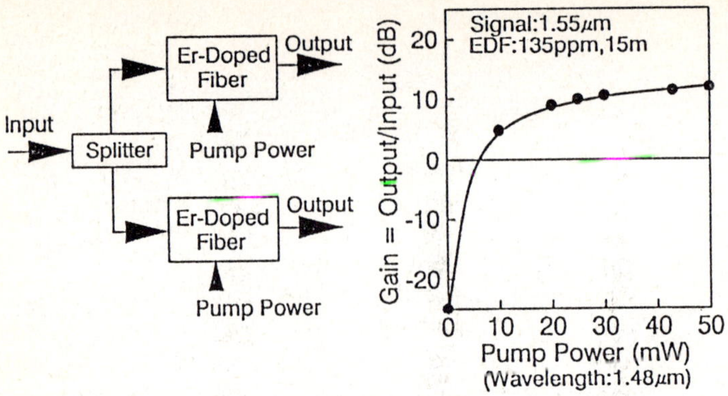

Fig.4 Principle of Erbium-Doped Fiber Active Switch

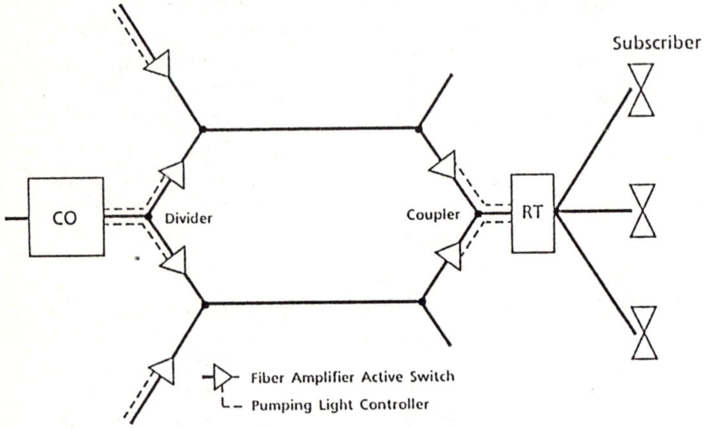

Fig.5 Optical Amplifiers in Subscriber Loop System for Wide Service Area and Network Reconfiguration

4.3 Broadband Distribution Systems

An example of an EDF amplifier in a broadband distribution system such as CATV is shown in Fig. 6. In existing coaxial networks, the service area is limited to 7 km from the head end, many electrical amplifiers must be installed, and cable length is critical to the video quality. EDF amplifiers can increase transmission length to about 40 km by developing a 1.5 micron waveband technology[2]. The effect of applying EDF amplifiers not only increases the transmission length, but also offers the possibility of optical branching. An FM-FDM signal format will effectively increase the number of branchings to 1000 or more in the future.

In a multi-repeatered digital transmission system experiment, an in-line repeater gain of 85.5 dB was achieved at 5 Gb/s. This means that a numerous optical digital distribution of 10^8 branchings is theoretically

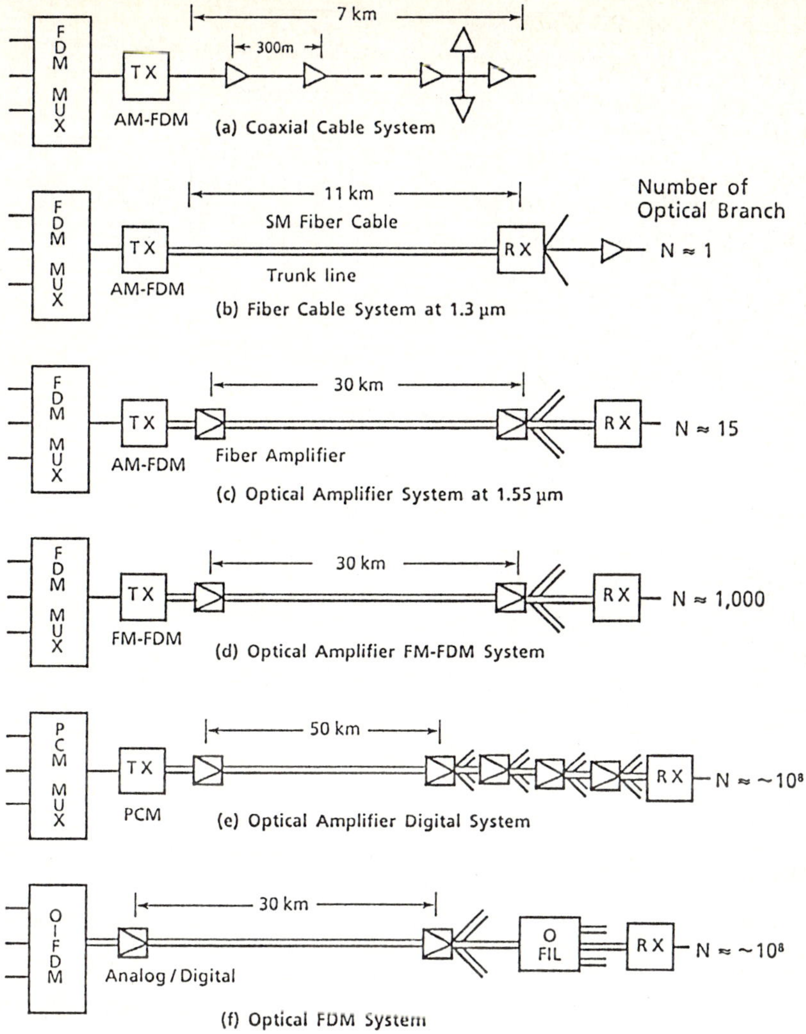

Fig.6 Present and Future Broadband Distribution Systems

feasible, even if a 25.5 dB line loss is included, for a 50 Mb/s 100 video channel service distribution.

4.4 Transport Node Systems

All optical network configurations are receiving attention because switching and cross-connections as well as transmission can be performed without converting optical signals into electrical ones, because the naturally wide bandwidth characteristics of light are fully utilized.

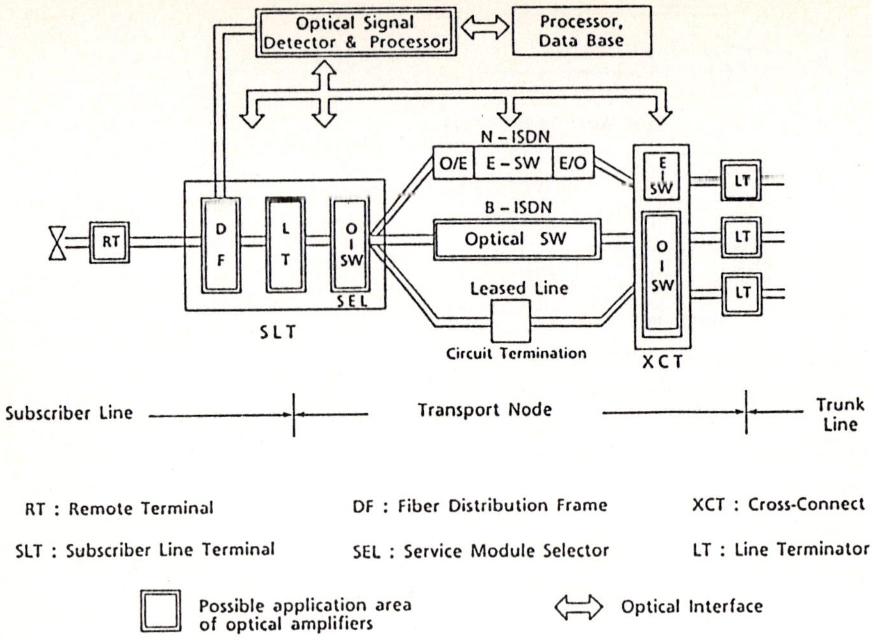

Fig.7 Role of Optical Amplifiers in Transport Nodes

However, the problem of optical loss exists in present switching circuits. In most switching systems, electronic switching is preferred over optical to electrical conversion, because electronic signal processing is better at present than optical processing in terms of integration and higher functionality, even if electronic bandwidths are fairly narrow. If optical amplifiers become feasible, optical loss will no longer be a major problem and the research goal of these applications will be changed.

An example of a node system using optical switching is shown in Fig.7. Optical amplifiers are applicable in various transport node systems. The main application modules are broadband switches, cross-connects and service module selectors as well as optical signal detectors and processors. Optical amplifiers are directly applicable to compensate for the attenuation caused by connections and branchings. Fiber active switches[3] can be used for cross-connections, service module selectors or fiber distribution frames when the block-mode switching shown in Fig.8 is allowed. Bit-mode switching devices, however, will be needed for broadband switches. The potential devices and switching speeds necessary to realize the applications being considered for all optical networks are indicated in table 2. High-speed switching circuits will become a reality once optical amplifiers have been functionally integrated with the circuits for absorption, coupling or feedback among others.

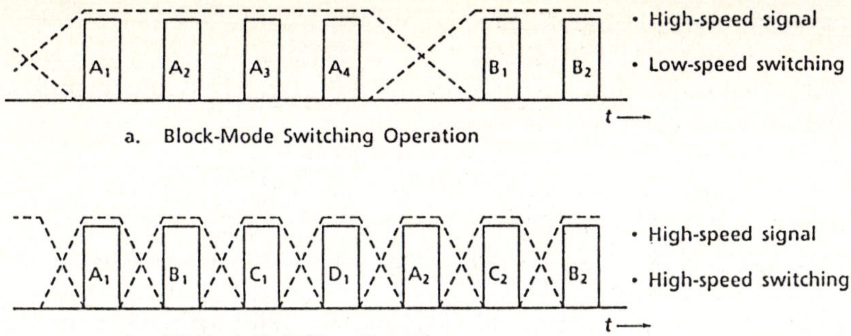

a. Block-Mode Switching Operation

b. Bit-Mode Switching Operation

Fig.8 Optical TD-Switch Operation Format

Table.2 Optical Switching Devices for Transport Nodes

Main Module		Broadband Switch	Cross-Connect	Service Module Selector
Main Function		Circuit (B-call) switching	Path configuration	Interface path between modules
Signal SW Mode		Bit-mode preferable	Block-mode allowed	Block-mode allowed
Proposed optical switching circuits	SD	Fiber active switch; AO, MO, Mechanical switch; EO ($LiNbO_3$) switch; Liquid crystal switch		
	TD	Gate (WG, QW), Bistable diode switch; OEIC		
	FD	Nonlinear switch; Tunable LD switch; Silica-based waveguide switch		
	CD	Code division multiple access		
Switching speed		fs, ps, ns ⟵⟶ μs ⟵⟶ ms		
Role of optical amplifiers		· Attenuation compensation for connection and branching · Functional integration of absorption, coupling, feedback etc. with amplifier		

5. Conclusion

Recent research on EDF amplifiers, semiconductor laser amplifier/functional-devices and their future applications are described in this paper. It is clarified that optical amplifiers will have a great influence on the development of future optical communication systems. Research into optical amplifier applications has just started and many research items still remain for future study. It is conceivable that a new revolution in optical communication systems will lead towards all-optical networks being brought about by the employment of these optical amplifiers.

Acknowledgement:

The author wishes to thank Dr. S. Shimada, Dr. H. Kimura and his colleagues for their guidance and cooperation

References:

[1] H.Hagimoto, K.Iwatsuki, A.Takada, M.Nakazawa, M.Saruwatari, K.Aida, K.Nakagawa and M.Horiguchi: "250km nonrepeated transmission experiment at 1.8Gb/s using LD pumped Er^{3+}-doped fiber amplifier in IM/direct-detection system", Electron. Lett., 25-10, p.662-664, 1989.

[2] E.Yoneda T.Tuchiya, K.Suto and Kikushima : "Erbium-doped fiber amplifier for video distrihbution networks", ITE of J, ROFT 89-31, 1989 (in Japanese).

[3] Y.Sato, K.Aida, and K.Nakagawa : "An Erbium-doped fiber active switch controlled by coded pump light", PS '90, 13 C-4, 1990.

Nonlinear Optical Devices for Photonic Switching

K. Kubodera

NTT Opto-electronics Laboratories, 3-1 Morinosato Wakamiya,
Atsugi-shi, Kanagawa-ken 243-01, Japan

Abstract. Recent developments in nonlinear optical switching devices with fast response times are reviewed. Examples are etalon-type semiconductor devices of superlattice and doped glass materials, ultrafast glass fiber switches, and organic switches.

1. Introduction

Optical devices utilizing nonlinear optical effects have attracted much attention as key devices for photonic switching. Wavelength conversion effects such as second-harmonic generation have been studied from the beginning because of their practical use. Other nonlinear optical effects, such as refractive index changes or absorption coefficient changes depending on the applied electric field or the incident light intensity (optical Kerr effect), have been extensively investigated recently. These effects are applied to light controlling devices such as electrically-controlled optical switches or optically-controlled optical switches [1,2]. It is also important that these devices show nonlinear input-output characteristics, such as thresholding, optical memory, and triode logics.

In this paper, prospective applications of such devices to future optical systems are described. Additionally, recent developments in switching devices, especially all-optical switching devices with fast response times, are shown in the following sections.

2. Nonlinear Optical Devices in Future Systems

Table 1 summarizes several nonlinear optical devices for photonic switching and their prospective applications in future optical communication and information processing systems. Switch matrices are electrically-controlled optical switches utilizing the refractive index change of $LiNbO_3$ crystal or absorption spectrum changes of semiconductor materials. Such electrically-controlled optical switches are indispensable because they are compatible with electrically-controlled computers.

Thresholding gate devices and optical memory devices which exhibit abrupt jumps in their input-output characteristics, are key components in time-division multiplexing systems. In high-bit-rate time-division systems, beyond 100 Gb/s for example, all-optical switches are especially important because they are not constrained by the CR-constant limitation. Optical time switches can be constructed by combining these all-optical devices with electrically-controlled write and read gates.

Wavelength converters and filters are very important electrically-controlled optical switches in wavelength-division switching systems. Recently, such wavelength switches have been demonstrated using the nonlinear absorption effect in semiconductor lasers and their refractive index changes, caused by varying the injection current [2,3].

Springer Series in Electronics and Photonics, Vol. 29 Photonic Switching II
Editors: K. Tada · H.S. Hinton © Springer-Verlag Berlin, Heidelberg 1990

Table 1. Nonlinear optical devices and their prospective applications.

	Nonlinear Optical Devices & Functions			Control-Signal
Wired Sw	Space-div. Sw	Switch Matrix		E-O
	Time-div. Sw	Thresholding Gate	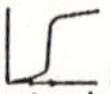	E-O, O-O
		Optical Memory	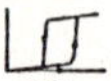	E-O, O-O
		Write/Read Gate	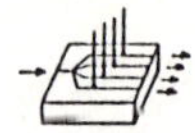	E-O
	Wavelength-div. Sw	Wavelength Converter/Filter		E-O
Beam Sw	Free-space Sw	Switch Array	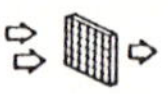	E-O O-O

Optical gate array devices consisting of integrated gated switches or bistable devices are indispensable in free-space beam switching architecture. Especially, parallel all-optical switches will be key devices in future optical signal processing systems, utilizing the free space's wide bandwidth and its image-forming feature.

Among the various nonlinear optical devices described above, the following sections are focused on all-optical switching devices with fast response times.

3. Studies on All-Optical Switching Devices

3.1. Classification of All-Optical Switches

Table 2 shows the classification of optically-controlled optical switches. Among them, O-E feedback devices such as SEED (self-electro-optic effect device) have been investigated intensively because of their advantages of high optical sensitivity and easy application to two-dimensional arrays. A large scale array of 2 kbits has already been fabricated [4].

Laser diode switches are regenerative devices, yielding large fan-out and wavelength controllability. Optical etalon devices consisting of an optical Kerr medium and etalons are very simple thresholding and optical bistable devices, and are investigated extensively using several semiconductor materials.

In order to obtain much faster response times, very fast nonlinear optical materials, such as glass or organic materials can be used. By combining these materials with an interferometer or a waveguide coupler, picosecond or femtosecond response devices are obtainable.

3.2. Semiconductor Multiple Quantum Well Etalons

Direct transition semiconductor materials are known to show efficient nonlinear refractive indices at the band gap wavelength due to the dynamic band filling effect. Among them, semiconductor superlattice materials (MQW; multiple quantum well materials) show the largest nonlinearity in the wavelength region of the exciton absorption tail, due to the screening effect of excitons, superimposed on the band filling effect. The structure of an InGaAs/InAlAs MQW etalon investigated in our group is shown in

Table 2. Classification of optically-controlled optical switches.

Classification	Structure	Waveguide	Vertical	Function	Material	Power-Speed (Reported)	Remarks		
O-E Feed Back	• SEED • VSTEP	—	O	• Gating • Bistability	• GaAs	6μW-100μs ~3.6mW-1ns	• High-sensitivity • 2-Dim Array		
Laser Osc. /Amp.	• LD-Osc. + Saturable Abs. • LD-Amp.	O	O	• Gating • Bistability • Wavelength Converter	• GaAs • InGaAsP	1μW-1ns	• Large fan-out • Wavelength Control	Semiconductor	
Optical Kerr switch	• Etalon + Kerr Medium	—	O	• Gating • Bistability	• GaAs, InSb • ZnS, ZnSe • MQW • Doped glass	4mW-20ns ~ -150ps	• 2-Dim Array		1ns
	• Fiber Kerr' Medium + Interferometer	O	—	• Gating	• SiO_2 glass	1.8W-<100ps ~32kW-100fs	• Picosecond Response	Glass	1ps
	• Waveguide Kerr Medium + Coupler	O	—	• Gating	• Organics	1MW- <400fs	• Ultrafast Response	Organics	

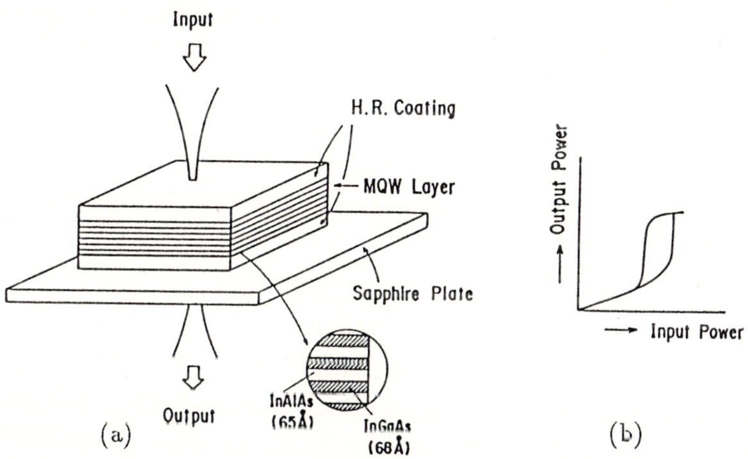

Fig. 1. 1.5 μm InGaAs/InAlAs superlattice etalon device. (a) Device structure. (b) Bistable input-output characteristics.

Fig. 1(a) [5]. The etalon consists of an MQW layer grown by molecular beam epitaxy, with high-reflection dielectric mirrors directly coated on both surfaces. The etalon is fixed with epoxy glue on a heat sink sapphire plate. Optical bistable operations of about 10 ns switching times were observed by the 20 mW color-center laser, pumping at a wavelength of 1.5 μm, as shown in Fig. 1(b).

Aiming at vertically-accessable optical parallel processing devices, several advanced etalon devices have been developed for GaAlAs systems. Semiconductor monolithic mirrors are being studied to prevent heating up, and to more precisely construct the etalon cavity [6]. Monolithic microresonator array devices of 3 x 3 μm^2 spacing have also been fabricated, where a much faster response time of 150 ps, caused by the acceleration of the surface recombination process, was confirmed [7].

3.3. Semiconductor-Doped Glass Etalons

Semiconductor-doped glass materials are effective for constructing fast switching devices [8]. As a result of the fast carrier recombination process in microcrystallites with diameters of less than 10 nm, a response time of about 10 ps has been reported by degenerate four-wave mixing experiments [9].

Optical bistability, making use of the 10 ps response time, was observed as shown in Fig. 2 [10]. An etalon was constructed with a 300 μm thick CdSSe-doped glass sample and two external mirrors. Optical pulses of a SHG light (0.532 μm) from a mode-locked Q-switched YAG laser was used as the input light. Nonlinear input-output characteristics, i.e. limiting, differential gain, and bistability were observed by varying the initial detuning of the cavity length from the off resonance to the on resonance, as shown in Fig. 2(b). The switching time of the bistable operation was 25 ps, which correlates well with the theoretical results, if the material lifetime of 10 ps and the cavity lifetime of 20 ps are considered. The switching energy density required to obtain a one-bit operation was 8.3 μJ, which is comparable to bulk semiconductor crystals.

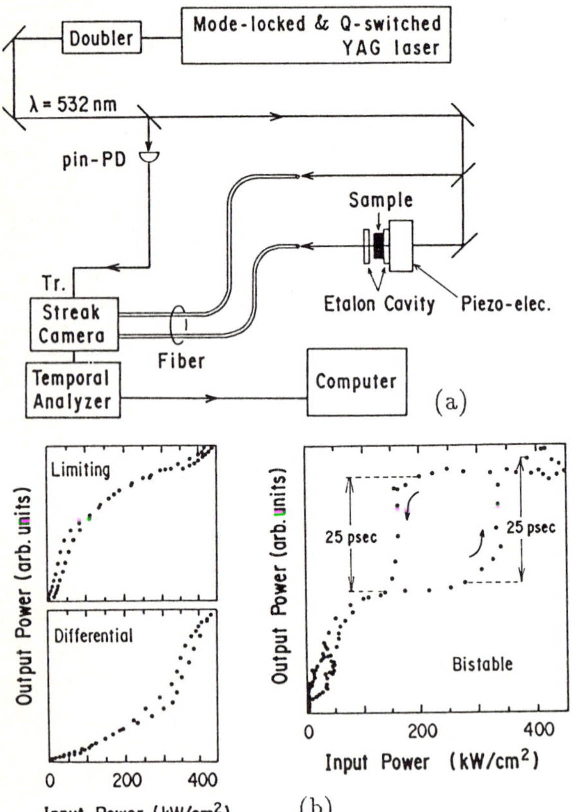

Fig. 2. Picosecond optical bistable operation using a semiconductor-doped glass etalon. (a) Experimental setup. (b) Nonlinear input-output characteristics.

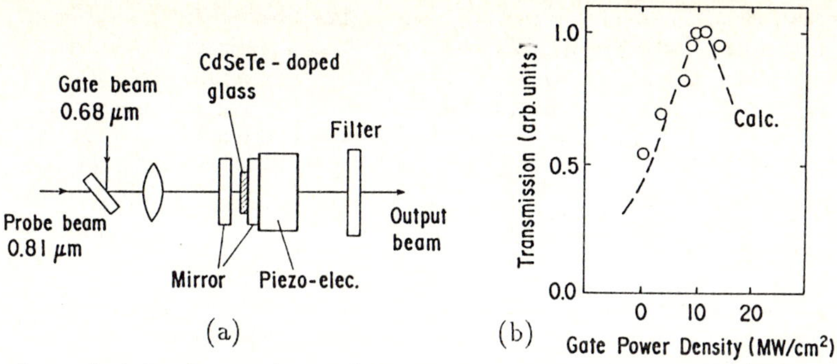

Fig. 3. Optical gating experiment using a semiconductor-doped glass etalon. (a) Experimental setup. (b) Signal transmission.

A demonstration of an efficient optical gating operation using a CdSeTe-doped glass etalon has been done using different wavelengths for the probe light and the gate light (non-degenerate case) [11]. A laser diode with a suitable wavelength was selected as the probe light. A pulsed dye laser which has a much higher absorption efficiency, was used as the gate light. The experimental setup and the results of the optical gating are shown in Fig. 3. The features of the varied material design and the fast response time of 10 ps would be profitable if applied to time-division multiplexing systems in the future.

3.4. Glass Fiber Switches

Nonlinearities of silica glass fibers are fairly high because of the small cross sectional areas and long lengths, although the nonlinear refractive index of the material is very small. Several kinds of ultrafast optically gated optical switches have been demonstrated using silica glass fibers. Concepts of these devices, such as a Mach-Zehnder interferometer type, a directional coupler type, an optical Kerr shutter type, and a loop mirror type are shown in Fig. 4. An efficient laser-diode-pumped, all-optical gate switch has been demonstrated using a 38 m long Mach-Zehnder interferometer. The interferometer consists of an ordinary silica glass fiber and a highly GeO_2 doped glass fiber with high nonlinearity [12]. Ultrafast switching of 100 fs response time has been demonstrated using a dual core fiber directional coupler [13]. A successful operation of 5 Gb/s time-domain switching has also been demonstrated using a fiber loop mirror switch [14].

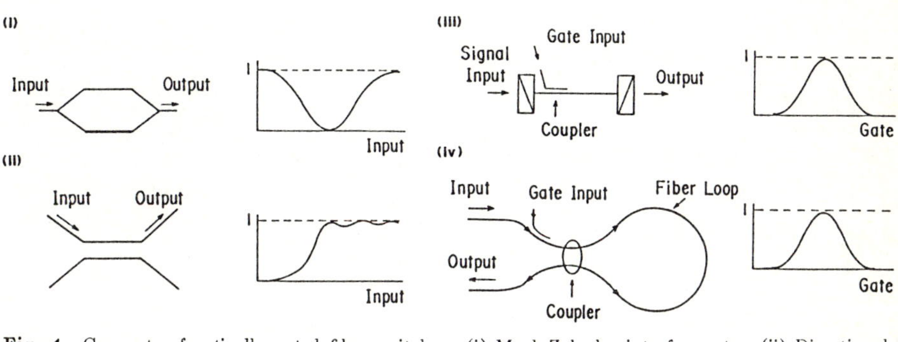

Fig. 4. Concepts of optically gated fiber switches. (i) Mach-Zehnder interferometer, (ii) Directional coupler, (iii) Optical Kerr shutter, (iv) Loop mirror.

The mechanism of the nonlinearity in the silica glass fiber is the generation of nonlinear (third order) electric polarization. Since nonlinear polarization gives the refractive index change without absorption, it makes it possible to use long fiber lengths. Recently, a very efficient third order polarization effect has been detected in high-index chalcogenide glass [15]. The chalcogenide glass, because of its easy fabrication and low loss property over long distances, is expected to aid in the development of efficient nonlinear optical fiber switches.

3.5. Organic Nonlinear Optical Switches

Similarly to glass materials, some organic materials exhibit efficient refractive index changes with fast response times due to nonlinear polarization. Representative materials are diacetylene polymers and some monomer materials such as diphenyl compounds, where conjugated π electrons effect the nonlinearity [16]. For example, four orders of magnitude larger nonlinearity than that of silica glass, has been reported for the polydiacetylene.

A demonstration of an organic optical Kerr shutter experiment is shown in Fig. 5 [17]. The experiment was carried out using a solution of stilbazolium salt in a 1 mm thick glass cell, with 0.7 μm dye laser gate pulses, and a 0.83 μm laser diode probe beam. The transmission efficiency was four times that of the conventional CS_2 material.

An ultrafast switching experiment that exhibited a response time of less than 400 fs has been reported using a solid state polymer film and a grating coupler to the polymer waveguide [18]. Studies on organic devices are well under way, with importance being placed on material design and device optimization, such as film formation and waveguide construction.

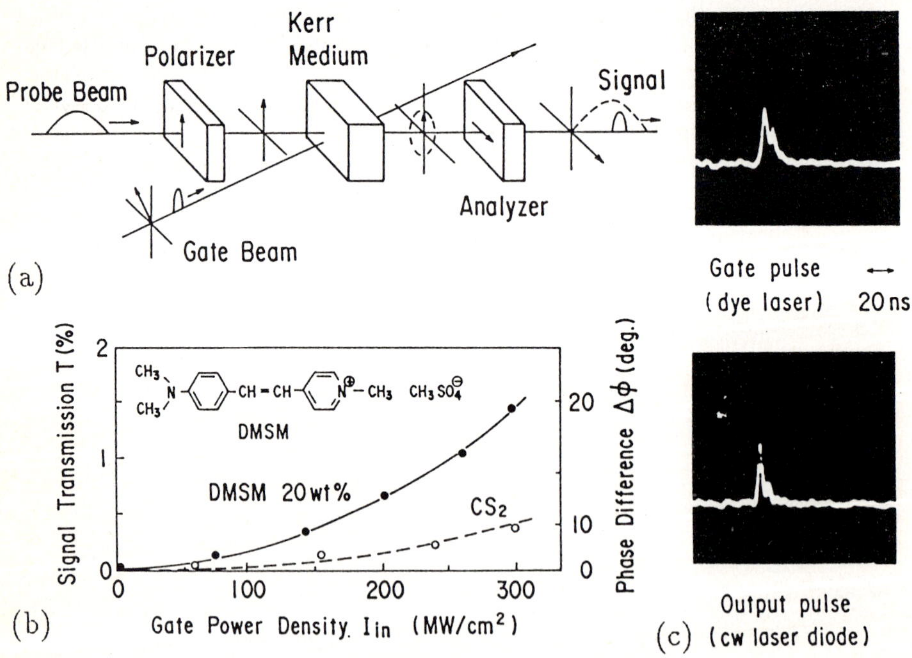

Fig. 5. Optical Kerr shutter experiment using organic material. (a) Experimental setup. (b) Signal transmission. (c) Gate and output waveforms.

4. Conclusion

Recent developments in nonlinear optical switching devices with fast response times, such as optical bistable devices and optically-gated optical switches, were reviewed. Low-power devices consisting of semiconductors and ultrafast devices consisting of glass or organic materials are expected to be key components in tomorrow's optical systems. In addition to the continuous effort in development, major breakthroughs in the search for new materials, as well as novel device structures are earnestly desired.

Acknowledgements. The author would like to thank Dr. Tetsuhiko Ikegami and Dr. Takayuki Sugeta for their continuous encouragement.

References

1. P. W. Smith: IEEE Circuits and Devices Magagine **May**, 9 (1987).
2. T. Ikegami and H. Kawaguchi: IEEE J. Selected Areas in Commun. **6**, 1131 (1988).
3. H. Kawaguchi, et al.: Electron. Lett. **23**, 1088 (1987).
4. A. L. Lentine, et al.: IEEE Photonics Tech. Lett. **2**, 51 (1990).
5. K. Nonaka, et al.: IEEE Photonics Tech. Lett. **1**, 55 (1989).
6. R. Kuszelewicz, et al.: Appl. Phys. Lett. **53**, 2138 (1988).
7. J. L. Jewell, et al.: Appl. Phys. Lett. **51**, 94 (1987).
8. G. I. Stegeman and R. H. Stolen: J. Opt. Soc. Am. **B-6**, 652 (1989).
9. M. Mitsunaga, et al.: J. Opt. Soc. Am. **B-5**, 1448 (1988).
10. J. Yumoto, et al.: Opt. Lett. **12**, 832 (1987).
11. H. Kobayashi, et al.: Topical Meeting on Photonic Switching, Kobe, Japan, 13C-34 (1990).
12. I. H. White, et al.: Electron. Lett. **24**, 340 (1988).
13. S. R. Friberg, et al.: Opt. Lett. **13**, 904 (1988).
14. M. Jinno and T. Matsumoto: Topical Meeting on Photonic Switching, Kobe, Japan, 13C-16 (1990).
15. H. Nasu, et al.: J. Non-Cryst. Solids **110**, 229 (1989).
16. K. Kubodera and T. Kaino: "Nonlinear Optics of Organics and Semiconductors" ed. T. Kobayashi (Springer, 1989) p.163.
17. H. Kanbara, et al.: IEEE Photonics Lett. **1**, 149 (1989).
18. R. Burzynski, et al.: Appl. Phys. Lett. **53**, 2011 (1988).

Photonic Switching Fabrics Based on S-SEED Arrays

H.S. Hinton, F.B. McCormick, T.J. Cloonan, F.A.P. Tooley, A.L. Lentine, and S.J. Hinterlong

AT&T Bell Laboratories, 200 Park Plaza, Naperville, IL 60566, USA

Abstract. This paper will review the hardware required for photonic switching fabrics based on S-SEED arrays and free-space digital optics. These fabrics are based on diffraction limited optics, regenerative digital logic devices, differential digital signals, 2-D pixelated images, a fan-in and fan-out of two, and single wavelength operation.

1. Introduction

The optical domain offers the strengths of both temporal and spatial bandwidth to be exploited in the design of new switching and computing systems. The majority of the research has focused on the efficient utilization of the available temporal bandwidth of optical fiber. This includes switching fabrics based on time-division multiple access, code-division multiple access, and wavelength-division multiple access. The second strength of the optical domain, and least understood, is the spatial bandwidth provided by imaging systems. This spatial bandwidth can be viewed as connections or optical pin-outs.[1] By utilizing these "connections" through the use of free-space digital optics new switching fabrics can be built.

2. Free-Space Digital Optics Hardware

This section will discuss the basic hardware components that are required to implement a switching fabric based on free-space digital optics.[2] It assumes that the devices are limited to a fan-in and fan-out of two and that the two input signals to any device are different polarizations. The four basic components required for a free-space digital optical fabric are: 1) 2-D arrays of optical logic devices, 2) a spot array generator to provide the optical power for the devices, 3) an optical interconnect to provide the device-to-device connectivity, and 4) a beam combiner to get information into and out of the devices. Each of these components will be discussed in more detail below.

2.1 Devices

There are several devices that have been proposed for use in systems based on free-space digital optics. One of the leading contenders is the symmetric-SEED.[3] As shown in Figure 1 this device is made by electrically interconnecting two MQW pin diodes in series. When the diodes are connected in this fashion, they become complementary, in that, when one of the diodes is "on" the other will be "off." Thus, one of the diodes will be in the absorbing state while the other is in the transmissive state. Perhaps the greatest strength of these devices is that changing states is a function of the *ratio* of the two input powers and not the function of

Springer Series in Electronics and Photonics, Vol. 29 **Photonic Switching II**
Editors: K. Tada · H.S. Hinton © Springer-Verlag Berlin, Heidelberg 1990

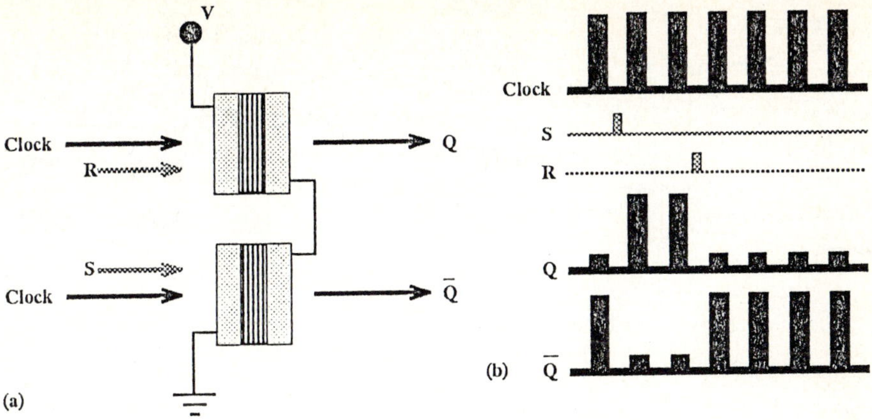

Figure 1. Symmetric SEED as an S-R latch

the absolute power of the input beams. The importance of this is that the allowable noise on the signal inputs can be much greater than for the case of a critically biased device such as a resistively loaded SEED.

The S-SEED can be configured such that it can operate as an S-R latch as illustrated in Figure 1. In part (a) of this figure, the inputs are separated into an S (Set) input, and R (Reset) input, and a clock input, where the clock has the same intensity for both inputs. The S and R inputs are also separated in time from the clock inputs as shown in part (b) of this figure. The S or R inputs are used to set the state of the device. When the S input is illuminated, the S-SEED will enter a state where the upper MQW pin diode will be transmissive while the lower diode will be absorptive. When the R input occurs, the opposite condition will occur. Low switching intensities are able to change the device's state when the clock signals are not present. After the device has been put in its proper state, the clock beams are incident on both inputs. Since the two clock beams are equal in intensity, the ratio between the incident beams should be one, which will prevent the device from changing states. These higher energy clock pulses will be used to transmit the state of the device to the next stage of the system. Since the S or R inputs are low intensity pulses and the clocks are high intensity pulses, a large differential gain may be achieved. This type of gain is referred to as time-sequential gain.

The S-SEED is also capable of performing logic functions (such as NOR, OR, NAND, and AND) in addition to memory functions (i.e., set-reset latch). The inputs will also be differential, thus still avoiding any critical biasing of the device. A method of achieving logic gate operation is shown in Figure 2. The logic level of the inputs will be defined as a function of the power of the signal on the set input relative to the power of the signal incident on the reset input. For example, when the power of the signal incident on the set input is greater than the power of the signal on the reset input, a logic "1" will be present on the input. On the other hand, when the power of the signal incident on the set input is less than the power of the signal on the reset input, a logic "0" will be incident on the input. For the noninverting gates, OR and AND, we can represent the output logic level by the power of the signal coming from the Q output relative to the power of the signal coming from the $\overline{Q}$ output. As before, when the power of the signal leaving the Q output is greater than the power of the signal leaving the $\overline{Q}$ output, a logic "1" will be represented on the output. To achieve AND operation, the device is initially set to its "off" or logic "0" state (i.e., Q low and $\overline{Q}$ high) with preset pulse, $Preset_b$, incident on only one p-i-n diode as shown in Figure 2. If both input signals have logic levels of "1" (i.e., set = 1, reset = 0), then the S-SEED AND gate is set to its "on" state.

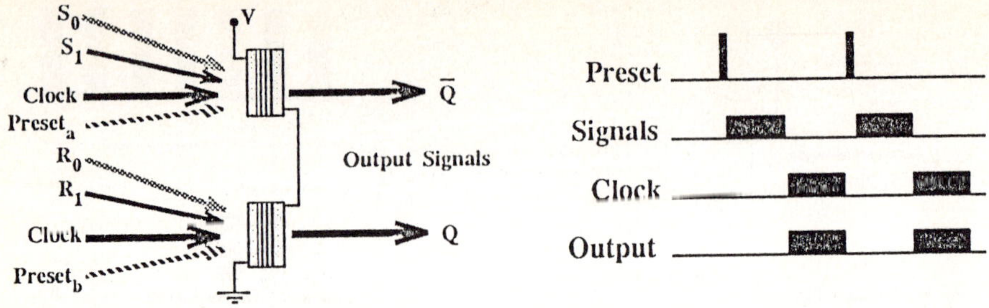

Input₀ Input₁

S_0	S_1	R_0	R_1	$Preset_a$	$Preset_b$	Q	$\bar{Q}$	
0	0	1	1	0	1	0	1	
0	1	1	0	0	1	0	1	AND / NAND
1	0	0	1	0	1	0	1	
1	1	0	0	0	1	1	0	
0	0	1	1	1	0	0	1	
0	1	1	0	1	0	1	0	OR / NOR
1	0	0	1	1	0	1	0	
1	1	0	0	1	0	1	0	

Figure 2. Optical Logic Based on S-SEED Devices

For any other input combination, there is no change of state, resulting in AND operation. After the signal beams determine the state of the device, the clock beams are then set high to read out the state of the AND gate. For NAND operation, the logic level is represented by the power of the $\bar{Q}$ output signal relative to the power of the Q output signal. That is, when the power of the signal leaving the $\bar{Q}$ output is greater than the power of the signal leaving the Q output, a logic "1" is present on the output. The operation of the OR and NOR gates is identical to the AND and NAND gates, except that preset pulse, $Preset_a$, is used instead of the preset pulse $Preset_b$. Thus, a single array of devices can perform any or all of the four logic functions and memory functions with the proper optical interconnections and preset pulse routing.

2.2 Spot Array Generation

The two-dimensional arrays of optical logic gates that have been previously discussed will require an optical power supply to either clock the devices, as in the case of S-SEED structures, or provide the optical bias power necessary to place the devices into the nonlinear region. These optical power supplies need to be 2-D arrays of uniform intensity spots. Binary phase gratings are one solution to this problem. These gratings are made by etching glass with a repetitive pattern of different thicknesses. For the case of a binary phase grating (BPG), there would be two thicknesses of glass. This grating is illuminated by a plane wave from a laser source as illustrated in Figure 3. The light transmitted through the grating is transformed at the back focal plane of a lens which is the output plane of the spot array generator. One approach in designing the BPGs assumes an even number of transition points and a phase difference of π between the two thickness.[4] [5] A second approach uses an odd number of transition points and does not constrain the phase difference to π.[6] Multi-level phase gratings have also been demonstrated.[7]

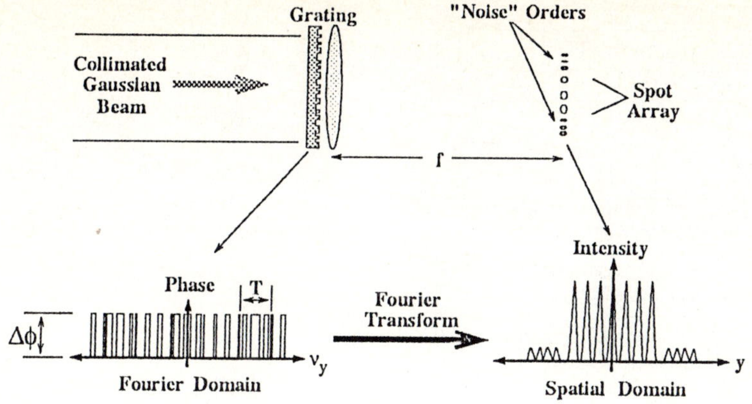

Figure 3. Spot Array Generation

2.3 Optical Interconnects

To create a large dimensional switching fabric a multi-stage interconnection can be used. These networks are typically composed of multiple stages of smaller switches. Perfect shuffle networks and their topological equivalents are examples. They may be composed of stages of 2x2 switches with perfect shuffle interconnects (or the topological equivalent) between each stage. An example of a topological equivalent is the crossover network,[8][9] which is illustrated in Figure 4. Part (a) in the figure illustrates the topology of the interconnect. The output of each node is split into two components: a straight-through and a crossed interconnect. To convert to a 3-D network, the 2-D network can be fan-folded on the fold lines shown in part (a) to create the 3-D network of part (b).[10] Note that the interconnect between each stage of the 3-D network is still a 2-D interconnect. Part (c) shows the operation of the interconnect system. An input image enters the system from the left. Each pixel or spot is circularly polarized. The perpendicular component of the input light will be imaged onto the mirror. The reflected light from the mirror will have its polarization rotated, allowing it to pass through the polarizing beamsplitter where it can be imaged onto the output image plane. The parallel component of the input light will pass through the beamsplitter and be imaged onto the prism grating. The light incident on the prism grating will be reflected to the opposing mirror where it will be redirected back into the beamsplitter. This reflection process shifts the position, in space, of the light being imaged onto the grating corresponding to the cross-over patterns of the crossover interconnection topology. The light leaving the prism grating will have its polarization rotated allowing it to be redirected and focused onto the output image plane. Thus, both the straight through and crossed interconnects have been achieved. The amount of shift in space is related to the period of the prism grating.

2.4 Beam Combination

Free-space photonic switching systems based on S-SEED arrays require that each device must be able to receive two input signals plus a clock.[11] [12] In addition, the reflected output signal must be directed from the device to the stage of the switching fabric. The major constraints of this problem are that the spots must be small ($<5\mu m$), requiring the entering signals to use the full aperture of an imaging lens, and that the signals must not interfere at the device's optical window.

An example of a beam combination system using space-multiplexing is illustrated in Figure 5. This beam combiner is composed of a polarizing beamsplitter that is surrounded on

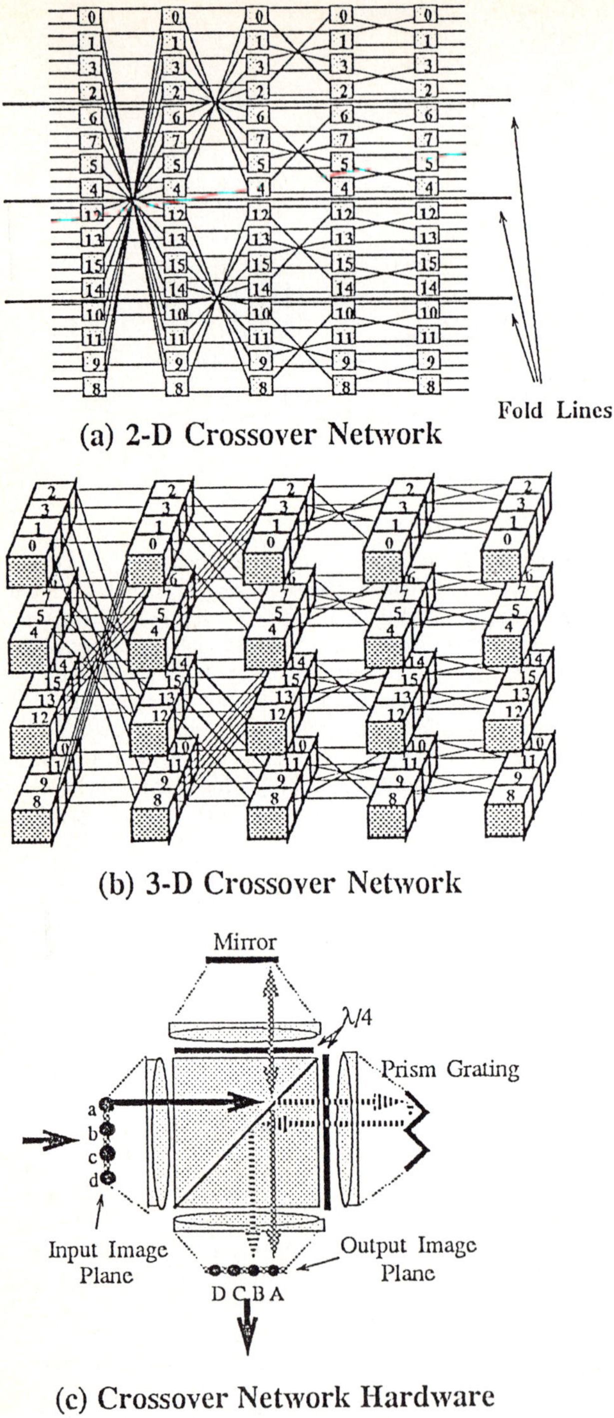

Figure 4. Crossover Interconnection Network

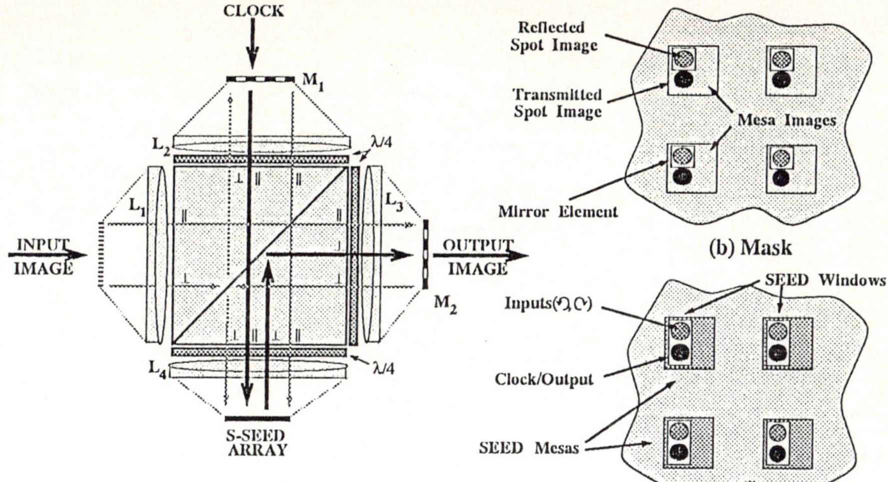

(a) Beam Combiner

(b) Mask

(c) S-SEED Array

Figure 5. Space-Multiplexed Beam Combination

three sides by λ/4 plates. The lenses, L_{1-4}, are used in the infinite conjugate mode. At the bottom of the beam combiner is the S-SEED array. The location of the beams incident on the S-SEED array are shown in part (c). As an input image enters the beam combination system on the left (this image is composed of an array with each of the two input signals being associated with one of the two linear polarizations) the perpendicular component is directed upward while the parallel component passes through the beamsplitter. Both components travel through λ/4 waveplates where the linear polarization is converted to circular. The light traveling upward is imaged onto patterned mirrors of mask M_1 as shown in part (b). This mask consists of an array of small mirrors that are located in the image plane of L_2 such that light reflected from them will be imaged onto the upper half of the rectangular window of the S-SEED array. Therefore, the image traveling upward reflects off M_1, passes again through a λ/4 waveplate which changes the polarization from circular to parallel allowing the light to be passed through the polarization beamsplitter onto the upper half of the S-SEED array.

The parallel component of the input image passes though the polarization beamsplitter, is imaged onto the small patterned mirrors of mask M_2, reflected by the polarization beamsplitter, and then imaged onto the upper half of the S-SEED array. Since these two inputs will have orthogonal circular polarizations they will not interfere.

The optical clock will enter the beam combination unit by being imaged onto the transparent region below the small mirrors of mask M_1. This array of high power spots will be imaged onto the bottom half of the S-SEED array. The reflected output signal of the S-SEED array, derived from the clock, will have its polarization rotated allowing it to be reflected by the polarization beamsplitter and then imaged onto the transparent region below the small mirrors of mask M_2. This output image can then be collected and used by an optical interconnect or another beam combination unit.

3. Optical Hardware Module

By combining all of the piece parts that have been discussed, an optical hardware module (OHM) can be created to provide the optical hardware necessary for each stage of a multi-

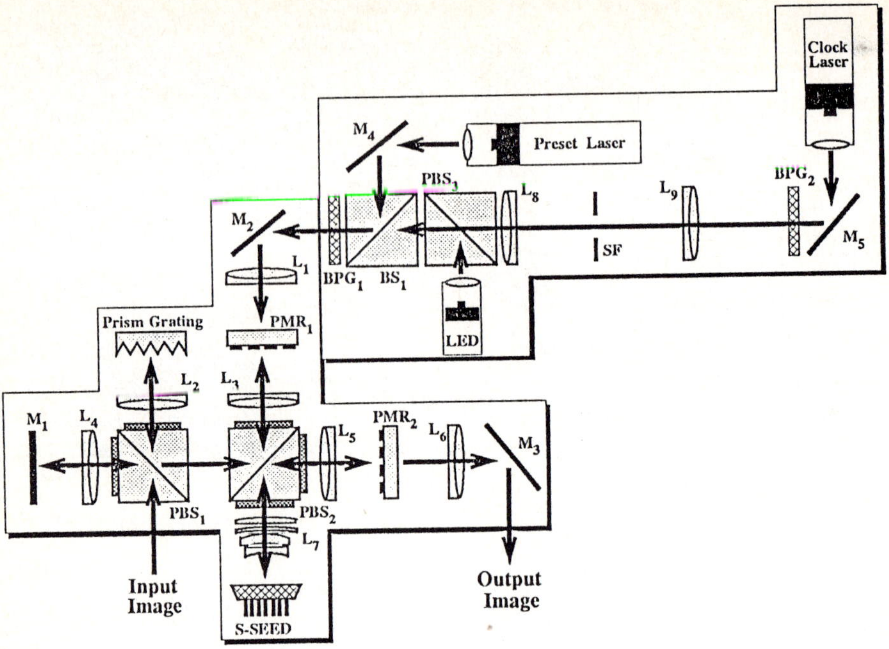

Figure 6. Optical Hardware Module

stage network. Figure 6 is an example of a possible OHM.[13] In this figure M_1, L_4, the polarizing beam splitter PBS_1, L_2, and the prism grating are the basic parts of the optical interconnect. The beam combination unit is composed of L_7, PBS_2, L_5, L_3, the patterned mirror reflectors PMR_1 and PMR_2, L_1, L_6, M_2, and M_3. The preset signals required for optical logic operation of the S-SEEDs are generated by the preset laser and then directed to the beam combination unit by M_2, the beamsplitter BS_1, and the binary phase grating BPG_1. The clock signal is generated by the clock laser and then distributed to the beam combiner by M_4, BPG_2, L_9, the spatial filter SF, L_8, PBS_3, BS_1, and BPG_1.

4. Free-Space Photonic Switching Fabrics

Using the OHMs that have been previously discussed, photonic switching fabrics can be implemented. Examples include blocking networks, such as the omega[14] and baseline networks,[15] and rearrangeably nonblocking networks such as the Ofman network.[16][17]

A free-space digital optical version of an Ofman rearrangeable non-blocking fabric based on (2,1,1) nodes and crossover interconnects is illustrated in part (a) of Figure 7. The (2,1,1) triplet represents the following: (Number of inputs, Number of outputs, Capacity of the node).[18] The first two parameters of the triplet represent the number of inputs and outputs, thus a (2,1,1) node has two inputs and one output. The third parameter indicates the number of channels that can be passed through the node at a given time. A logic diagram of a (2,1,1) node is shown in part (b) of Figure 7. The Ofman network was chosen since it guarantees that a node will never have two input signals present at the same time.[17] These networks require $2\log_2 N-1$ (2,1,1) nodes plus an initial fan-out array or $4\log_2 N - 1$ S-SEED arrays (OHMs).

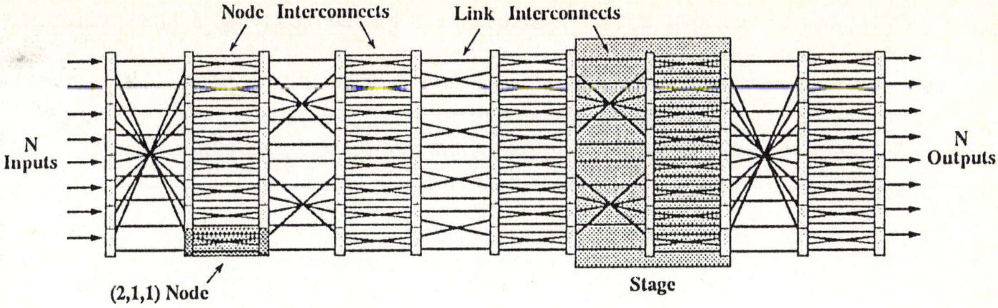

(a) Ofman Interconnection Network

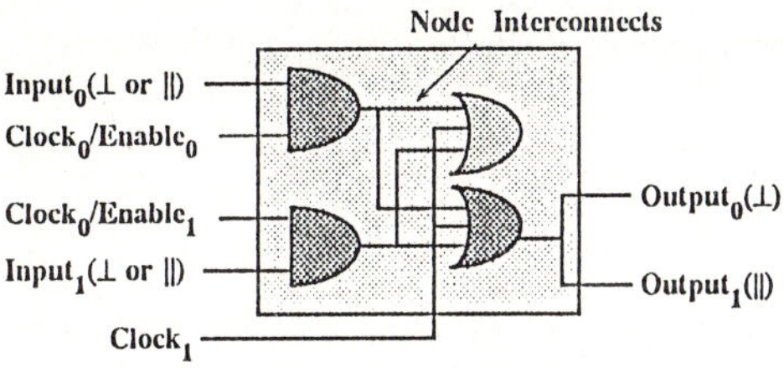

(b) Logic Equivalant of (2,1,1) Node

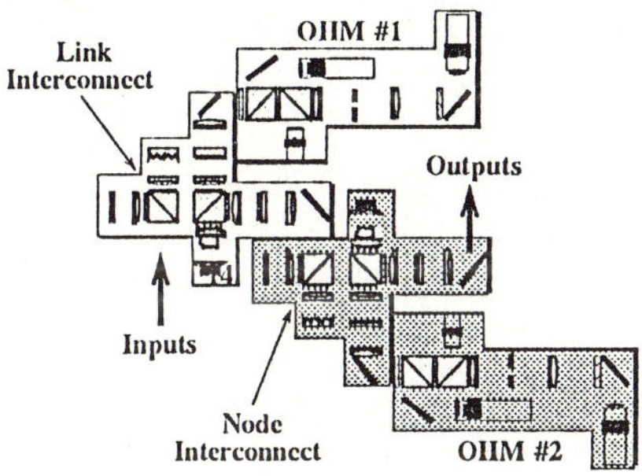

(c) Hardware Implementation of Stage

Figure 7. Ofman network using (2,1,1) nodes and crossover interconnects

In this example, there are eight inputs and eight outputs. The network is composed of six stages, where each stage contains eight nodes composed of two arrays of optical logic gates. The first logic gate array in each node will logically select which of the two input channels will be allowed to pass its traffic to the node output. The second logic gate array provides an OR function. The output of the node will have a fan-out of two, half of its energy in each of the two linear polarizations. A two-dimensional array of signals will enter each stage through a link interconnect. This interconnect will have a different period for each stage and provide the node-to-node connectivity required by the network topology. The crossover interconnect will implement this function providing the straight through connection to one of the polarizations while providing the crossover function for the other. This permuted 2-D array of signals will then pass through the beam combination unit onto the AND logic gates. The enable signal will allow only one of the two possible signals to pass through the node interconnect to the OR gate. Note that the node interconnect is a fine period crossover network. The output of the OR gate will then pass through the beam combination unit to the next stage of the fabric. A means of implementing the AND function is to use a S-SEED as a latch to capture and store the information for both inputs. By placing a SLM in one of the image planes of the spot array generator, the S-SEEDs receiving a clock pulse can be selectively controlled.[19] Using this clock control method, only the input AND gate containing the information to be moved to the output would be clocked. Thus, the paths through the network are controlled through the state of the SLM.

The chip-to-chip connections provided by the free-space digital optics will allow a 1024 x 1024 Ofman fabric to be created with only 39 S-SEED arrays (OHMs).

5. Conclusions

Free-space digital optics offers the opportunity for new advances in switching fabrics through the exploitation of the connections available using optical pin-outs. A small number of optical hardware modules consisting of S-SEED arrays, crossover interconnects, spot array generators, and beam combiners, can be concatenated to create large dimensional photonic switching fabrics.

REFERENCES

1. H. S. Hinton,"Architectural Considerations for Photonic Switching Networks," *Journal of Selected Areas of Communications,* Vol. 6, pp. 1209-1226, August 1988.

2. H. S. Hinton, "Relationship between the hardware required for photonic switching and optical computing based on free-space digital optics," *OSA Proceedings on Photonic Switching,* J. E. Midwinter and H. S. Hinton, eds. (Optical Society of America, Washington, D. C., 1989), Vol. 3, pp. 184-191.

3. A. L. Lentine et al. "Symmetric self-electro-optic effect device: Optical set-reset latch, differential logic gate, and differential modulator/detector" *IEEE Journal of Quantum Electronics,* Vol. 25, No. 8, pp. 1928-1936, August, 1989.

4. H. Dammann and K. Gortler, "High-efficiency in-line multiple imaging by means of multiple phase holograms," *Optics Communications,* Vol. 3, No. 5, July 1971, pp. 312-315.

5. H. Dammann and E. Klotz, "Coherent optical generation and inspection of two-dimensional periodic structures," *Optica Acta,* Vol. 24, No. 4, 1977, pp. 505-515.

6. U. Killat, G. Rabe, and W. Rave, "Binary phase gratings for star couplers with high splitting ratio," *Fiber and Integrated Optics,* Vol. 4, No. 2, 1982, pp. 159-167.

7. J. Jahns, N. Streibl, and S. J. Walker, "Multilevel phase structures for arrays generation," *OE/LASE'89,* Los Angeles, CA, 1989.

8. J. Jahns and M. J. Murdocca, "Crossover Networks and their Optical Implementation," *Applied Optics,* Vol. 27, pp. 3155-3160, August 1, 1988.

9. T. J. Cloonan, "Topological equivalence of optical crossover networks and modified data manipulator networks," *Applied Optics,* Vol. 28, No. 13, 1 July 1989, pp 2494-2498.

10. T. J. Cloonan et al. "An all-optical implementation of a 3-D crossover switching network," *To be published in IEEE Photonics Technology Letters.*

11. M. E. Prise et al. "Design of an optical computer," *Journal de Physique,* Colloque C2, Supplement au n°6, Tome 49, juin 1988.

12. F. B. McCormick and M. E. Prise, "Optical circuitry for free-space interconnections," *Applied Optics,* Vol. 29, May 1988.

13. F. B. McCormick et al. To be published at CLEO'90.

14. D. H. Lawrie, "Access and alignment of data in an array processor," *IEEE Transactions on Computers,* Vol. C-24, No. 12, December 1975, pp. 1145-1155.

15. C. Wu and T. Feng, "On a Class of Multistage Interconnection Networks," *IEEE Trans. Computers,* Vol. 29, No. 8, Aug.1980, pp. 694-702.

16. J. P. Ofman, "A universal automaton," *Trans. Moscow Math Soc., Vol. 14, 1965 (translation published by Amer. Math Soc., Providence, RI, 1967, pp. 200-215).*

17. C.-T. Lea, "Bipartite graph design principle for photonic switching systems," *IEEE Transactions on Communications,* April 1990.

18. E. Kerbis et al. "An all-optical realization of a 2 x 1 free-space switching node," *Submitted to IEEE Photonics Technology Letters,*

19. T. J. Cloonan et al. "A 3-D crossover switching network based on S-SEED arrays," *Proceedings of 1990 Topical Meeting on Photonic Switching,* Kobe, Japan, April 12-14, 1990.

Part II

Electrically Controlled Switches

Bandfilling or Stark Effect for Photonic Switching – A Comparison

O. Sahlén[1], *L. Thylén*[1], *A. Karlsson*[2], *and U. Olin*[3]

[1]Ericsson Telecom AB, Fibre Optic Research Centre, S-12625 Stockholm, Sweden
[2]Royal Institute of Technology, Microwave Technology, S-10044 Stockholm, Sweden
[3]Institute of Optical Research, S-10044 Stockholm, Sweden

Abstract. Bandfilling in bulk semiconductors is an attractive alternative to Stark shift in quantum wells for moderate-speed, space and wavelength waveguide switches, requiring large index changes.

1. Introduction

In waveguide devices relying on electrically-controlled refractive index changes the refractive index change Δn that can be induced in the material should be large while the absorption coefficient α is low. We compare the two mechanisms known to give the largest Δn in semiconductors at wavelengths close to the bandgap: Bandfilling via carrier injection and Quantum Confined Stark Effect (QCSE). The calculations are done for bulk GaAs and MQWS-GaAs/$Al_{0.6}Ga_{0.4}As$ (100Å/100Å) respectively, but the conclusions are qualitatively applicable also to InGaAsP and MQWS InGaAs/InP. Finally, experimental data on bandfilling in 1.55 μm bulk InGaAsP are presented.

We require that $\Gamma \cdot \alpha \cdot L \leq 1$ for all applied voltages/currents (L = the waveguide length, Γ = irradiance confinement factor). The waveguide could be a building block in space switches (e.g. directional coupler) or wavelength switches (tuneable Bragg filter or reflector). For dense integration, L should be kept small. However, L=100-200 μm is the minimum conceivable length in most cases.

2. Theoretical models and results

QCSE is computed using a model [1] similar to [2]. However, the first "forbidden" transition (hh2 -> e1) is also included, and the exciton lineshape is described by a function that is Gaussian near the peak, but exponential at small energies (far from the exciton peak) in order to describe the Urbach tail (found also in MQW material [3]). The polarisation is parallel to the wells.

Carrier injection effects are computed with a model [4] based on the theory in [5]. It includes bound exciton states, continuum enhancement, carrier-dependent line broadening, screening, bandfilling and bandgap renormalisation. Since we only are interested in the behaviour close to the bandgap, the free-carrier intra-band effect (about 10 times smaller [6]) is neglected. The material is assumed to be undoped which is essential in order to minimize band tailing and maximize $\Delta n / \alpha$.

Springer Series in Electronics and Photonics, Vol. 29 **Photonic Switching II**
Editors: K. Tada · H.S. Hinton © Springer-Verlag Berlin, Heidelberg 1990

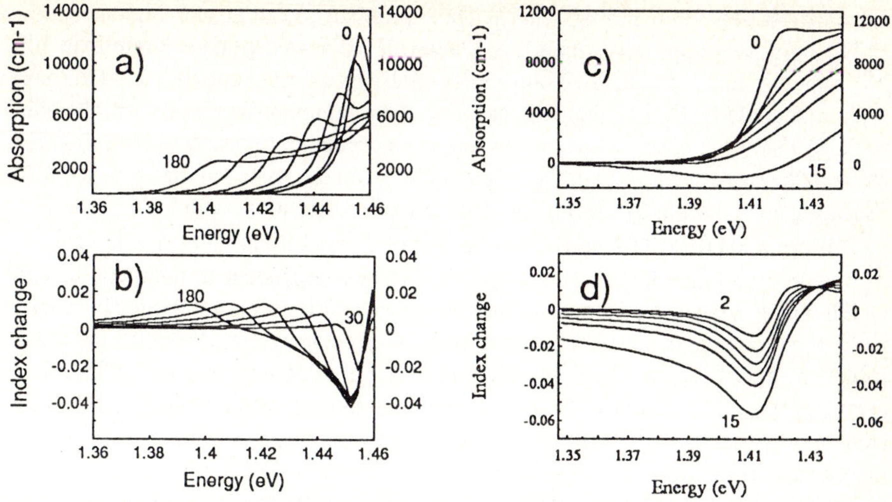

Figure 1a: Stark shift of absorption spectrum for different applied electric fields in MQW. The field increases from right to left as 0, 30, 60, 90, 120, 150 and 180 kV/cm. ***b****: Corresponding index changes.* ***c****: Absorption spectra in bulk for different carrier densities in the current injection case. For decreasing absorption the density varies as 0, 2, 4, 6, 8, 10 and 15 • 10^{17} cm^{-3}.* ***d****: Corresponding index changes.*

Figure 1 shows calculated QCSE and bandfilling data. The mode absorption is $\alpha_{mode} = \Gamma \cdot \alpha$, and confinement factors of $\Gamma = 0.5$ allows one to write $\Delta n_{eff} \approx \Gamma \cdot \Delta n$. A detailed examination of the data reveals that the maximum Δn is about 0.007 if $\alpha <$ 100 cm^{-1} in the QCSE case, and can be increased to 0.010 if $\alpha <$ 500 cm^{-1} is allowed. (Above 150 kV/cm, $\Delta n / \alpha$ starts to saturate, so further increase in field beyond 180 kV/cm will not help.) In the bulk bandfilling case, Δn = 0.01 can be achieved at a density of 8•10^{17} cm^{-3} while $-35\,cm^{-1} < \alpha < 100\,cm^{-1}$. Increasing the current slightly, the maximum Δn is about 0.01 at a carrier density of 1•10^{18} cm^{-3} under the condition $|\alpha| < 100\,cm^{-1}$. If, however, a somewhat larger gain is allowed, $-150\,cm^{-1} < \alpha < 100\,cm^{-1}$, an index change $\Delta n = 0.02$ can be obtained at this density by operating with wavelengths closer to the bandgap.

3. Discussion

The index changes that can be induced under the condition that neither the gain nor the absorption is larger than about 100 cm^{-1} are comparable, $\Delta n \approx 0.01$, though somewhat larger for bandfilling. To make e.g. a directional coupler, a phase shift $\Delta\Phi = 2 \cdot \pi \cdot \Delta n \cdot \Gamma \cdot L / \lambda = \sqrt{3} \cdot \pi$ is sufficient, so that with $\Gamma = 0.5$ and L = 200 μm, $\Delta n \approx$ 0.007 is required. To make wavelength tuneable filters and lasers, the maximum possible value of $\Gamma \cdot \Delta n$ is desired. A target value is $\Delta n_{eff} / n_{eff} \approx 0.01$, allowing a wavelength range of over 10 nm to be continuously accessed.

For such large $\Delta n > 0.02$ only bandfilling is appropriate. Larger index changes are possible if the gain is allowed to increase ($\Delta n \approx 0.05$ at a gain of about 1000 cm^{-1}

and a carrier density of $1.5 \cdot 10^{18}$ cm^{-3}). Three problems occur at such high densities: 1) The noise output power from a $L = 100$ μm, $\Gamma = 0.5$ waveguide is larger than 100 μW, which is unacceptable. This noise is spectrally broad and centred around the gain peak, and could be filtered and further reduced by detuning the signal from the gain peak. 2) For some specific switch types it is unacceptable to decrease the refractive index in the same arm as one has gain, particularly for the so-called digital switch [7]. 3) High gain can bring the device into oscillation if reflections occur.

If Δn = 0.005 - 0.01 suffices, what is best? Bandfilling can provide gain to overcome waveguide losses. It is also polarisation independent in bulk, contrary to QCSE which is intrinsically polarisation sensitive. Stark shift is not limited by carrier recombination time (a few ns) but rather by RC time constants, and hence allows much faster operation. If picosecond switching is not required, bandfilling is preferable. In many applications a few ns set-up time is tolerable. (Faster operation using bandfilling can be obtained in barrier reservoir and quantum well electron transfer structures, at the expense of reduced Δn compared to current injection in a similar quantum well structure, and of not achieving gain nor polarisation independence [8].)

4. Experimental results of bandfilling in InGaAsP

The comparison was made for GaAs/AlGaAs, since most material parameters are well known. (Some uncertainty lies in the Urbach parameters. We have probably underestimated the absorption tail in the QCSE case at high electric fields, by using a constant, field-independent Urbach parameter of 7 meV. This means that carrier injection could be even more favourable than shown by our results.) However, the conclusions are qualitatively also applicable to InGaAsP/InP. We have measured index changes up to 100 nm above the bandgap wavelength in a 1.53 μm laser, by determining the shift in the Fabry-Perot peaks for different injection currents and wavelength regions, measuring the luminescence with a high-resolution, high-sensitive and averaging optical spectrum analyser. Figure 2 shows that a material index change of 0.025

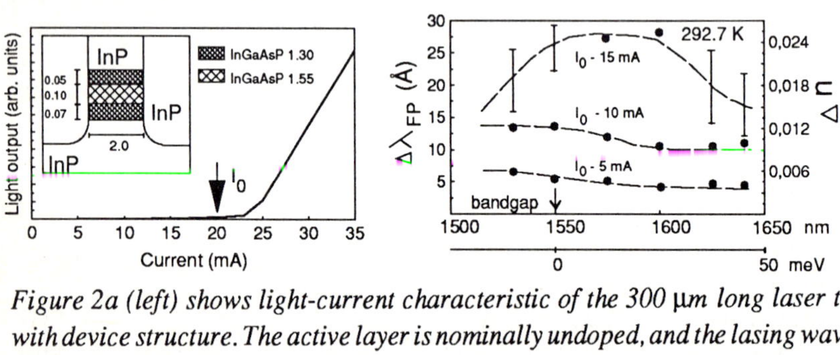

Figure 2a (left) shows light-current characteristic of the 300 μm long laser together with device structure. The active layer is nominally undoped, and the lasing wavelength is 1.53 μm. Figure 2b (right) shows Fabry-Perot shifts ($\Delta\lambda_{FP}$) when the current is decreased from the reference current I_0 = 20 mA. The right axis shows the corresponding (material) refractive index change $\Delta n \approx \Delta n_{eff} / \Gamma$. The dashed lines are guides for the eye only. (The laser was fabricated by the Swedish Institute of Microelectronics in Kista, Sweden.)

($\Delta n \approx \Delta n_{eff} / \Gamma$, $\Gamma = 0.23$) is found when changing the current from 5 mA to 20 mA for wavelengths 30 - 50 nm above the bandgap. Large index changes are also found up to 100 nm (50 meV) above the gap, the longest wavelength at which we could measure any luminescence.

5. Conclusion

Bandfilling through carrier injection in bulk is superior to QCSE in MQW in many moderate-speed applications requiring large $\Delta n / \alpha$. Large $\Delta n \approx 0.025$ due to bandfilling have been measured in 1.55 μm InGaAsP waveguides for wavelengths 50 nm above the bandgap wavelength.

6. References

1 . A. Karlsson, unpublished.

2 . P. J. Stevens, M. Whitehead, G. Parry and K. Woodbridge, IEEE J. Quant. Electron., **QE-24**, 2007 (1988).

3 . A. Kost, H. C. Lee, Y. Zou, P. D. Dapkus and E. Garmire, Appl. Phys. Lett., **54**, 1356 (1989).

4 . U. Olin, Institute of Optical Research, Stockholm, Technical Report TR 207 (1989).

5 . L. Banyai and S. Koch, Z. Physik B. **63**, 283 (1986).

6 . J. G. Mendoza-Alvarez, L. A. Coldren, A. Alping, R. H. Yan, T. Hausken, K. Lee and K. Pedrotti, J. Lightwave Techn., **LT-6**, 793 (1988).

7 . P. Granestrand, B. Lagerström, P. Svensson, L. Thylén, B. Stoltz, K. Bergvall, J.-E. Falk and H. Olofsson, Electron. Lett., **26**, 4 (1990).

8 . J. E. Zucker, T. Y. Chang, M. Wegener, N. J. Sauer, K. L. Jones and D. Chemla, IEEE Photon. Techn. Lett., **2**, 29 (1990).

Electrorefractive Waveguide Devices Using Tunable Electron Density Quantum Wells

J.E. Zucker, M. Wegener, K.L. Jones, T.Y. Chang, N. Sauer, and D.S. Chemla

AT&T Bell Laboratories, Room 4F 319, Holmdel, NJ 07733, USA

Abstract. We demonstrate a novel tunable-electron-density electrorefractive multiple-quantum-well waveguide with high speed capability. 5.4 V produces 180° phase shift in a 650 μm long device.

Most bulk semiconductor waveguide switches have employed either current injection [1,2] or the linear electro-optic Pockels effect [3] to produce a refractive index change in the material. In the case of current injection, large index changes Δn can be achieved. Thus the device length needed to produce 180° phase shift $L \equiv \lambda / 2\Delta n$ can be quite compact. However, the speed of response is limited to $\sim$ 50 MHz by the carrier recombination time [2]. Pockels effect devices have a speed advantage in that response time is limited only by the RC time, which can allow $>$ 20 GHz operation. However the magnitude of the Pockels effect in semiconductors such as InP is relatively weak, forcing device lengths to be on the order of several millimeters.

In this paper we present a novel semiconductor waveguide heterostructure that has large electrorefraction associated with free electrons but avoids speed limitations imposed by electron-hole recombination. In the Barrier Reservoir And Quantum Well Electron Transfer (BRAQWET) device, we accurately control the number of electrons inside multiple InGaAs quantum wells by means of an applied voltage. In the current structure, five BRAQWET periods form the core of an n-i-n InAlAs waveguide. Each period contains a 500 Å InGaAlAs electron reservoir, a 210 Å InGaAlAs spacer, and a 90 Å InGaAs quantum well (QW) followed by a 600 Å InAlAs barrier. At zero bias each well is empty and hence has high absorption. As electrons are synchronously transferred in, the absorption is bleached yielding an effective blue-shift of the absorption edge. The *intrinsic* response time is under 20 psec and is limited by the speed of electron transfer across the spacer between well and electron reservoir layers. In practice, the actual device speed is determined by the RC time. Refractive index changes associated with the electron transfer process are linear in the applied voltage and large: the index change per carrier is of the same order as that obtained by current injection [4,5,6].

Fig. 1 shows Δn as a function of field measured with incident light propagating normal to the QW layers. The field shown is the applied

Springer Series in Electronics and Photonics, Vol. 29 **Photonic Switching II**
Editors: K. Tada · H.S. Hinton © Springer-Verlag Berlin, Heidelberg 1990

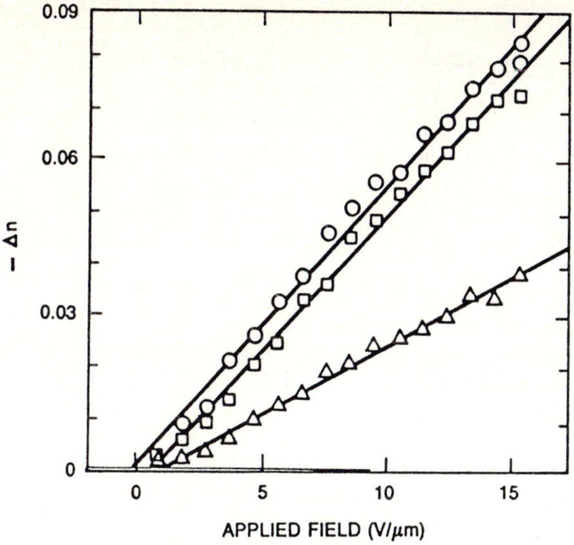

Fig. 1 Refractive index change as a function of applied field measured in normal incidence at wavelengths 1.56 μm (circles), 1.575 μm (squares), and 1.59 μm (triangles).

voltage divided by the thickness of the undoped region, 0.51 μm. Δn is obtained from the field-induced optical phase change $\Delta\phi = 2\pi\Delta nl/\lambda$ that is produced by the total thickness of active QW material l=450 Å and measured with the sample in one arm of a Mach-Zehnder interferometer. The index change per applied field $\Delta n/E$ for the BRAQWET structure is significantly larger than $\Delta n/E$ associated with the Quantum Confined Stark Effect (QCSE) in InGaAs QWs. For example, at 1.59 μm (a wavelength detuned $\sim$ 30 meV from the bandedge) Fig.1 shows constant $\Delta n/E = 2.7\times10^{-7}$cm/V. This value is 3.8 times that obtained via the QCSE in InGaAs QWs at the same detuning and a field 5 V/μm.

In addition to larger $\Delta n/E$, a significant advantage of BRAQWET over QCSE for electrorefractive devices is that a large ratio of phase to intensity modulation can be obtained at wavelengths just below the bandedge. This feature is a consequence of the the blue-shift of the absorption edge with applied bias in the BRAQWET structure. As shown in Fig.2, at 1.59 μm the ratio of Δn to Δk, the field-induced change in extinction coefficient, is $>$ 10 for field 10 V/μm. By contrast, the corresponding $\Delta n/\Delta k$ associated with the red-shifting QCSE is only $\sim$2.

For waveguiding devices, the large index change of the QW material is diluted by the optical confinement factor Γ that describes the overlap between the mode and the 5 QWs. We calculate $\Gamma = 1.3\times10^{-2}$ for the central well, and 60 % smaller for the outermost wells. Therefore the index change in the waveguide is only five percent of that measured in normal incidence. Nevertheless, we obtain efficient phase modulation in very

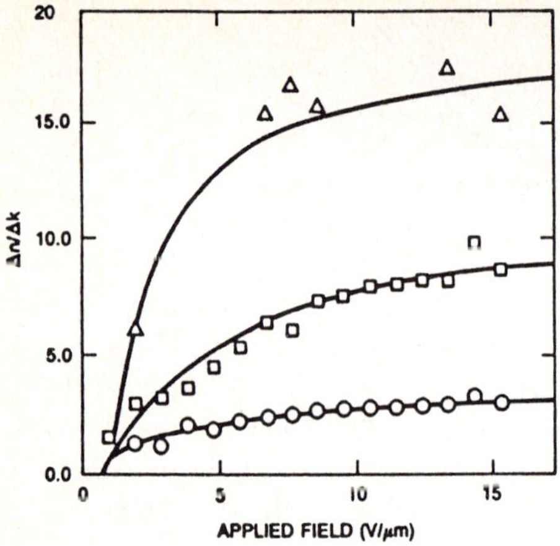

Fig.2 Ratio of change in refractive index to change in extinction coefficient as a function of applied voltage measured in normal incidence for same wavelengths as in Fig.1.

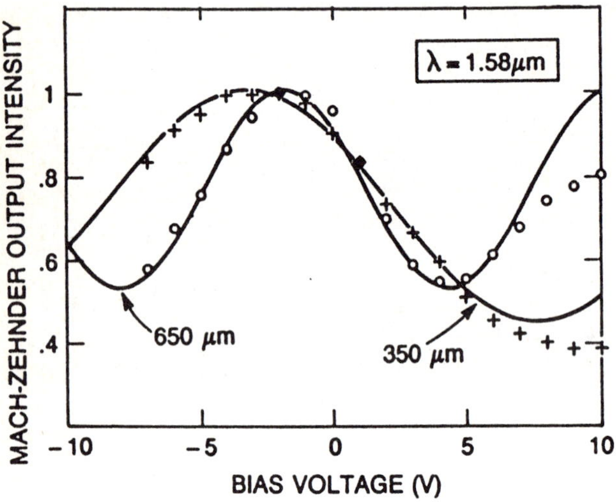

Fig. 3 Output intensity of Mach-Zehnder rib waveguide interferometer as a function of applied voltage at 1.58 μm wavelength for two different active lengths, 650 and 350 μm.

compact waveguide device lengths. To monitor the phase modulation in the waveguide geometry, we formed rib-waveguide Mach-Zehnder interferometers by wet chemical etching of the wafer. Fig. 3 shows the output intensity at 1.58 μm as a function of applied voltage for interferometers with active device lengths L=650 and 350 μm. These show the cosine-like behavior that is expected for a linear electro-optic effect, i.e. the product $V_\pi \times L$ = constant, where V_π is the half-wave voltage. For

L=650 μm, V_π=5.4 V for single arm drive voltage. When both sides of the interferometer are operated in push-pull configuration, the voltage is decreased by a factor of 2. No damping of the output characteristic with increased bias is observed, as is the case for Mach-Zehnder devices that utilize the QCSE, indicating that electroabsorption effects are negligible.

In summary, we present a novel multiple QW waveguide structure that has both high-speed potential and large electrorefractive effects. The voltage-length product in the structure described here is $V_\pi \times L$=3.5 V-mm. By increasing the filling factor of the waveguide, we can achieve $V_\pi \times L \sim 1$ V-mm. In addition to the suitability for high-density monolithic integration, the compact device length is also an advantage in terms of device speed: with device capacitance under 1 picofarad, we have attained multi-gigahertz modulation bandwidths in the BRAQWET waveguide.

REFERENCES

[1] K. Ishida, H. Nakamura, H. Matsumara, T. Kadoi, and H. Inoue, Appl. Phys. Lett. 50, 141-142 (1987).

[2] F. Hernandez-Gil, T. L. Koch, U. Koren, R. P. Gnall, and C. A. Burrus, Electron. Lett. 25, 1271 (1989).

[3] K. Komatsu, M. Sugimoto, A. Ajisawa, and A. Suzuki, in *OSA Proceedings on Photonic Switching,* J. E. Midwinter and H. S. Hinton, eds. (Optical Society of America, Washington, D. C., 1989), Vol.3,pp. 92-95.

[4] J. E. Zucker, T. Y. Chang, M. Wegener, N. J. Sauer, K. L. Jones and D. S. Chemla, Photonics Technol. Lett. 2, 29 (1990).

[5] M.Wegener, J. E. Zucker, T. Y. Chang, N. J. Sauer, K. L .Jones, and D. S. Chemla, Phys. Rev. B 41, 3097 (1990).

[6] J. E. Zucker, M. Wegener, K. L. Jones, T. Y. Chang, N. Sauer, and D. S. Chemla, Appl. Phys. Lett. May 14, 1990.

GaAs/AlGaAs Reflection-Type Optical Switch Using Heterojunction Bipolar Transistor Waveguide Structure

Y. Okada and K. Tada

Department of Electronic Engineering, University of Tokyo,
7-3-1 Hongo, Bunkyo-ku, Tokyo 113, Japan

Abstract. Theoretical analysis of the optical switching characteristics expected for a GaAs/AlGaAs carrier-injected reflection-type optical switch employing a double-heterojunction bipolar transistor waveguide structure is presented on the basis of the experimentally determined results. It has been shown that an extinction ratio of better than 12 dB could be obtained at the input base current of as small as 1.5 mA for a X-crossing optical switch with an intersecting angle of 4°.

1. Introduction

For the past few years, research activities on efficient semiconductor optical modulators and switches have been increasing as these devices are fundamentally important components in the areas of optical transmission, photonic switching and signal processing, and photonic integrated circuits. We have been exploring possible applications of bipolar transistor waveguide structures for realization of high-speed, compact size, and monolithic integration of semiconductor optical modulators and switches operated by free carrier injection [1], [2].

Recently, we have reported the first experimental results obtained with a GaAs/AlGaAs double-heterojunction bipolar transistor (DHBT) waveguide structure carrier-injected optical intensity modulator grown by molecular beam epitaxy (MBE) [3]. In this work, the experimental data obtained with the above device have been applied to design and analyze the optical switching characteristics expected for a DHBT waveguide structure reflection-type X-crossing optical switch [4]–[6] by using a beam propagation method (BPM).

2. Design and Analysis

A schematic illustration of the DHBT waveguide structure carrier-injected X-crossing optical switch considered is shown in Fig. 1. The

Springer Series in Electronics and Photonics, Vol. 29 **Photonic Switching II**
Editors: K. Tada · H.S. Hinton © Springer-Verlag Berlin, Heidelberg 1990

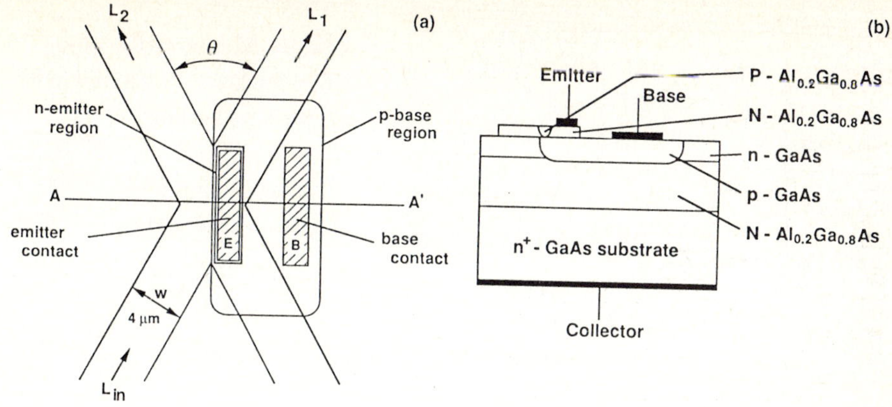

FIG. 1 A schematic of GaAs/AlGaAs DHBT structure reflection-type optical switch; (a)overview, and (b) A-A' cross-section.

layer parameters as shown in Fig. 1(b) are almost identical to the structure previously reported in Ref. 3, and the 0.25 μm-thick p^+-GaAs layer doped to 1×10^{18} cm^{-3} functions as the base of an N-p-N HBT structure, as well as a single-mode optical waveguide. When the free carriers are injected into the base/waveguide of this transistor, the refractive index in this region reduces for a properly selected wavelength, primarily because of the carrier-induced band-filling and plasma dispersion effects [2]. As a final consequence, the incident light power can be switched from the transmission port L_1 to the opposite reflection port L_2 by total internal reflection.

As is reported in Ref. 3, typical common-emitter dc current gain h_{FE} of the HBT measured was 38. Also, the estimated transit time τ_T of the diffusing minority carriers (electrons) across the 0.25 μm-thick base was about 9 ps, and the switching rise time t_r of the DHBT structure in the common-emitter mode for a 50 Ω load circuit was about 1.5 ns, which had been limited by the parasitic impedance, the main contribution of which was due to the emitter contact resistance. Free-carrier injection has modulated the absorption coefficient α in the waveguide and consequently, an optical on/off modulation ratio of up to 2.1:1, corresponding to a modulation efficiency of $\Delta\alpha / \alpha \approx 89\%$ has been demonstrated at the injected base current of 3 mA as shown in Fig. 2, for a device with a waveguide width of 7 μm and a length of 190 μm. Assuming Stern's absorption model for the base/waveguide layer [2], the variation of the real refractive index $\Delta\mu$ with injected current was then determined by taking a Kramers-Kronig transformation of the absorption variation $\Delta\alpha$. The results are plotted in Fig. 3, and it can be

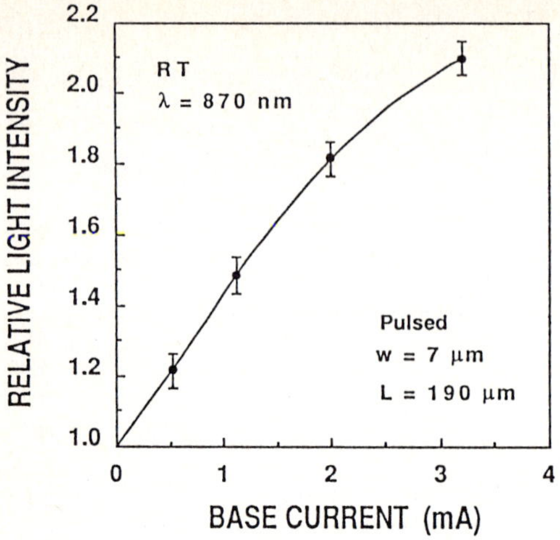

FIG. 2 Measured optical output power variation with current injection.

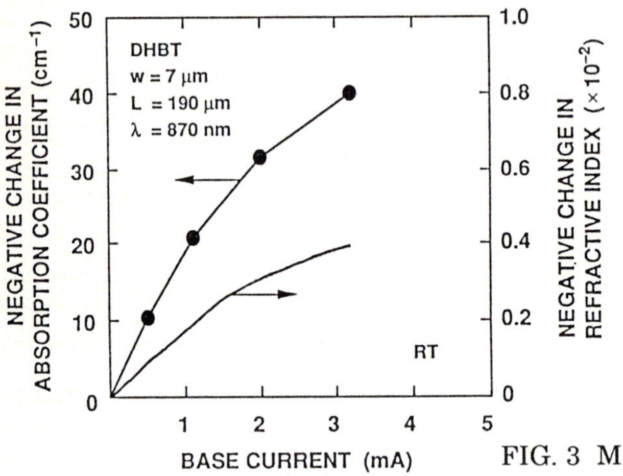

FIG. 3 Measured $\Delta\alpha$ and $\Delta\mu$.

observed that $\Delta\mu$ varies in relative proportion with $\Delta\alpha$, and $\Delta\mu \approx -0.4 \times 10^{-2}$ was obtained at the base current of $I_b = 3$ mA or equivalently, at the injected carrier density of $\Delta n \approx 3 \times 10^{17}$ cm^{-3}.

The above results were then used to calculate the optical switching characteristics expected for the reflection-type X-crossing optical switch as in Fig. 1 by using beam propagation method (BPM) analysis. For the computation, the following was assumed; the waveguide width of w = 4 μm, light wavelength of λ = 870 nm, switch length of L = 110 μm, and the identical electrical transistor characteristics as discussed in Ref. 3. The result obtained for an intersecting angle of $\theta = 4°$ is shown in Fig. 4. It can be observed that the amount of the reflected optical power L_2 could be controlled by the base current, and an extinction ratio of better than 12 dB could be achieved at $I_b \approx$ 1.5 mA. The common-emitter mode

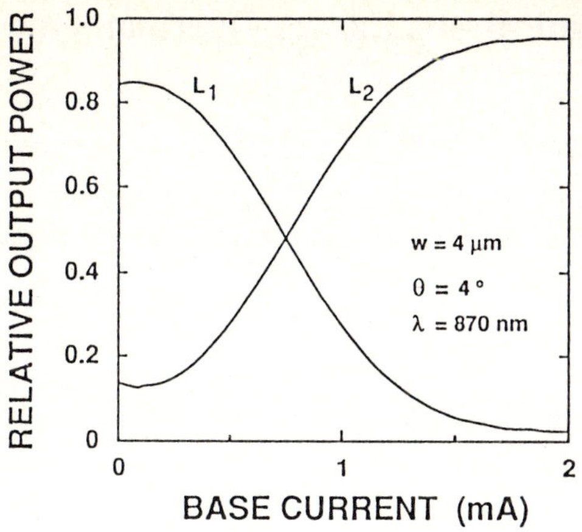

FIG. 4 BPM simulation results.

switching time for the given geometry could be estimated as about 0.9 ns had the parasitic impedance been reduced to a minimum.

The results are therefore, very promising for realizing compact size DHBT waveguide structure carrier-injected reflection-type optical switches with fast switching speeds and high extinction ratios.

References

[1] K. Tada and Y. Okada, IEEE Electron Dev. Lett., **EDL–7**, 605 (1986)
[2] Y. Okada, R-H. Yan, L. A. Coldren, J. L. Merz, and K. Tada, IEEE J. Quantum Electron., **25**, 713 (1989)
[3] Y. Okada, T. Ishikawa, and K. Tada, Appl. Phys. Lett., **55**, 2591 (1989)
[4] O. Mikami and H. Nakagome, Electron. Lett., **20**, 228 (1984)
[5] K. Ishida, H. Nakamura, H. Matsumura, T. Kadoi, and H. Inoue, Appl. Phys. Lett., **50**, 141 (1987)
[6] T. Kikugawa, K. G. Ravikumar, K. Shimomura, A. Izumi, K. Matsubara, Y. Miyamoto, S. Arai, and Y. Suematsu, IEEE Photonics Tech. Lett., **1**, 126 (1989)

Insertion Loss Reduction of InP Based Carrier-Injection-Type Optical S^3 Switch

T. Kirihara[1], *S. Kashimura*[2], *H. Inoue*[1], *H. Sano*[1], *Y. Sasaki*[1], *and K. Ishida*[1]

[1]Central Research Laboratory, Hitachi Ltd., Kokubunji, Tokyo 185, Japan
[2]Cable Research Laboratory, Hitachi Cable Ltd., Hitachi, Ibaraki 319-14, Japan

Abstract. An 11dB insertion loss in a carrier-injection type 4 x 4 optical single-slip structure (S^3) switch array is demonstrated by optimizing the waveguide structure and anti-reflection coating. To reduce further insertion loss, a novel switch structure that integrates an optical traveling-wave amplifier in the optical S^3 switch is also proposed.

1. Introduction

The optical switch is one of the key components for future photonic switching systems. We have been developing an InP based carrier-injection type optical single-slip structure (S^3) switch which has suitable features for large scale integration, polarization-independence and wavelength-insensitivity. However, the insertion loss previously reported for this switch was more than 20dB [1]. This value is not sufficient when applied to practical photonic switching systems.

Here, recent results of insertion loss reduction are presented. An 11dB insertion loss in a carrier-injection type 4 x 4 optical S^3 switch is obtained by optimizing the waveguide structure and anti-reflection coating. A carrier-injection type optical S^3 switch with traveling wave amplifier (COSTA) is also proposed for further insertion loss reduction.

2. Insertion loss reduction of 4 x 4 optical switch array

Schematic views of a 4 x 4 optical switch array and a magnified unit cell of an S^3 optical switch are shown in Fig. 1. Sixteen S^3 optical switches are integrated on one chip. Insertion loss of this switch can be classified into 5 factors ; waveguide propagation loss, mode-mismatching loss, Fresnel reflection loss, bending loss and branching loss. A loss reduction effort was primarily targeted for the first 3 factors. The fabrication process and the device structure are almost the same as was previously reported [1]. Improvements of the device are as follows. 1) A 1μm buffer InP layer (n~$2\times10^{16}cm^{-3}$) was inserted between the InP substrate (n~$2\times10^{18}cm^{-3}$) and the InGaAsP guiding layer to decrease free carrier absorption in the InP substrate. 2) The InGaAsP guiding layer thickness was increased from 1μm to 2μm to improve waveguide confinement and mode-mismatching between the fiber and waveguide 3) Waveguide facets received an anti-reflection coating.

Obtained switching characteristics of the 4 x 4 optical switch array are summarized in Table 1. A 1.31μm BH laser diode input light source and a

Springer Series in Electronics and Photonics, Vol. 29 **Photonic Switching II**
Editors: K. Tada · H.S. Hinton © Springer-Verlag Berlin, Heidelberg 1990

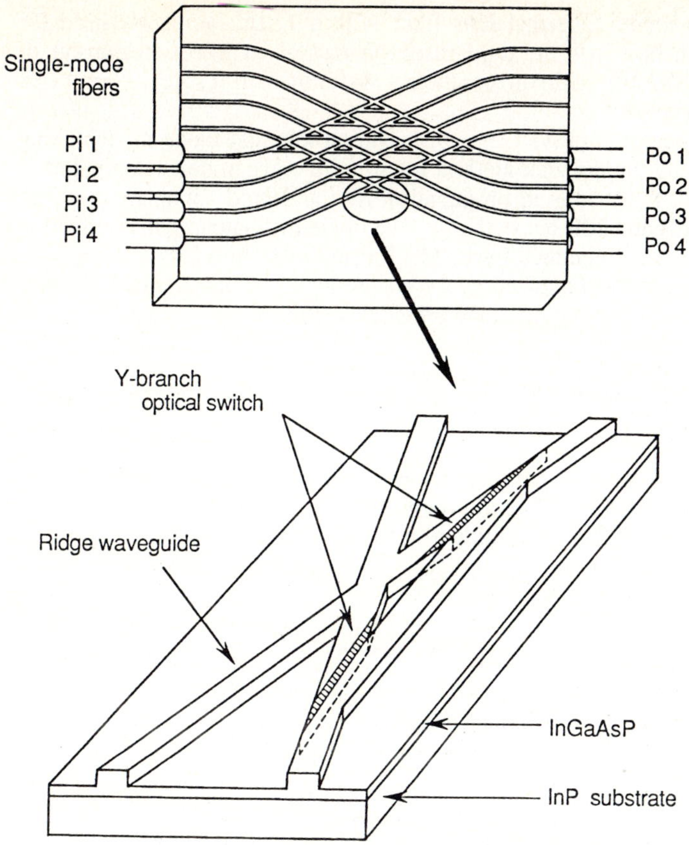

Fig. 1. Schematic view of the single-slip structure (S^3) optical switch as a unit cell of nonblocking 4 x 4 optical switch array.

TABLE 1 SWITCHING CHARACTERISTICS OF 4 X 4 OPTICAL SWITCH

in \ out	Po 1	Po 2	Po 3	Po 4
Pi 1	11.0 (23.2)	15.4 (8.7)	16.8 (12.5)	16.9 (16.8)
Pi 2	14.2 (15.7)	15.6 (15.9)	14.6 (20.6)	18.7 (23.6)
Pi 3	16.1 (15.5)	16.5 (17.5)	15.7 (15.8)	16.2 (14.7)
Pi 4	13.6 (16.2)	15.7 (20.8)	17.7 (24.7)	17.1 (21.0)

Insertion loss ,dB
(ON/OFF ratio ,dB)

conventional single-mode spherical lens fiber with a 12μm radius are used for measurement. Each box lists the total insertion loss at an injection current of 200mA, with the ON/OFF ratio in brackets. As shown in Table 1 , an 11dB insertion loss is achieved.

To classify the above improvements, propagation loss and mode-mismatching loss were measured by the cutback method for straight waveguides. Compared to the previous sample, reductions in propagation loss and mode-mismatching loss were confirmed at 4dB and 2dB respectively. The mode-mismatching loss reduction is consistent with results calculated by BPM. Fresnel reflection loss, estimated at 1.5dB for each facet, is almost avoided by the anti-reflection coating. Finally, a 9dB loss reduction is attained comparing to the previous result.

3. Carrier-injection type optical S^3 switch with traveling-wave amplifier(COSTA)

For further insertion loss reduction, a carrier-injection type optical S^3 switch with traveling wave amplifier (COSTA) is proposed [2] .

A schematic view of COSTA is shown in Fig. 2. A traveling-wave amplifier is integrated to the slip waveguide of the S^3 switch. This structure has the

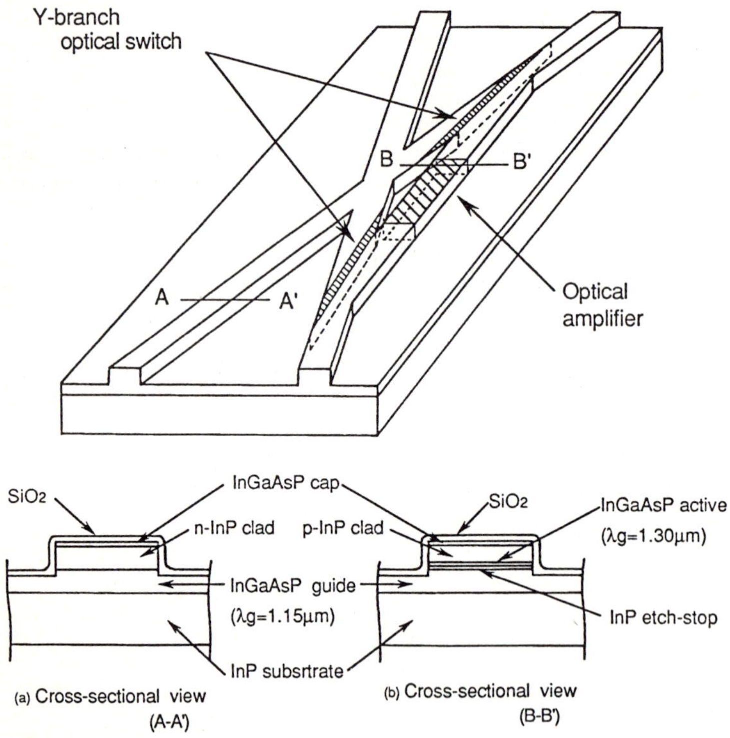

Fig. 2. Schematic view of Carrier-injection type Optical S^3 switch with Traveling-wave Amplifier (COSTA). (a) Schematic cross-section of waveguide region (A-A'). (b) Schematic cross-section of traveling-wave amplifier region (B-B').

advantages of not only insertion loss reduction but also ON/OFF ratio improvement. When both optical amplifier and switch are in the ON state, the switched signal is amplified and transmitted to the output port. On the contrary, when both of them are in the OFF state, the cross talk due to the signal leakage via the slip waveguide is absorbed by the integrated amplifier. Another advantage of the integrated amplifier is compensation of loss fluctuation in switch array characteristics.

Evanescent coupling is used for amplification. This traveling-wave amplifier structure is shown in Fig. 2.(b). An InGaAsP (λg=1.30μm) active layer is loaded on the common guiding layer. This structure has the following advantages. 1) Polarization-independence in amplification. 2) Coupling loss reduction at the waveguide and amplifier interface. 3) Reflection reduction in the active layer, acting as a traveling wave amplifier. Since optical field is almost confined in guiding layer, the Γ-factor of the active layer is small. However, a gain of more than 10dB is expected from calculations where amplifier length is estimated at 500μm, injection current at 150mA, and Γ=0.065.

To confirm the fundamental function, an amplifier was integrated into the 5μm wide ridge waveguide. A preliminary fabricated sample shows as much as a 4dB gain, compared to the structure without an amplifier, a 29dB of ON/OFF ratio, and less than 1dB polarization-dependance for a 1.3μm incident light. These results suggest the COSTA has the potential to realize a small size but large capability optical switch array for photonic switching.

4. Conclusion

In conclusion, an 11dB insertion loss in a 4 x 4 carrier-injection type optical switch array was obtained. A 0dB insertion loss 4 x 4 optical switch array is expected to integrate with a traveling wave amplifier.

References.

1. H.Inoue, H.Nakamura, K.Morosawa, Y.Sasaki, T.Katuyama, and N.Chinone, An 8 mm Length Nonblocking 4 x 4 Optical Switch Array, IEEE J. Sel. Areas in Comm., vol. 6, No. 7, pp. 1262-1266, 1988.
2. H.Inoue, T.Kirihara, Y.Sasaki, and K.Ishida, Carrier-Injection Type Optical S^3 Switch with Traveling-wave Amplifier, IEEE Photonics Tech. Lett., Vol. 2, No. 3, pp. 214-215, 1990.

Design and Performance of Monolithic LD Optical Matrix Switches

S. Oku, K. Yoshino, M. Ikeda, M. Okamoto, and T. Kawakami

NTT Opto-electronics Laboratories, 3-1 Morinosato Wakamiya, Atsugi-shi, Kanagawa-ken 243-01, Japan

Monolithic integrated LD optical matrix switches are fabricated in the form of a fiber pigtailed module. Half-mirror type large angle optical branching/combining circuits are found to be suitable for their construction. The LD matrix switches are fabricated using a Br_2 dry etching process on an InGaAsP/InP wafer. The 2×2 full switching operation is demonstrated.

1. Introduction

Optical matrix switches are key devices for future optical communication systems. Laser diode optical switches (LDSWs), proposed by the authors[1,2], have a gain in the ON state and a large isolation level even for a small device length[3]. A combination of the LDSWs and large angle optical branching/combining circuits[4] enable one to construct considerably compacter integrated optical matrix switches. In this paper, we describe the network designs and the large angle branching/combining circuitries for the configuration of the LD matrix switches. The full switching operation of a monolithic half-mirror type 2×2 LD matrix switch is demonstrated for the first time.

2. Network Design of LD Optical Matrix Switches

Optical switch networks are required to have several functions, such as a non-blocking[5] and 1-to-N connection. In constructing LD matrix switches having these functions, the total number of LDSWs and chip size must be considered. There are two types of network design. One is a link connection type and the other is a tree connection type, as shown in Figs. 1 (a) and (b).

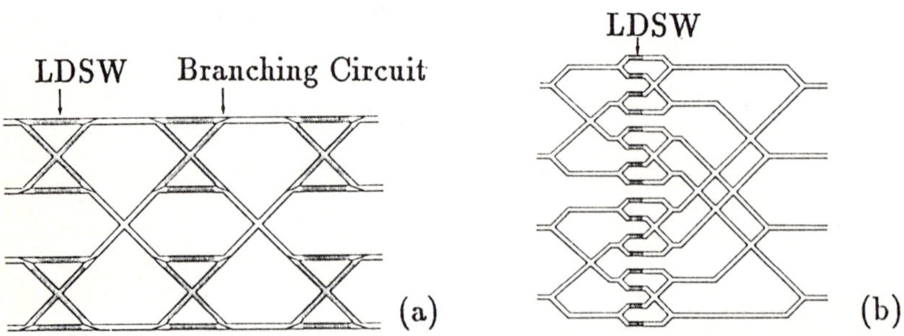

Fig. 1 4×4 non-blocking network. (a) Link connection type and (b) tree connection type.

Springer Series in Electronics and Photonics, Vol. 29 **Photonic Switching II**
Editors: K. Tada · H.S. Hinton © Springer-Verlag Berlin, Heidelberg 1990

In both networks, the routing operation is carried out by a splitting or combining function of the branching circuit and gate operation of the LDSW. The link-type network has plural routes between an arbitrary input and an output port. A large gain is obtained because an input signal is amplified through 3 LDSWs. The signal-to-noise ratio (S/N), however, deteriorates due to the accumulated spontaneous emission of the LDSWs[6]. In the tree-type network, an input signal is gated through only one LDSW. A good S/N is anticipated, but, this network has only one route between an input and output port. The link network is advantageous for sizes larger than 16×16 because the total number of LDSWs (224 for link type and 256 for tree type) and chip size (3.2 ×6.0 mm for link type and 26×26 mm for tree type) can be reduced compared to those of the tree type.

3. Optical Branching/Combining Circuit

Branching/combining circuits are essential for gate-type matrix switches. A large splitting angle, 45°in this experiment, can be used in the LDSWs because of a large ON state gain. Two types of branching circuits, a half-mirror and a Y junction as shown in Figs. 2 (a) and (b), were investigated. The propagating light is split vertically by an etched groove in the half-mirror and horizontally by a wedge-shaped reflection mirror in the Y junction (Figs. 2). Two transversely propagating modes are supported in the 5 μm wide ridge waveguides used in this experiment. The light distribution is narrow in the vertical direction and somewhat spread in the horizontal direction. This indicates that an etching control with high accuracy is required for the fabrication of the half-mirror grooves compared with that required for the Y junction mirrors. However, the light transmittance measurement for the Y junctions revealed the following serious problems. One is that a loss larger than 10 dB/one branching circuit occurred because of the roundness of the wedge apex, which was produced in the actual fabrication processes. The other is that the branching ratio of the Y junction was influenced by the light launching conditions due to the multimode waveguides. These results indicate that the half-mirror circuits are suitable for constructing large scale matrix switches using multimode waveguides.

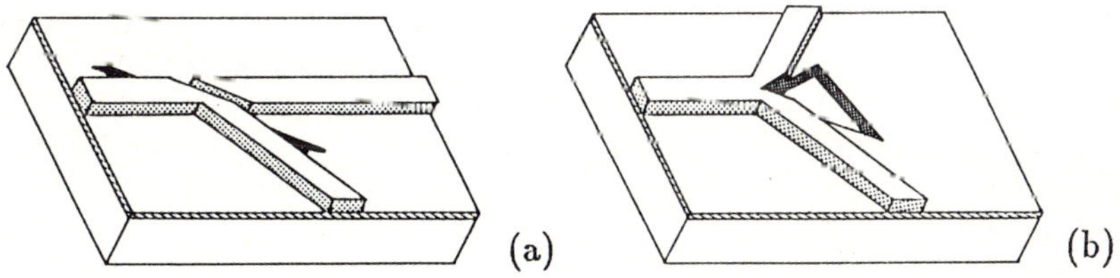

Fig. 2 Optical branching circuits. (a) Half-mirror type and (b) Y junction type.

4. Fabrication and Performance of 2×2 LD Switches

Half-mirror type 2×2 LD matrix switches were fabricated using 1.54 μm InGaAsP/InP DH wafers grown by MOCVD. Ridge waveguides and half-mirror grooves were etched by Br_2 reactive ion beam etching. The 1 μm wide grooves for the branching circuits were etched down to just above the active layer. The etching rate was controlled at 0.1 μm/min. The irregularity of the side wall surfaces was

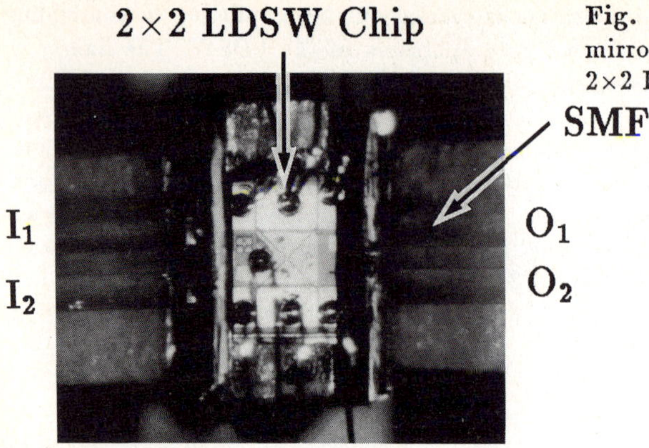

Fig. 3 Photograph of the half-mirror type monolithic integrated 2×2 LD matrix module.

less than 0.1 μm. The 2×2 LD matrix switch module is shown in Fig. 3. The LD matrix switch chip size is 400 μm square. The single mode fiber interval is 200 μm.

Switching performance was measured using single mode fibers as input and output ports. The signal light source was a 1.54 μm DFB laser. A gate pulse of 350 ns duration and 80 mA for each LDSW was applied in the ON state. The net loss and cross-talk level for each path in the chip are summarized in Table 1. The net loss was evaluated as due to the splitting loss from the half-mirrors and the fiber coupling loss. The half-mirror loss was about 10 dB/mirror. The coupling loss between the LDSW chip and the single mode fiber was about 10 dB for one port. Because of the lack of anti-reflection coating on each cleaved surface, an injection current higher than 80 mA led to the laser oscillation of the LDSW. Therefore, a large gain could not be effectively obtained. This indicates that anti-reflection coating for both facets is essential for constructing low loss (or gain) matrix switches.

Figure 4 shows a switching operation for an input pulse train of light signal whose repetition rate is 4.7 MHz. Two input pulses are gated at the cross state in this experiment. In the upper trace, the slightly raised base line is due to the spontaneous emission power from the LDSWs. The spontaneous emission power deteriorates the signal-to-noise ratio, however, it can be much improved by an optical filter which is inserted just before an APD[6].
The spontaneous emission power was effectively reduced by an optical filter having a 4 nm band width and an 8 dB insertion loss. The trace with this optical filter is shown as the lower trace in Fig. 4.

Table 1 Insertion loss and crosstalk level (dB)

Loss (dB) / Cross-talk	O_1	O_2
I_1	-6.5 / <-45	-3.2 / <-45
I_2	-5.6 / <-45	-5.2 / <-45

Without Optical Filter

With Optical Filter

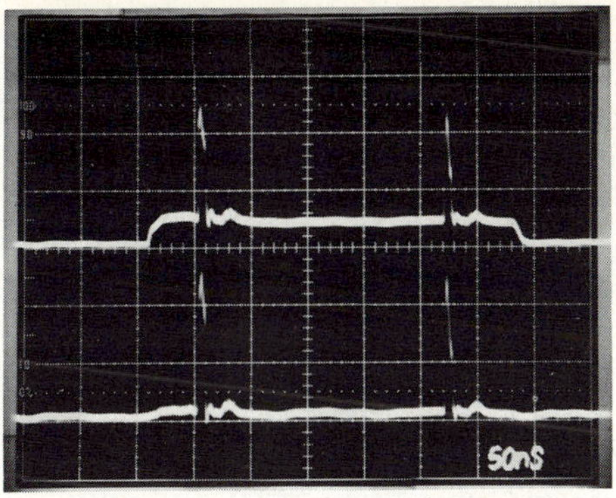

Fig.4 Gated output signal waveform at the cross state. Upper trace: without an optical filter, lower trace: with it.

Optical switches are used in the form of single mode fiber pigtailed modules in optical communication systems. All components of the module must be fixed by a metal for long term stability. Both sides of the two fiber arrays[7] were simultaneously fixed by a Pb-Sn solder using the terminal voltage detection method[8]. For the positioning of the four port fibers, four modulated lights with different frequencies were injected into each port. The insertion loss increased about 5 dB/port compared with that before fixing through the cooling process.

5. Conclusion

In conclusion, monolithic integrated LD optical matrix switches were fabricated using a Br_2 dry etching process on an InGaAsP/InP wafer. The full switching operation of the 2×2 LD matrix switch was demonstrated for the first time. The half-mirror type branching circuit was found to be more suitable for the link connection switching network than the Y junction type. The net insertion loss was about 5~6 dB/port and the cross-talk level was smaller than -45 dB.

Acknowledgement

The authors would like to thank K. KURUMADA and H. KATO for their helpful discussions.

References

[1] M. Ikeda, Electron. Lett.,**17**, 899 (1981).
[2] M. Ikeda, ECOC 87', p227.
[3] M. Ikeda, J. Appl. Phys.,**56**, 1319 (1984).
[4] P. Albert et al., ECOC 88', p235.
[5] C. J. Smyth, IEEE, J. Select. Area Commun.,**6**, 1052 (1988).
[6] M. Ikeda, Opt. Quantum Electron.,**20**, 515 (1988).
[7] H. Inoue et al., OSA Proceedings on Photonic Switching,**3**, 241 (1989).
[8] K. Yoshino et al., Electron. Lett., **25**, 62 (1989).

Polarization and Wavelength Independent Four-Port Optical Routing Switch with Semiconductor Y-Junction Optical Cross-Points

K. Ueki, Y. Kamata, and H. Yanagawa

Opto-technology Laboratory, The Furukawa Electric Co., Ltd.
6, Yawata-kaigandori, Ichihara, Chiba 290, Japan

Abstract. We propose a polarization and wavelength independent four-port routing switch with semiconductor Y-junction optical switching cross-points, and report the performance of the fabricated routing switch.

1. Introduction

As optical communication develops, the significance of switching system for the optical communication network increases, and a routing switch which can change the communication route dynamically can be one of the key components in the system. As such it is preferable to have one more switching state, that is, a loop-back state, in addition to the straight-through and cross-over states of a conventional 2-by-2 switch.

The optical fiber used in the present optical communication network is non-polarization-maintaining type, so the propagating light-wave will not be always linearly polarized. Consequently, polarization independent switches are required. Furthermore, for wavelength division multiplexing systems, wavelength independent switches will be necessary.

In this paper, we propose a polarization and wavelength independent routing switch using a semiconductor Y-junction switch, and report the performance of the fabricated routing switch.

2. Optical Routing Switch

There are three switching states as shown in Fig.1, that is, straight-through (a), cross-over (b) and loop-back (c). To realize this switch, we must add a loop-back mechanism to the conventional 2-by-2 matrix switch. A conceptional diagram of a routing switch using a 2-by-2 matrix switch with loop-back paths is shown in Fig.2. The loop-back mechanism was realized by attaching the ends of two fiber coils to a switch chip. A cross-point switch was composed of a pair of Y-junction switches which will

Springer Series in Electronics and Photonics, Vol. 29 Photonic Switching II
Editors: K. Tada · H.S. Hinton © Springer-Verlag Berlin, Heidelberg 1990

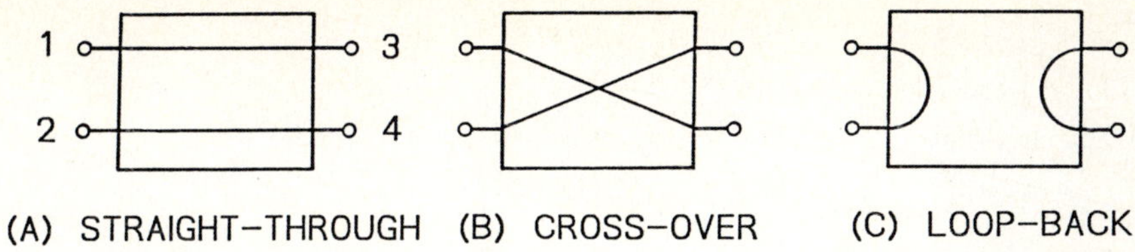

Fig.1 Switching states of the routing switch. (a) Straight-through, (b) cross-over, (c) loop-back

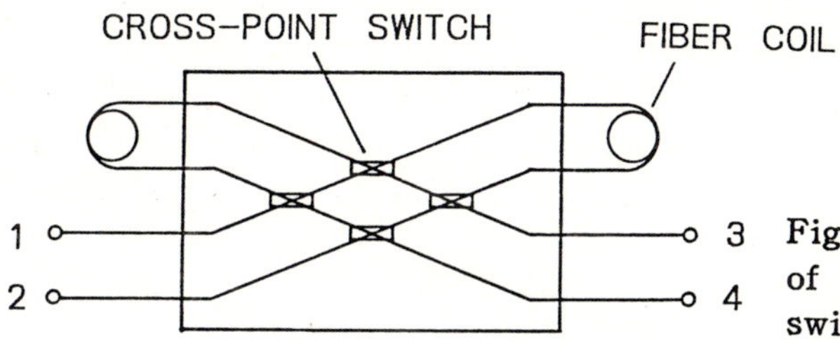

Fig.2 Conceptional diagram of the proposed routing switch

be described next. The number of monolithically integrated Y-junction switches was eight in total.

3. Semiconductor Y-junction Optical Switch [1]

Figure 3 shows a conceptional view of the semiconductor Y-junction optical switch. In this switch, a switching operation is achieved by mode cut-off of one of the branching waveguides caused by current injection into the current injection area formed on both the branching waveguides. As a result, this switch has excellent characteristics of polarization and wavelength independence. Since this switch is usually used in the mode cut-off

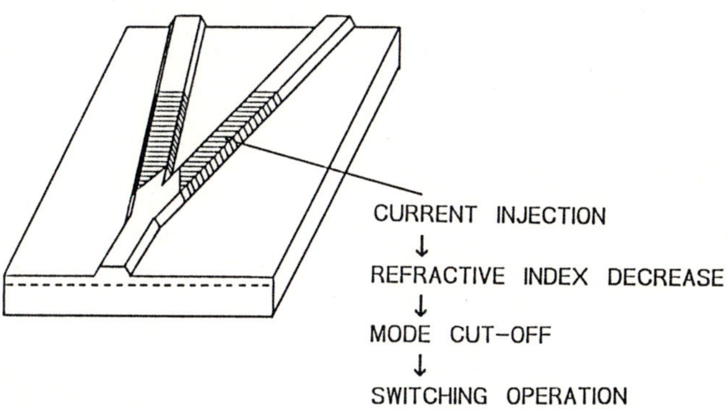

Fig.3 Conceptional view of the semiconductor Y-junction optical switch

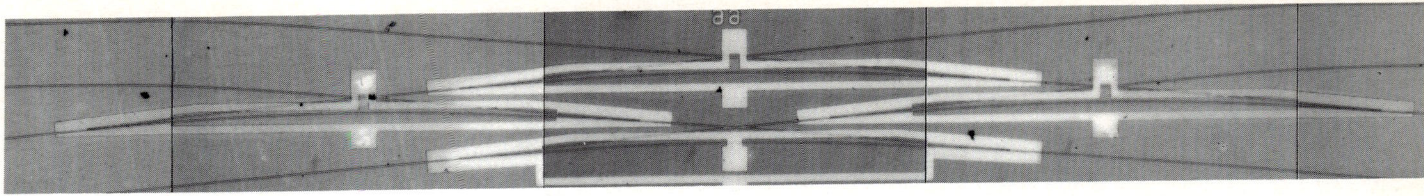

Fig.4 Top view of an experimental sample of the routing switch

state, it can operate even if the branching angle is rather large, which is effective in reducing device length.

4. Experiment

Figure 4 shows a top view of an experimental sample of the routing switch chip. Each Y-junction switch has a branching angle of 2 degrees, a waveguide width of 6 μm, and crossing angle of 10 degrees at the waveguide intersection. A standard single-mode fiber was used as the coil fiber.

Figure 5 shows the cross-sectional diagram of the current injection area of a Y-junction switch. After several n-type epitaxial layers were grown on a GaAs substrate using a MOCVD technique, a ridge was formed in the upper cladding layer by wet etching to obtain a ridge loaded waveguide structure. On the current injection area, Zn was diffused to make a p-n junction. The waveguide structure was chosen because it is compatible with very low propagation loss [2].

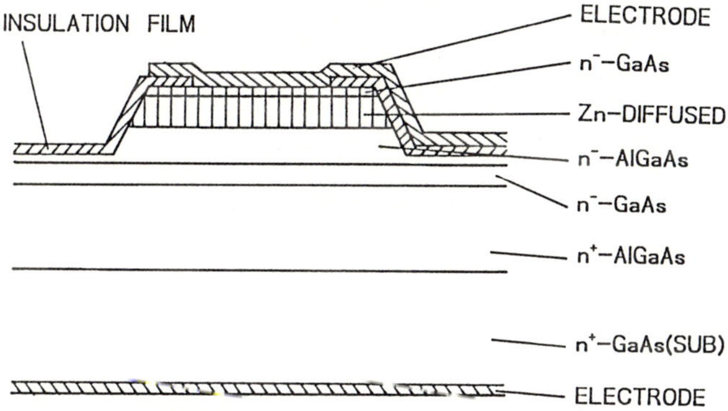

Fig.5 Cross-sectional diagram of current injection area of the Y-junction switch

5. Measured Switching Characteristics

Before attaching fiber coils, the near-field pattern was observed by using an infrared TV camera in order to confirm the operation of a constituent 2-by-2 matrix switch. After that, fiber coils were attached to both facets of the switch. Figure 6 shows the switching characteristics in a loop-back state which was observed through each port. It is clearly seen in this figure that this switch operated as a routing switch.

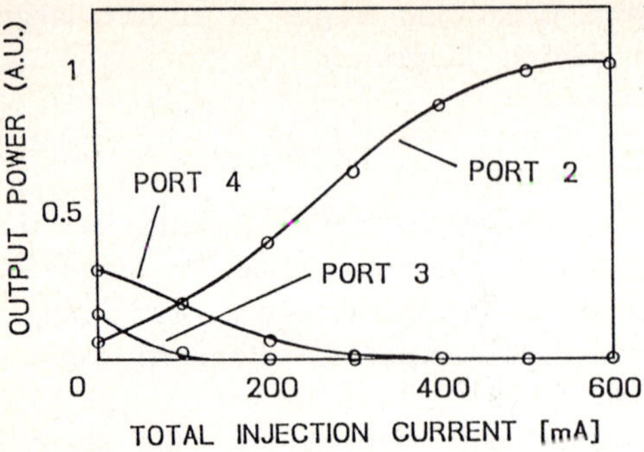

Fig.6 Switching characteristics in a loop-back state

6. Conclusion

In conclusion, we proposed a four-port routing switch which has three states. This switch was constructed by adding a loop-back mechanism to a conventional 2-by-2 matrix switch, and we reported the principal operation characteristics experimentally observed.

References

1. H.Yanagawa, Y.Kamata, K.Ueki, H.Miyazawa, and M.Ogai : ECOC'89 WeP-5
2. Y.Kamata, K.Ueki, H.Yanagawa, and H.Miyazawa : OEC'88 2C1-2

4 × 4 GaAs/AlGaAs Optical Matrix Switches with Electro-Optic Guided-Wave Directional Couplers

K. Komatsu, M. Sugimoto, A. Ajisawa, K. Hamamoto, Y. Kohga, and A. Suzuki

Opto-Electronics Research Laboratories, NEC Corporation,
4-1-1 Miyazaki, Miyamae-ku, Kawasaki 213, Japan

Abstract. 4x4 GaAs/AlGaAs optical matrix switches constructed from 12 integrated electro-optic directional couplers have been realized. They have a simplified tree architecture, which is suitable for achieving uniform device characteristics. Matrix switches with much more uniform characteristics than ever reported before have been realized.

1. Introduction

There has been a growing interest in optical switches made of III-V semiconductor compounds for photonic switching systems, because of their large scale integration capability and their ability to be integrated with other semiconductor devices. A GaAs/AlGaAs electro-optic directional coupler (EODC) is attractive as a cross-point element for integrated matrix switches, because of its low absorption loss in the long wavelength region [1], fast switching speed, low electric power consumption and wavelength independent operation capability. Although many reports, including reference [2] published by the authors, have been made on discrete EODCs, few have been made on semiconductor integrated matrix switches [3],[4].

This report describes 4x4 matrix switches with a simplified tree architecture constructed from 12 integrated GaAs/AlGaAs EODCs. 4x4 matrix switches with much more uniform device characteristics than ever reported before have been realized. The switching voltage deviation of all 12 cross-point elements of a matrix switch was very small, 9.0 ± 0.5V for ⓧ-state and 21.9 ± 1.5V for ⊜-state, and the crosstalk deviation of the 12 cross-points was also very small, -21.3 ± 3.4dB for ⓧ-state and -20.9 ± 3.8dB for ⊜-state.

2. Matrix Switch Design and Fabrication

A schematic drawing of the 4x4 GaAs/AlGaAs matrix switch is shown in Fig.1. Twelve guided-wave EODCs with pin structure, and straight and bent waveguides connecting EODCs are integrated on the same chip.

Springer Series in Electronics and Photonics, Vol. 29 **Photonic Switching II**
Editors: K. Tada · H.S. Hinton © Springer-Verlag Berlin, Heidelberg 1990

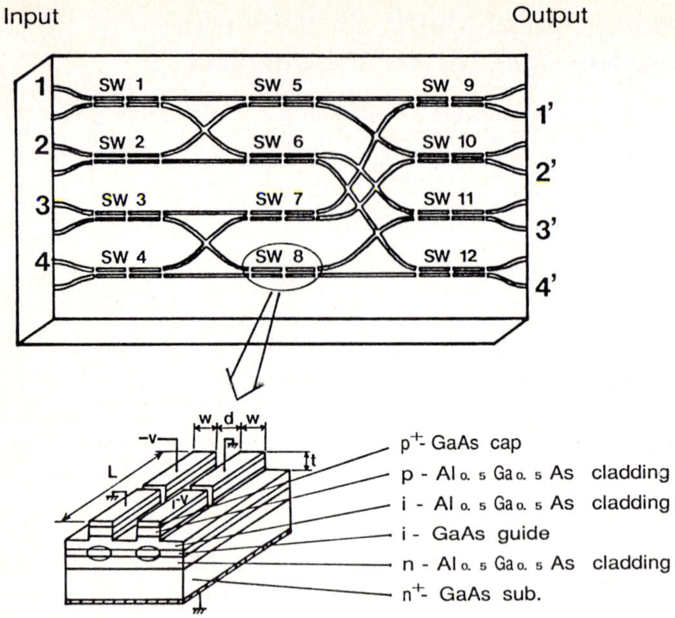

Fig 1. The 4x4 GaAs/AlGaAs optical matrix switch with electro-optic directional couplers using a simplified tree structure

The p-side electrode of each EODC is divided into two equally long sections in order to enable alternating $\Delta\beta$ operation, which does not require an exact coupler length to coupling length ratio. A simplified tree architecture [5],[6] is used as the matrix switch architecture because the architecture requires fewer cross-points, fewer ranks, and thus a smaller chip size than the conventional one. Moreover, the architecture has the advantages of low crosstalk nature and smaller path-dependence of propagation loss. These features of the architecture are very attractive from the viewpoint of uniformity of device characteristics.

Each EODC is 3mm long. The radius of each waveguide bend was chosen to be 4mm to prevent a radiation loss. Total chip length of the matrix switch is 15mm. Waveguide widths and waveguide spacings in the directional coupler region are 2 μm and 2.5 μm, which were chosen by considering the single-mode condition and coupling length.

The matrix switches were fabricated on MBE (Molecular Beam Epitaxy) grown wafers using a conventional photolithography technique and Cl_2 RIBE (Reactive Ion Beam Etching). Here, MBE and RIBE were chosen from the viewpoint of uniformity of device characteristics, because MBE and RIBE can provide uniformity of crystal layer thickness and etching depth, respectively.

3. Device Characteristics

The matrix switch characteristics were measured by coupling 1.3 μm wavelength TE polarized light into an input waveguide using lens systems. Typical switching characteristics of a directional coupler cross-point under alternating $\Delta\beta$ operation are shown in Fig.2. Switching voltage deviation of all 12 cross-point elements of the matrix switch was very small, 9.0 ± 0.5V for ⊗ -state and 21.9 ±1.5V for ⊜ -state. Measured crosstalk deviation of the 12 cross-points was also very small, -21.3 ± 3.4dB for ⊗ -state and -20.9 ±3.8dB for ⊜ -state.

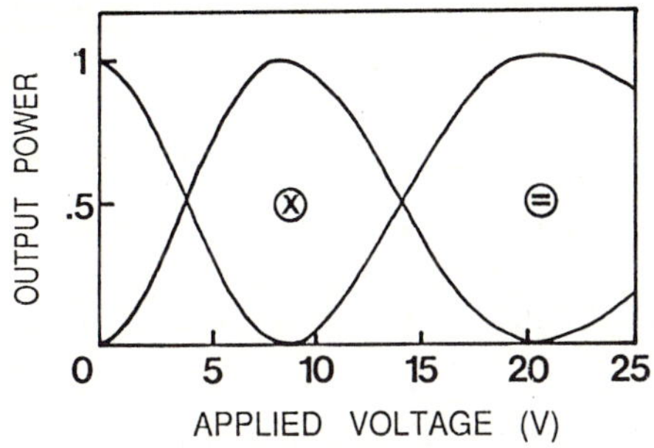

Fig 2. Measured typical switching characteristics of a directional coupler cross-point at λ =1.3 μm

An example of the switching experiments using the matrix switch is shown in Fig.3. In this experiment a variable 1x4 connection between an input channel and output channels was tested. Here, the light is coupled into input channel #2, and 7 cross-points, SW 2, SW 5, SW 6, SW 9, SW 10, SW 11 and SW 12, are operated. Figure 3 clearly shows the desired connections are successfully achieved with low crosstalk. Loss and crosstalk characteristics were measured during the 1x4 connection experiments. Results are summarized in Table 1.

Table 1. Measured propagation loss and crosstalk of the matrix switch

Connected Path	[Propagation Loss], Crosstalk			
	1'	2'	3'	4'
2 → 1'	[10.5 dB]	< -30 dB	< -30 dB	< -30 dB
2 → 2'	< -30 dB	[11.3 dB]	< -30 dB	< -30 dB
2 → 3'	< -30 dB	< -30 dB	[10.2 dB]	< -30 dB
2 → 4'	< -30 dB	< -30 dB	< -30 dB	[10.5 dB]

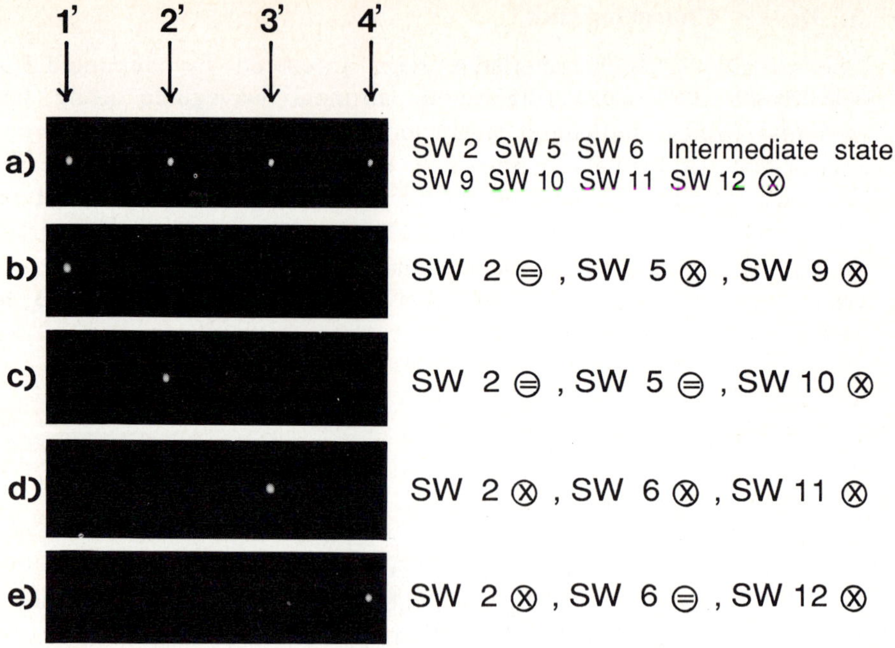

Fig 3. Variable 1x4 connection experiment using the matrix switch
a) input #2 → outputs #1', #2', #3' and #4'
b) input #2 → output #1'   c) input #2 → output #2'
d) input #2 → output #3'   e) input #2 → output #4'

Little path-dependence of propagation loss of ±0.5dB and very low crosstalk level, less than -30dB at output channels, were confirmed. This little path-dependence proved that there is little loss in intersecting waveguides, as expected, because different paths experience a different number of intersecting waveguides.

4. Conclusion

We report 4x4 GaAs/AlGaAs matrix switches constructed from 12 integrated EODCs for $\lambda = 1.3 \mu$ m with a simplified tree architecture. Very uniform characteristics were realized by choosing MBE and RIBE as the crystal growth and waveguide fabrication techniques, in addition to a simplified tree architecture as the matrix switch architecture.

References.

[1] E.Kapon, and R.Bhat, Appl. Phys. Lett. **50**, 1628(1987)

[2] K.Komatsu, M.Sugimoto, A.Ajisawa, and A.Suzuki, Proceedings of 2nd Topical Meeting on Photonic Switching, Salt Lake City, 1989, FE4

[3] H.Inoue, H.Nakamura, K.Morosawa, Y.Sasaki, T.Katsuyama, and N.Chinone, IEEE J.S.A.C. **SAC-6**, 1262(1988)
[4] T.Ushikubo, I.Asabayashi, and T.Ishida, Technical Digest of Optical Fiber Communication Conference (OFC), Houston, 1989, THB5
[5] H.Nishimoto, S.Suzuki, and M.Kondo, Electron. Lett. **24**, 1122(1988)
[6] A.Matoba, H.Okayama, R.Shibuya, and T.Ishida, Electron. Lett. **25**, 165(1989)

Photonic Spatial Logic Operations Using a Beam Scanning Laser Diode

H. Itoh, S. Mukai, M. Watanabe, M. Mori, and H. Yajima

Optical Information Section, Electrotechnical Laboratory,
Umezono 1-1-4, Tsukuba, Ibaraki 305, Japan

Abstract. A novel logic operation technique using a photonic spatial switching device is reported. An optical spatial-code encoding unit and fundamental logic units, NOT, XOR, and AND gates have been implemented. This technique has advantages for a parallel integration and cascade connection.

1. Introduction

Spatial modulation of a light beam has been proved to be a useful technique for optical parallel computing [1]. A logic operation system using direct photonic spatial modulation by semiconductor lasers is more attractive than using discrete light sources and passive elements for modulation. A promising element for this purpose is a beam scanning laser diode (BSLD) whose output beam can be deflected by changing the injected current distribution[2,3].

In this paper, we propose a method to implement binary logic operations using a BSLD and present experimental results.

2. Beam scanning laser diode

Fig. 1(a) shows a schematic device structure of the BSLD. The laser has two parallel electrodes (250 μm long, 2 μm wide, and 3 μm separated) on a 10 μm-wide mesa on a AlGaAs/GaAs double-heterostructure. A 2 μm-wide

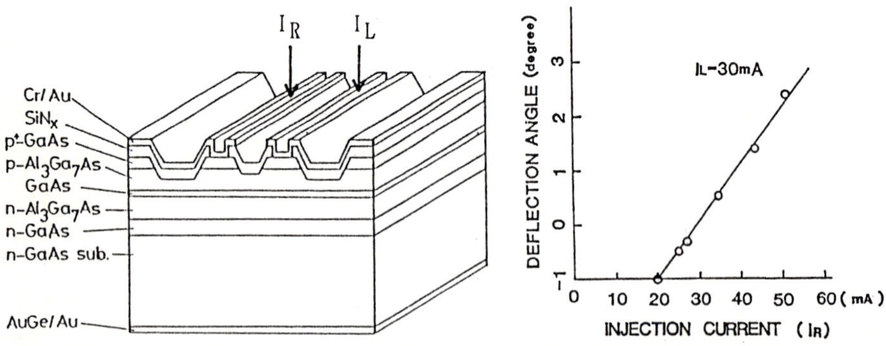

Fig. 1(a)Schematic structure of the BSLD (b)Beam deflection characteristics cavity length=250 μm, pulsed(1 μs, 1kHz)

Springer Series in Electronics and Photonics, Vol. 29 Photonic Switching II
Editors: K. Tada · H.S. Hinton © Springer-Verlag Berlin, Heidelberg 1990

trench between the electrodes was etched to separate carriers and optical fields.

Fig. 1(b) shows a typical dependence of the deflection angle of the laser beam on the injection current to the right electrode, I_R, at the injection current to the left electrode, I_L, of 30mA. The peak shifts from -2° to +1.5° continuously as I_R increases from 20 to 50mA.

3. Beam scanning binary logic system

Fig. 2 shows a schematic diagram of the beam scanning binary logic (BSBL)[4] system using the BSLD. The BSBL is composed of two input photodetectors (D_a and D_b), two amplifiers (A_a and A_b), and a BSLD. $Z_{+\theta}$, Z_0, and $Z_{-\theta}$ are the photodetectors of the next stage and are set in the deflection angle of $+\theta$, 0, and $-\theta$, respectively. Signal currents of D_a and D_b are amplified by A_a and A_b, and injected to the laser.

Table 1 shows the functions of input and output detectors for each logic gate. In the case of 2-input-1-output and 2-input logic gates, two data X and Y are spatially encoded to (X_a, X_b) and (Y_a, Y_b). X_a+Y_a and X_b+Y_b are introduced to D_a and D_b, respectively.

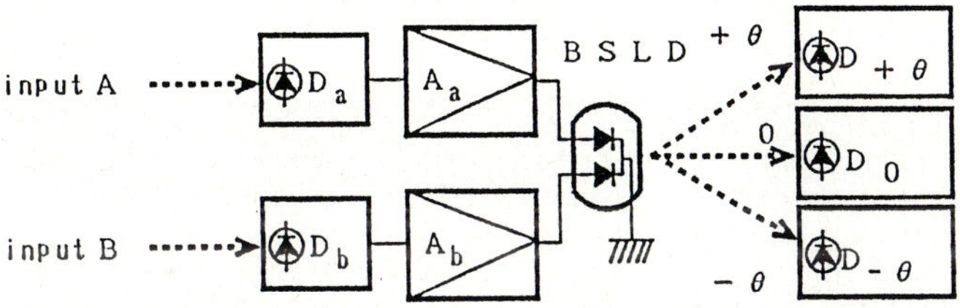

Fig. 2 Schematic setup of beam scanning binary logic system

Table 1 Function selection table of photodetectors

	ENCODER	NOT	XOR	AND
D_a	BIAS	X_a	X_a+Y_a	X_a+Y_a
D_b	X	X_b	X_b+Y_b	X_b+Y_b
$D_{+\theta}$	Z_a	Z_a	Z_{b1}	Z_{b1}
D_0			Z_a	Z_{b2}
$D_{-\theta}$	Z_b	Z_b	Z_{b2}	Z_a

4. Results and discussion

Experiments were executed by pulse current injections to the electrodes of the BSLD at room temperature. The output beam was detected by PIN photodiode array which was 40mm away from the BSLD. The width of the input pulse signal was 1μs and the repetition rate was 1 kHz.

Fig. 3 shows a result of the encoder from an amplitude to a spatial code in the case of X=1. The four curves show the injection current to the right electrode (I_R), the bias current (50 mA) to the left electrode (I_L), and the spatial coded outputs (Z_a and Z_b), respectively.

Fig. 4 shows the result of NOT gate in the case that the coded input pair (X_a, X_b) is (1, 0) (X=1). Z_a and Z_b are set in the angles of the deflected beams, $+\theta$, and $-\theta$, respectively. The output beam is detected at $D_{-\theta}$ and the spatial output (Z_a, Z_b) is (0, 1), (Z=0).

Fig. 5 shows the result of XOR gate in the case that the input (X, Y) is (1, 0). Z_a, Z_{b1}, and Z_{b2} are set in the angle of deflected beams, 0, $+\theta$, and $-\theta$, respectively, and Z_{b1} and Z_{b2} are connected in parallel. The output beam is detected at D_0. The spatial output (Z_a, $Z_{b1} \cup Z_{b2}$) is (1, 0), (Z=1).

Fig. 6 shows the result of **AND** gate in the case that the input (X, Y) is (1, 0). Z_a, Z_{b1}, and Z_{b2} are set in the angles of the output beams, $-\theta$, 0, and $+\theta$, respectively, and Z_{b1} and Z_{b2} are connected in parallel. In this case, the laser beam is detected at D_0. The spatial output (Z_a, $Z_{b1} \cup Z_{b2}$) is (0, 1), (Z=0).

The BSBL has several advantages as follows:

・The BSBL executes multi-functional logic operations simultaneously. The BSBL can realize the logic operation by selecting the position of detectors with the same condition of the operating unit.

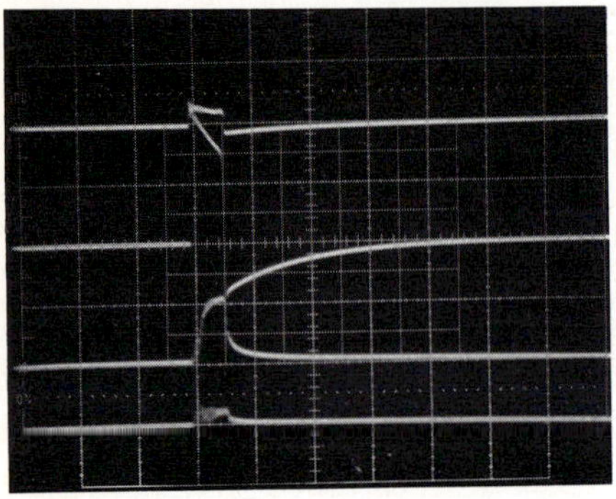

Fig. 3 Result of spatial encoding

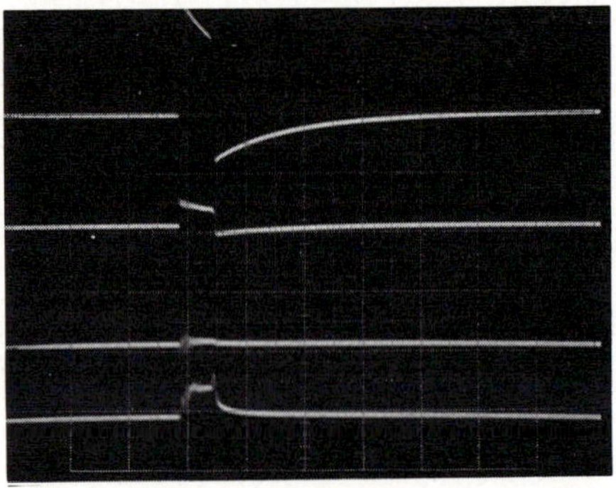

Fig. 4 Result of NOT gate

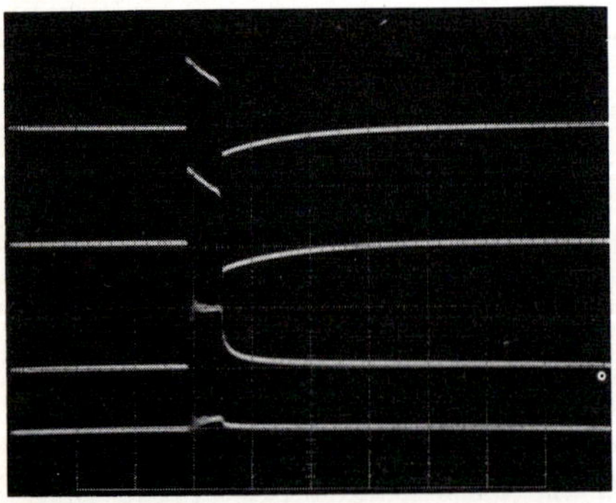

Fig. 5 Result of XOR gate

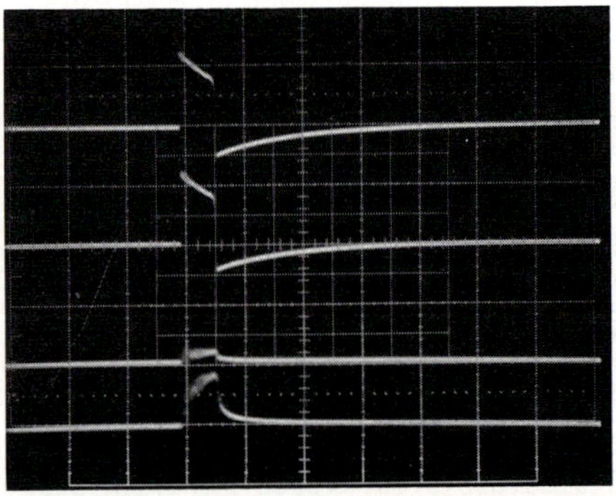

Fig. 6 Result of AND gate

A 2-input-2-output gate (half adder) which is composed of XOR gate and AND gate is easily realized by using two stacked pairs of output-detectors, which detect the output beam independently. The half adder of the electronic registor-transistor logic needs at least 7 active devices, while the element of the BSBL is 5. Replacing a pair of a photodetector and an amplifier by a photoconductive device or a phototransistor reduces the number of active devices to 3. As the space resolution of beam deflection increases, more complicated logic operations become possible.

・Conventional electronic logic gates have different dimensions and gate delay times, i.e. the AND gate is made by the cascade connection of NAND gate and NOT gate. On the other hand, the hardware configurations and delay times of the BSBL for each operation are almost the same. The estimation of the operational timings for the parallel logic operations is much easier. It is advantageous for designing and debugging a large scale integrated BSBL.

・The cascade connection and monolithic parallel integration are possible.

・The interconnection between gates is flexible because the output power and the direction of the laser beam can be electrically controlled. The operational speed is much faster than the other beam scanning devices.

5. Summary

A novel logic operation technique using beam scanning laser diodes has been implemented. This logic system has more advantages for reducing the number of active devices than ordinary binary logic operations. Implementation of arithmetic operations using these devices is under development.

References

1. P. Chavel, A. A. Sawchuk, T. C. Strand, A. R. Tanguay, Jr., and B. H. Soffer, Opt. Lett., 5, 398 (1980).

2. S. Mukai, M. Watanabe, H. Itoh, H. Yajima, Y. Hosoi, S. Uekusa, Opt. Quantum Electron., 17, 431 (1985).

3. S. Mukai, Y. Kaneko, M. Watanabe, H. Itoh, and H. Yajima, Appl. Phys. Lett., 51, (25), 2091 (1987).

4. H. Itoh, S. Mukai, M. Watanabe, M. Mori, and H. Yajima, Technical Digest of Int. Topical Meeting on Optical Computing, Kobe, Japan, 9D18 (1990).

Cross-Coupled Lateral Modes in a Twin-Stripe Laser: Analysis Leading to the Mechanism and Bistability

M. Watanabe[1], *I.H. White*[2], *and J.E. Carroll*[2]

[1]Electrotechnical Laboratory, Optical Information Section, Umezono, Tsukuba, Ibaraki 305, Japan

[2]Cambridge University, Engineering Department, Trumpington Street, Cambridge CB2 1PZ, UK

Abstract. Mechanism of the cross-coupled lateral mode in a twin-stripe laser is clarified, which leads to bistability between the mode and its mirror image.

1. Introduction

A cross-coupled lateral mode which couples from one stripe to the other in a gain-guided twin-stripe laser was observed by White and Carroll [1]. This mode is particularly useful when the switching of optical signals between different waveguide channels is required. However, the mechanism for generating the mode has not been clarified although it has been attributed intuitively to directional coupling [1] or a longitudinal perturbation [2].

In this work, cross-coupled resonant lateral modes in a twin-stripe four-contact laser with diagonal carrier injection (where each diagonal pair of stripes are equally pumped) are analysed. The analysis clarifies the mechanism and the conditions for generating cross-coupled modes. It is shown that a cross-coupled mode can be generated even in a two-contact twin-stripe laser, which leads to bistability between the mode and its mirror image.

2. Calculation procedure

The structure analysed here is illustrated in Fig.1. This shows a laser with twin buried-heterostructure waveguides made of AlGaAs material. This structure was adopted to make the calculation more straightforward than for gain-guided structure. The dielectric constants of the active regions, the lateral claddings, and the transverse claddings are ε_a=13.1-0.023i, ε_l=12.9, and ε_t=11.4, respectively. The refractive-index change Δn due to the carrier density N is taken into account by the relation $\Delta n=(-5.4+1.5i)\times 10^{-20}cm^3\times N$. The average carrier density N_0 is fixed throughout this work, for simplicity, at a value $1.8\times 10^{18}cm^{-3}$ which allows the modal gain to compensate the facet losses when the cavity length L is equal to the coupling length Lc ($\sim$510 μm). The carrier-density difference between the adjacent waveguides is represented by Δ_c. For carrying out the analysis, Maxwell's equation is solved to get the lateral modes, first, along the transverse (y) direction and then along the lateral (x) direction. In the lateral direction, after solving the 3-layer problem to get the fundamental mode for each waveguide, coupling between the modes is considered. Then we get an eigen-value problem by setting the condition that the mode to be observed should not change its field distribution after a round trip. By solving the problem, we get two lateral resonant modes which satisfy the condition. One of these modes which has the higher gain than the other is to be observed.

Springer Series in Electronics and Photonics, Vol. 29 **Photonic Switching II**
Editors: K. Tada · H.S. Hinton © Springer-Verlag Berlin, Heidelberg 1990

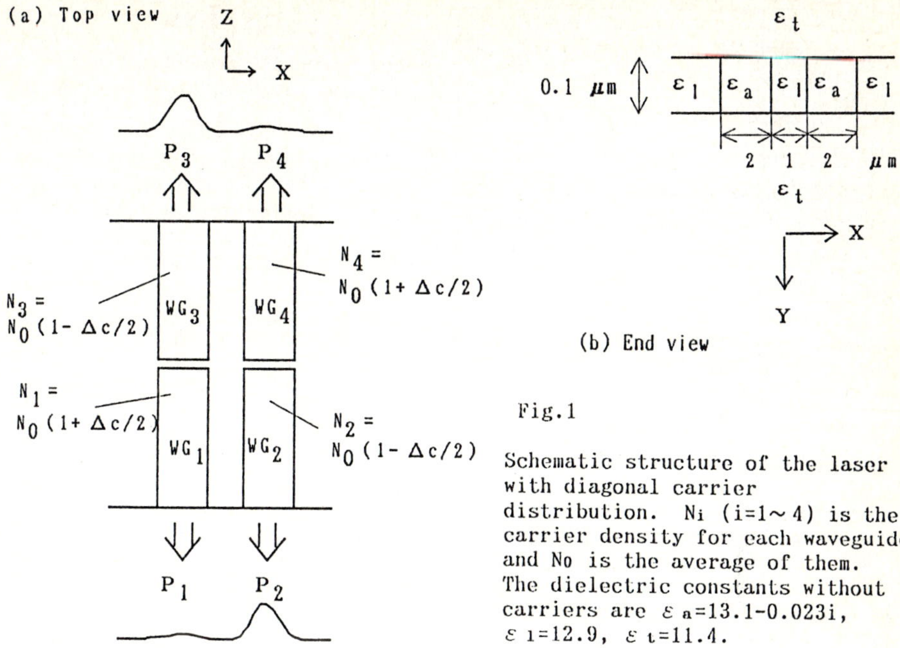

Fig.1

Schematic structure of the laser with diagonal carrier distribution. N_i ($i=1\sim4$) is the carrier density for each waveguide and N_0 is the average of them. The dielectric constants without carriers are $\varepsilon_a=13.1-0.023i$, $\varepsilon_1=12.9$, $\varepsilon_t=11.4$.

3. Results

First, we solve the problem with the carrier density in each waveguide fixed as shown in Fig.1 and Sect. 3.1. Second, we calculate the currents which are necessary to maintain the carrier density and the optical power distribution, supposing that rapid diffusion ensures local uniformity of the carriers but that the laser structure ensures good lateral confinement (section 3.2).

3.1 Optical power ratio vs cavity length

Fig.2(a) shows a typical dependence of P_1/P_2 on L/Lc with small Δc ($=6\times 10^{-3}$) where Lc is the coupling length ($\sim 510\mu m$) for $\Delta c=0$. Fig.2(b) shows the difference between the phases of the modal field in each waveguide, and Fig.2(c) shows the round-trip gain defined by (initial amplitude)/(final amplitude after a round trip) excluding the facet losses. Due to the diagonal distribution of the carrier density, $P_4/P_3=P_1/P_2$ (the field is coupled from one stripe to the other) is always satisfied. Referring to (c), the solid and the broken lines are of the dominant and the subordinate resonant lateral modes. If the carrier distribution is fixed and the gain of these modes is not identical, only the dominant mode appears because any pattern can be expressed as a superposition of the two resonant modes, and the weight of the dominant mode increases with every round trip. The most interesting behavior of the dominant mode in Fig.2(a) is that the highest power appears in the waveguide with the least carriers ($P_1 < P_2$) when $L < Lc$, which is explained as follows.

Fig.2(c) indicates that the dominant mode is odd-like (the phase difference is near to 180°). There are two factors which determine the resonant mode gain. One is the local gain and the other is the confine-

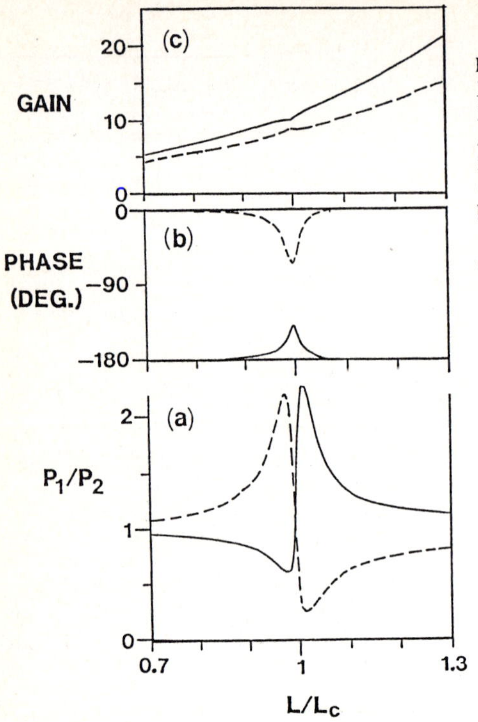

Fig.2

Results for the dominant (solid line) and the subordinate (broken line) resonant lateral mode with $\Delta c=6\times 10^{-3}$; (a) shows the power ratio between the waveguides, (b) shows the phase difference between the waveguides, and (c) shows the round-trip gain excluding the facet losses.

ment factor of the lateral field in the gain region. When Δc is as small as this case, the contribution of the latter is much larger than the former, which makes the gain of the odd-like mode larger than the even-like mode. It can be proved theoretically that the odd-like resonant mode has the highest power in the waveguide with the least (most) carriers when $L<(>)\ Lc$, which is consistent with the result of simulation shown in Fig.2(a). The proof of this interesting behavior of the field pattern is complicated and cannot be described here, but we just note that the behavior comes from the change of the twin waveguide at the center of the device (from WG_1 and WG_2 to WG_3 and WG_4).

3.2 Optical power ratio vs current ratio - bistability between cross-coupled modes

Fig.3 shows a typical result of P_1/P_2 vs Cu_1/Cu_2 with $L/Lc=0.9$ and the output power=1.3mW, where the solid and the broken lines are the stable and unstable solutions. The curve shows bistability with fixed currents. In particular, even if all the four contacts are driven with equal currents (equivalent to the symmetrically pumped two-contact twin-stripe laser), the output power is not symmetrical, giving $P_1/P_2=3$ or 1/3. (Note that the carrier densities are not equal.) A brief physical picture of the behavior follows.

Suppose that $L<Lc$ and all the contacts are equally driven but a slightly asymmetric field (e.g. $P_1<P_2$) appears at a facet for some reason. This field propagates to the other facet where it becomes nearly the mirror image ($P_3>P_4$) due to the coupling between the waveguides. This internal field distribution causes $N_1>N_2$ and $N_3<N_4$ (i.e. $\Delta c>0$). Since $L<Lc$, this carrier distribution supports the asymmetric light distribution ($P_1<P_2$), and hence a cross-coupled mode results. The mirror

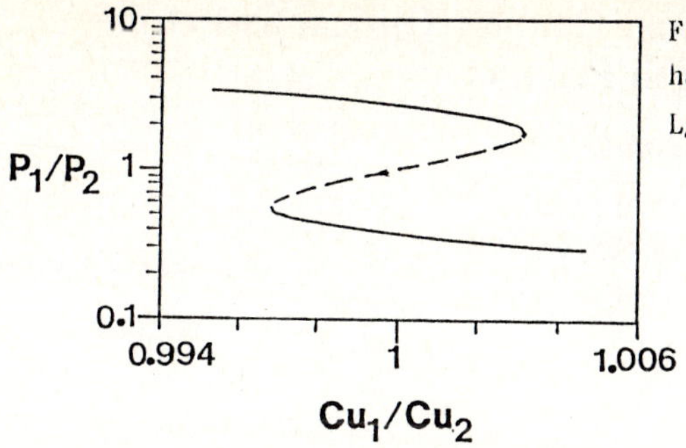

Fig.3 Power ratio vs current ratio having bistable region, with $L/L_c=0.9$ and $P_{out}=1.3mW$.

image of this field can also be supported so that bistable switching is predicted. The state can be switched to its mirror image by injecting a pulsed light or current, and hence the device is useful for spatial photonic switching. The theory assumes perfect symmetry in the laser structure so that the operation is sensitive to fabrication tolerances.

4. Summary

The paper has reported a mechanism within a closely coupled twin-stripe laser with index guiding whereby a cross coupled mode can be generated even when there is a uniform current flow. Bistability has also been shown to be possible.

References

[1] I.H. White and J.E. Carroll, Electron. Lett., 19, 337 (1983).
[2] K.A. Shore, Electron. Lett., 19, 874 (1983).

Low Drive Voltage Directional Coupler Optical Switch with Reduced Voltage–Length Product

H. Okayama, T. Ushikubo, and T. Ishida

R&D Group, Oki Electric Industry Co., Ltd.
550-5 Higashiasakawa-cho, Hachioji, Tokyo 193, Japan

Abstract. The phase change × interaction length product needed to switch a directional coupler is shown to be reduced by decreasing the coupling coefficient in the middle of the interaction region.

1. Introduction

In future high speed photonic switching systems, optical switch matrices with reasonable device length should be operable at a low drive voltage. Among many 2×2 optical switches, the directional coupler is used most widely. The simplest form of this switch is the uniform $\Delta\beta$ directional coupler in which an electrode is placed on each of the waveguides. The interaction length is set to an odd number of coupling lengths for the cross state at zero electrical bias. When an electrical bias is applied to the electrodes, phase mismatch $\Delta\beta$ is induced. The $\Delta\beta$ × interaction length (L) product ($\Delta\beta$L) required to drive the switch into the bar state is $\sqrt{3}\pi$, which is larger than $\Delta\beta$L of π in the case of the Mach-Zehnder modulator. Therefore the drive voltage of the directional coupler is inherently higher for an optimized electrode structure [1]. Some designs (including the balanced-bridge switch) can reduce the voltage-electrode length (Le) product ($\Delta\beta$Le) [2], but these designs do not fulfill the demand of reducing the total device length at a lower drive voltage. In this report we describe a new design which reduces the $\Delta\beta$ × device length (and the drive voltage × device length) product.

2. Device structure

In reference 2, the $\Delta\beta$Le product was reduced by using a smaller value of coupling coefficient K (longer interaction length) compared with $\Delta\beta$. In this report the value of K is reduced, at the middle of the device, without increasing the interaction length by weighted coupling. (The strong coupling effect takes place in the small section at both ends of the interaction region achieving the same overall coupling effect as the parallel waveguide device.) The structure we propose is shown in Fig. 1. To reduce the coupling coefficient at the middle of the interaction region, the gap between the waveguides is gradually enlarged from both ends to the center. The same effect can be realized by a structure in which the low index region, such as the etched groove, is

Springer Series in Electronics and Photonics, Vol. 29 Photonic Switching II
Editors: K. Tada · H.S. Hinton © Springer-Verlag Berlin, Heidelberg 1990

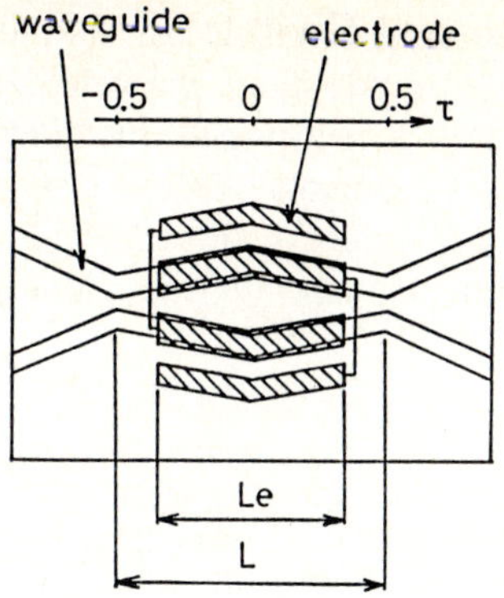

Fig. 1 Schematic configuration.

embedded between the parallel waveguides, at the expense of additional fabrication processes. To simplify the analysis, the gap between the waveguides is designed to be enlarged linearly in the direction of propagation. The coupling coefficient can be expressed as $K=K_0\exp(-\gamma d)$ with d as the gap between waveguides and K_0 and γ as constants. Using this relation, the conventional coupling equation [3] for this device is obtained as in Eq. 1.

$$\frac{\partial\rho}{\partial\tau} = -i\Delta\beta L\rho + i\pi m(-\ln R)R^{1-2|\tau|}(\rho^2-1)/[2(1-R)] \quad (1)$$

where $-0.5<\tau<0.5$ is the normalized position in the interaction length, ρ is the amplitude ratio of light in the two waveguides, L is the interaction length, m(=L/coupling length)=1, R is the min/max ratio of the coupling coefficient. In Fig. 2 calculated results are given showing that $\Delta\beta L/\pi$ reduces with decreasing R. In Fig. 3 the calculated switching digram is shown. The counter lines show the relative

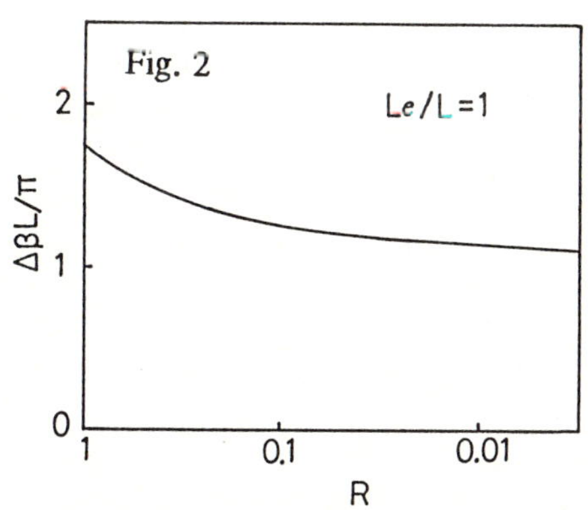

Fig. 3

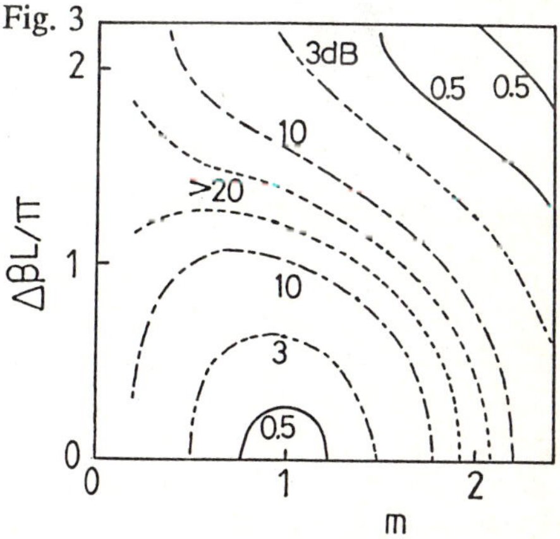

Fig. 2 Reduction of the phase mismatch ($\Delta\beta$) - interaction length (L) product by decreasing the coupling coefficient min/max ratio R.

Fig. 3 Calculated switching diagram for the device with R=0.1 and Le/L=1. The amount of crosstalk in the cross direction is shown.

crosstalk output power at the output port in the cross direction. The bar state is achieved at >20 and the cross state is achieved at <0.5. In this figure the value of R=0.1 and Le/L=1 was used. Similar to the device of Ref. 2, reducing the electrode length results in a decreased $\Delta\beta Le/\pi$ value [4]. The $\Delta\beta Le/\pi$ value reduces more effectively when R<1. Using the experimental values of $LiNbO_3$, the waveguide divergent angle was calculated to be less than 0.2 deg. for an interaction length longer than 5 mm when R=0.1.

3. Experiment

The single mode waveguides were fabricated on z-cut $LiNbO_3$ by in-diffusing 7 µm wide Ti. The thickness of the Ti was 750 Å for a 1.3 µm guide and 950 Å for a 1.55 µm guide. The diffusion temperature was 1050°C. The diffusion time was 7.5 hours for the 1.3 µm guide and 6 hours for the 1.55 µm guide. To obtain R<0.1, the gap between waveguides was enlarged at the middle of the interaction region for 5.7 µm. To obtain devices with one coupling length (m=1), the minimum waveguide gap of the device with 1 cm interaction length was 3.5 µm for 1.3 µm wavelength. The minimum waveguide gap of the device with 6 mm interaction length was 3.2 µm for 1.55 µm wavelength. With the minimum waveguide gap of 4.5 µm and with interaction length of 1.6 cm, a device with two coupling lengths (m=2) was obtained. An electrode structure proposed in Ref. 5 was used. The gap between electrodes was 4 µm and the width of the electrodes was 10 µm. The lengths of the electrodes were the same as the interaction lengths (L/Le=1). An SiO_2 buffer layer of 3000 Å thickness was placed between waveguide and electrode. A TM mode was launched into the waveguide. For 1.3 µm wavelength the swing voltage × interaction length product of 3.6 Vcm was achieved [4]. The measured response curves at 1.55 µm wavelengths for devices with m=1 and m=2 are shown in Fig. 4(a) and (b) respectively. The swing voltage of the devices shown in Fig. 4(a) was 9.3 V. For the device shown in Fig. 4(a), the crosstalk of the cross state was 17 dB and that of the bar state was 22 dB. The drive voltage was estimated to be reduced to 70 % by the improved electrode

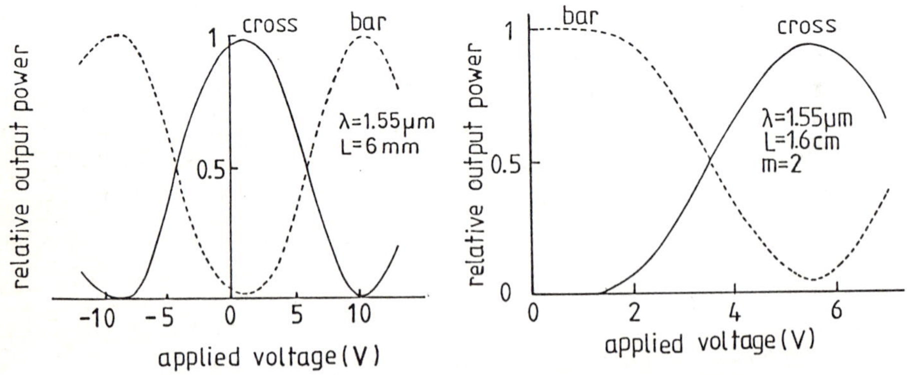

Fig. 4 Measured response curves of the devices with (a) m=1 and (b) m=2 at 1.55 µm wavelength.

structure and was reduced to about 70 % by the reduced coupling coefficient compared to conventional structure. The total reduction was more than 50 %. As shown in Fig. 4(b), for the device with m=2, the cross state was achieved by applying electrical bias to the electrode as predicted in Fig. 3. The electrical field / optical mode overlap parameter was 0.28. The fiber-to-fiber insertion loss was 4-7 dB.

4. Conclusion

The $\Delta\beta L/\pi$ value to switch the directional coupler can be reduced to almost one by decreasing the coupling coefficient at the middle of the device. The voltage-length products of 3.6-4 Vcm and 5-6 Vcm were obtained with $LiNbO_3$ at 1.3 μm and 1.55 μm wavelengths respectively. Further reduction of the drive voltage will be achieved by fabrication processes which yield a smaller optical mode or by using compound semiconductor materials.

References

1. Optical Fiber Telecommunications II, edited by S.E. Miller and J.P. Kaminow, (Academic Press, 1988).
2. L. McCaughan and S.E. Korotky, J. Lightwave Tech., LT-4, 1324, (1986).
3. T. Tamir (ed.): Guided-Wave Optoelectronics, 2nd ed., Springer Ser. Electron. and Photon. (Springer, Berlin, Heidelberg 1990).
4. To be published in Electronics Letters.
5. C. Duchet, M. DiMaggio and P. Fabre, Tech. Digest OFC/IOOC, Reno, Nevada, 1987, paper TUQ36.

A Broadband, Low Driving Voltage 2 × 2 Ti:$LiNbO_3$ Optical Switch

T. Nozawa, M. Yanagibashi, H. Miyazawa, K. Kawano, and H. Jumonji

NTT Opto-electronics Laboratories, 3-1, Morinosato Wakamiya, Atsugi-shi, Kanagawa-ken 243-01, Japan

Abstract A 20-GHz bandwidth, <5 V driving voltage at 1.53-μm wavelength improved thermal stability balanced-bridge Mach-Zehnder interferometric switch is demonstrated.

1. Introduction

Optical switches operating in the several 10's of GHz frequency region are attractive for optical communication, optical signal processing and fiber-optic test-instrumentation. Several different types of high-speed switches have been proposed and demonstrated(1-3). The switches proposed so far, however, have had high switching voltages, complex electrode structures, and severe fabrication tolerances. Furthermore, large thermal instability has been reported for the z-cut $LiNbO_3$ devices(4).

In this paper, we report on a 2x2 Ti:$LiNbO_3$ balanced-bridge Mach-Zehnder interferometric switch with 20-GHz bandwidth and 5 V driving voltage at 1.53-μm wavelength that also provides much improved thermal stability.

2. Design Concept

Figure 1 illustrates a waveguide balanced-bridge Mach-Zehnder interferometric switch. It consists of two directional couplers, an input splitter and interferometer output combiner, sets of coupler trimming electrodes and a CPW electrode. The CPW electrode is an 8-μm-wide center conductor separated from two grounded electrodes by 15 μm on each side. The interaction length is set at 27 mm to achieve a 5-V driving voltage.

To broaden the bandwidth, a thick buffer-layer and a shielding plane have been employed, which enable complete velocity matching of the electrical and optical waves(5). The shielding plane efficiently exploits the low dielectric constant of air laid over the CPW electrode. The characteristic impedance of the switch is about 50 ohms.

To improve thermal stability, dual oxidized-silicon films were used for the buffer layer of the switch. The first layer, the one closest to waveguides, is densified SiO_2 which effectively suppresses dc drift(6). However, this kind of film is known to have a tendency to permanent charge buildup. This tendency represents an undesirable driving point shift of the switch, as a result of a surface charge induced by the pyroelectric effect and other effects. Accordingly, a charge-carrier injector is required to compensate for the buildup charge in the first layer. Such an injector can be

Springer Series in Electronics and Photonics, Vol. 29 **Photonic Switching II**
Editors: K. Tada · H.S. Hinton © Springer-Verlag Berlin, Heidelberg 1990

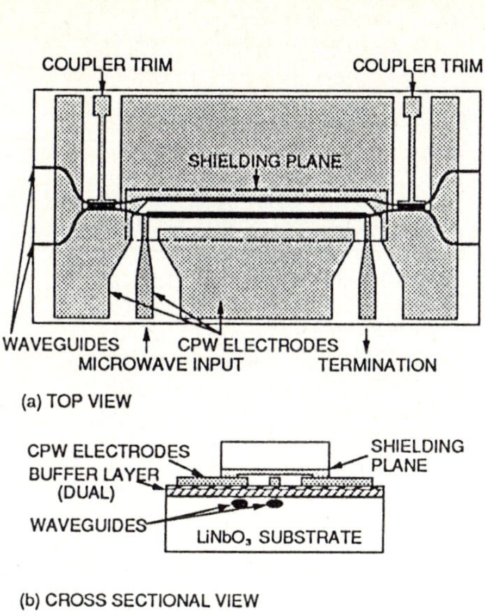

Fig.1 SCHEMATIC DIAGRAM OF THE 2x2 OPTICAL SWITCH

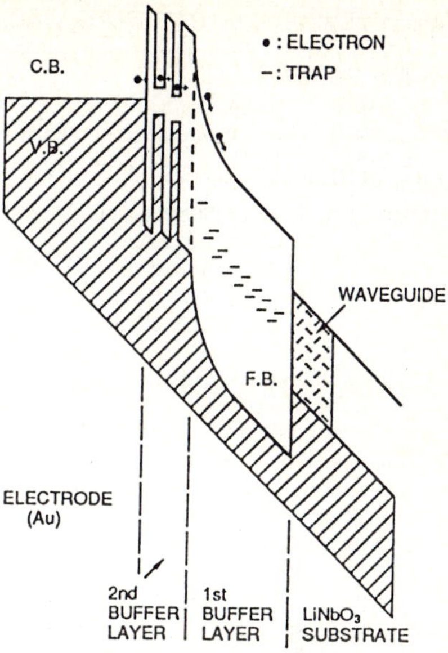

Fig.2 SCHEMATIC ENERGY BAND REPRESENTATION OF THE DUAL BUFFER-LAYER

obtained by inserting a two-phase mixture of Si and insulating oxidized-silicon between the SiO_2 and the electrode[7]. Figure 2 shows a schematic energy band representation of the dual-buffer layer. The layer should effectively suppress the instability because it produces charge-carrier injection from the contact electrode via the two-phase mixture into the SiO_2.

3. Device Fabrication

Table 1 summarizes the techniques used in fabricating the 2x2 optical switch. The optical waveguides were fabricated on the -z plane of optical-grade $LiNbO_3$ substrates using our standard process (1000°C·7hr. and wet Ar atmosphere). An oxidized-silicon dual-buffer layer was formed by ECR plasma CVD[8]. The thicknesses of the 1st and 2nd layers were 1.1 and 0.2 μm, respectively. The two-phase mixture of Si and insulating oxidized-silicon used for the 2nd layer was readily formed by operating at a certain condition[9]. The CPW electrodes were formed from 4-μm-thick electroplated Au. A metallized $LiNbO_3$ was used as a grooved shielding plane to prevent thermal expansion mismatch problems. PANDA (Polarization maintaining and absorption reducing) single-mode fibers were attached to the waveguides using the silicon V-grooved array technique. The fiber array was bonded to the crystal end-faces using a UV curable adhesive[10] closely index matched to the fiber.

Table 1 DEVICE FABRICATION TECHNIQUES

(1) WAVEGUIDE		
substrate	:	z-cut $LiNbO_3$
Ti-pattern	:	6-μm wide, 95-nm thick
diffusion	:	1000°C, 7 hr., wet Ar
(2) BUFFER LAYER		
apparatus	:	ECR plasma CVD
1st layer	:	densified SiO_2, 1.1-μm thick
2nd layer	:	two-phase mixture, 0.2-μm thick
(3) ELECTRODE		
electroplated Au, 4-μm thick		
(4) SHIELDING PLANE		
metallized $LiNbO_3$ with groove		
(5) PIGTAILING		
PANDA fiber array spaced on Si V-groove		

Table 2 PERFORMANCE OF FABRICATED OPTICAL SWITCH

WAVELENGTH	1.53 μm
BANDWIDTH	20 GHz
DRIVING VOLTAGE	<5 V
RISE TIME (10%-90%)	~15 ps (calculated)
INSERTION LOSS	4 dB
CROSSTALK	-20 dB
THERMAL STABILITY (-10—+60°C)	<±5% of V_π

4. Device Performance

Figures 3 and 4 show the measured small-signal optical frequency response and optical response for applied voltage at a through-port, respectively. From the figures, it is clear that the optical 3-dB-bandwidth for the switch is about 20 GHz and driving voltage is less than 5 V. The measured crosstalk at low frequency was -20 dB. The 10%-90% rise time of the optical signal was calculated to be about 15 ps. The fiber-to-fiber insertion loss was 4 dB. Figure 5 shows measured temperature characteristics of a driving point shift. The upper, (a), is that for constant heating and cooling rates (1° C/min) and the lower, (b), is that for abrupt temperature change between 20° C and 30° C. It is apparent that the dual-buffer layer effectively improves thermal stability. The performance of the fabricated optical switch is summarized in Table 2.

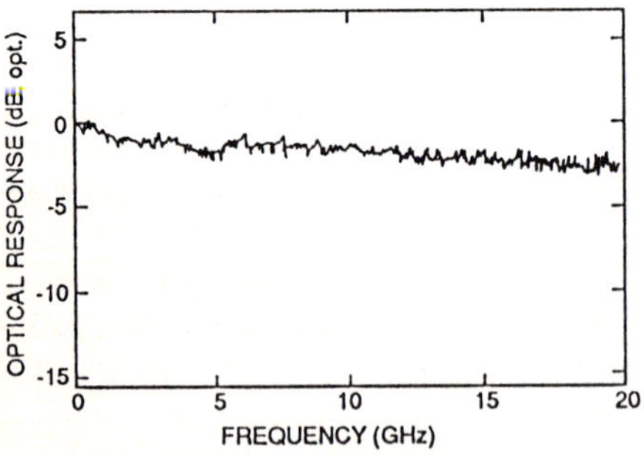

Fig.3 MEASURED FREQUENCY RESPONSE OF THE 2x2 OPTICAL SWITCH

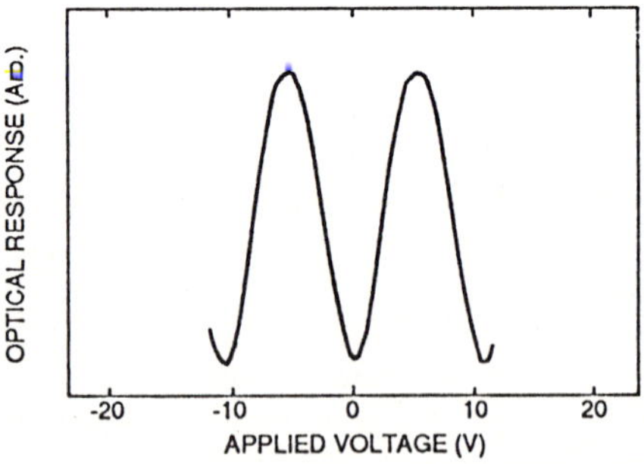

Fig.4 MEASURED OPTICAL RESPONSE vs APPLIED VOLTAGE

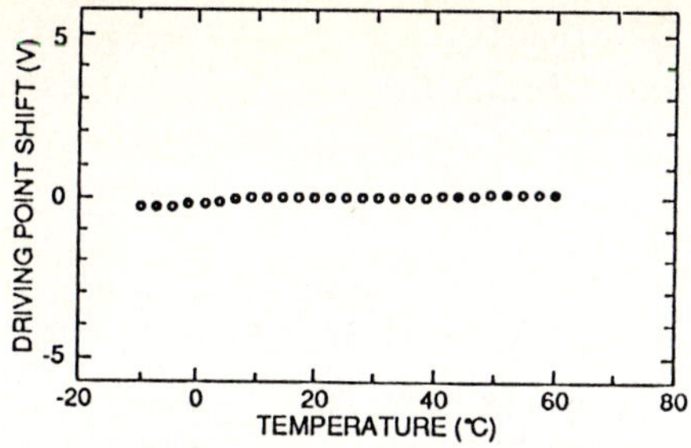

(a) CONSTANT RATE TEMPERATURE CHANGE
(1°C / min)

Fig.5 TEMPERATURE CHARACTERISTICS OF DRIVING POINT SHIFT

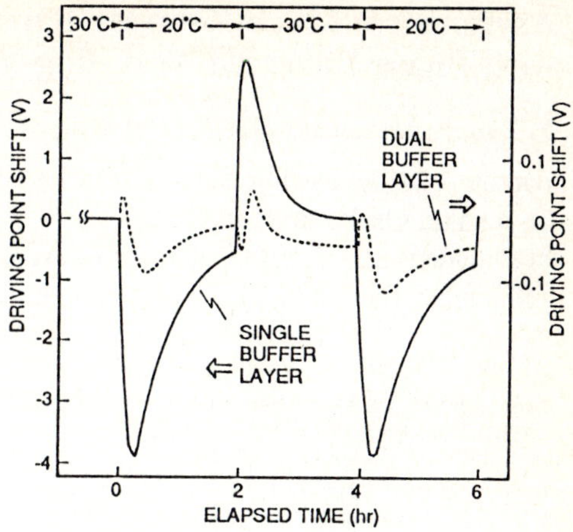

(b) ABRUPT TEMPERATURE CHANGE
(30°C→20°C→30°C→20°C→30°C)

5. Summary

We have fabricated a broadband (20 GHz) and low-driving voltage (<5 V) 2x2 optical switch that operates at 1.53-μm wavelength. We also proposed and experimentally demonstrated a dual-buffer layer which can improve thermal stability of z-cut $LiNbO_3$ optical devices.

Acknowledgments The authors wish to express their deep appreciation to Dr. T. Ikegami, Dr. T. Sugeta and Dr. K. Kurumada of NTT Opto-electronics Labs. for their encouragement.

References

(1) H. Kogelnik et al., IEEE J. Quantum Electron., **QE-12**, pp. 396-401 (1976).
(2) J. Ctyroky, J. Opt. Commun., **1**, pp. 139-143 (1986).
(3) S. K. Korotky et al., in Technical Digest, OFC'90, (1990) paper TUH2.
(4) C. H. Bulmer et al., Appl. Phys. Lett., **48**, pp. 1036-1038 (1986).
(5) K. Kawano et al., Electron. Lett., **25**, pp. 1382-1383 (1989).
(6) A. R. Beaumont et al., in Topical Meeting on IGWO'86, pp. 46-47 (1986).
(7) D. J. DiMaria et al., J. Appl. Phys., **51**, pp. 2722-2735 (1980).
(8) S. Matsuo et al., Jpn. J. Appl. Phys., **22**, pp. L210-211 (1983).
(9) T. Nozawa et al., to be submitted.
(10) K. Nakamura et al., Review of the ECL, **37**, pp. 127-132 (1989).

Asymmetric Mach-Zehnder Band Modulator with Phase Reversed Traveling-Wave Electrode

Yi Zhou, M. Izutsu, and T. Sueta*

Faculty of Engineering Science, Osaka University,
Toyonaka, Osaka 560, Japan
*Permanent address: Department of Applied Physics,
Shanghai Jiao-Tong University, Shanghai 200030, People's Rep. of China

Abstract: We propose a Ti:$LiNbO_3$ Mach-Zehnder band modulator with asymmetric X-junctions and a three sectional phase reversed electrode where the Mach-Zehnder interferometer has an asymmetric structure and the phase reversed effect was achieved by an asymmetric coplanar strip electrode. A 7 GHz 3-dB bandwidth centered at 12 GHz was experimentally obtained. The merit figure of $P/\Delta f$ is 3.9 mW/GHz.

1. Introduction

In order to exploit the very large bandwidth of single-mode fibers, light modulation at multigigahertz frequency is necessary. And, Ti:$LiNbO_3$ integrated optic modulators with traveling-wave structure have been regarded as promising candidates for this demand. To realize higher frequency modulation, required drive power increases in many cases. To reduce the drive power, several band limited modulation schemes have been proposed and experimentally demonstrated [1,2]. Here we report a novel 2 x 2 Mach-Zehnder(M-Z) band modulator using a phase reversed traveling-wave electrode. It is operated successfully in the frequency range from 8 to 16 GHz.

2. Device design and fabrication

An interferometric type modulator using Y branches is one of the most widely used structures for the broadband modulation, but the two arms of interferometer have to be separated wide enough to suppress coupling between the two divided guided modes. And therefore, the wider electrode gap is also required to employ push-pull operation. Using asymmetric X-junctions, the asymmetry of these two arms is led naturally. Waveguide and electrode with narrower separation can be used.

Fig.1 shows a schematic of the device. Asymmetric X-junctions, which have been known as 3-dB couplers [3], are employed instead of conventional Y branches so that the two interferometric arms of M-Z modulator become asymmetric. With the asymmetric M-Z modulator we can achieve higher modulator efficiency because the separation of parallel electrodes placed above two divided waveguides can be reduced. For instance, the coupling between guided modes in asymmetric interferometric arms of 6 μm separation is suppressed to achieve 20 dB extinction ratio if 1 μm difference of their widths are given [4].

On the other hand, to compensate for the inherent velocity mismatch between the modulating microwave and optical wave in traveling-wave modulator, periodic and aperiodic phase reversed structure has been adapted to realize flat base band frequency

Springer Series in Electronics and Photonics, Vol. 29 Photonic Switching II
Editors: K. Tada · H.S. Hinton © Springer-Verlag Berlin, Heidelberg 1990

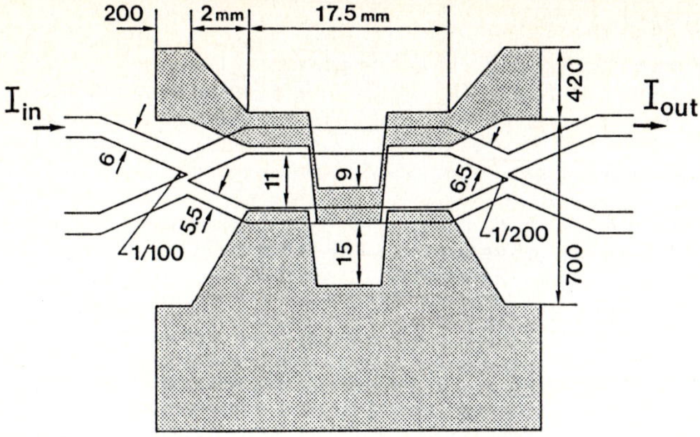

Fig.1. Asymmetric Mach-Zehnder band modulator and parameters of the waveguide and the electrode

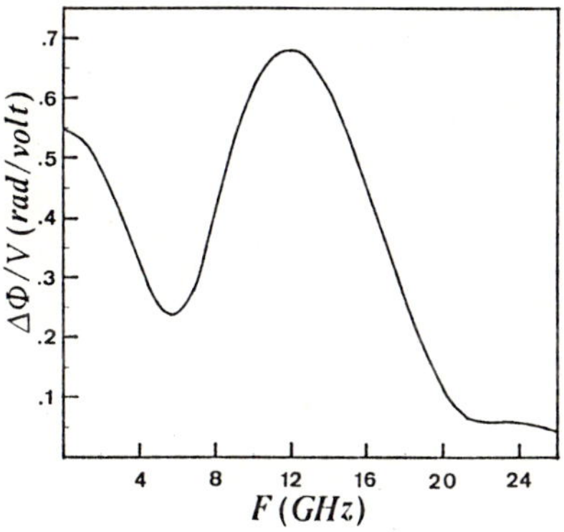

Fig.2. Calculated frequency response of the device

response [5]. The trade-off between bandwidth and drive power still exists. In our case, a phase reversed electrode was applied to obtain band operation [6]. As shown in Fig.1, an asymmetric coplanar strip line is chosen as three sectional phase reversed electrode because of its lower electrical transmission loss [7]. Fig.2 is a calculated example of frequency response for this structure of 2 μm thick electrode where transmission loss of electrode has been taken into account. Band limited operation can be expected in the frequency range from 6 GHz to 20 GHz by this structure with 1.65 cm active electrode length.

In Fig.1 the related parameters of waveguide and electrode are also shown. The waveguide was formed on a z-cut y-prop. $LiNbO_3$ substrate by diffusing an evaporated titanium film at 1050°C for 10 hours. A SiO_2 buffer layer of 110 nm thick was deposited on the surface by CVD technique. An aluminum electrode was then formed on the buffer layer by using the conventional dry etching process.

3. Modulation Experiment

The modulator was first tested at low frequency. The device has been evaluated using quasi-TM mode at 1.3 μm wavelength. The light was end-fire coupled into and out of the polished waveguide ends using microscope objectives. The measured switch or half-wave voltage is 10 V while the calculated value is 5.9 V. The extinction ratio is 17 dB.

The frequency response of the device in high frequency range was measured by utilizing an envelope detection method [5,8]. A dc voltage was applied to the modulator together with a microwave signal to obtain zero optical bias. The modulating microwave was further square-wave modulated at 1 kHz. A 1 kHz square-wave was, therefore, observed by detecting the output light with a photodiode. In Fig.3, the measured modulation depth with the drive power leveled at 27 mW is plotted versus frequency ranging from 8 to 18 GHz centered at 12 GHz where 1 rad phase retardation is estimated to be achieved. The switching voltage or half-wave voltage at this frequency is evaluated as 5.2 Vp-p. The measured 3-dB bandwidth is 7 GHz and the merit figure of $P/\Delta f$ is derived from these values as 3.9 mW/GHz for 88 percent intensity modulation. The solid line in Fig.3 is a frequency response calculated assuming 22 mW drive power.

4. Conclusion

We have proposed, fabricated and tested a high-speed, low drive voltage asymmetric M-Z band modulator. Asymmetric coplanar strip line has been applied to three sectional phase reversed electrode successfully to realize band limited operation. With the asymmetric X-junctions or the asymmetric M-Z modulator, an extinction ratio of 17 dB is achieved in spite of the narrow separation between two interferometric arms. High-speed modulation by low drive power has been realized with narrow electrode gap. It can be expected that the required drive power of modulator can be decreased dramatically if the separation between two arms, thus the electrode gap, is reduced further. This is useful in realistic system applications.

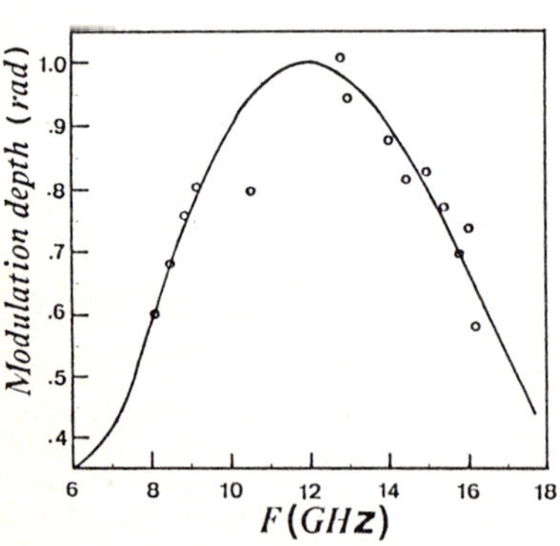

Fig.3. The measured and calculated frequency responses

References

1. M.Izutsu and T.Sueta, IOOC'89, 19D4-1, KOBE,July 1989.
2. G.E.Betts, IGWO'89, MAA4-4, Houston, Feb.1989.
3. M.Izutsu, A.Enokihara, and T.Sueta, Opt.Lett., Vol.7, pp549-551(1982).
4. J.L.Jacker and J.Johnson, J.Lightwave Technol., Vol.6, pp1348-1351(1988).
5. D.Erasme and M.G.F.Wilson, Opt. and Quantum Electron., Vol.18, pp203-211(1986).
6. D.Erasme, D.A.Humphreys, A.G.Roddie, and M.G.F.Wilson, J.Lightwave Technol., Vol.6, pp933-936(1988).
7. M.Izutsu, T.Itoh, and T.Sueta, IEEE J.Quantum Electron., Vol.QE-14, pp394-395(1978).
8. I.P.Kaminow, and W.M.Shapless, Appl.Opt., Vol.6, pp351-352(1967).

Polarization Beam Splitter Switch with Controlled Silica Waveguide Birefringence on Si Substrate

M. Okuno, N. Takato, M. Kawachi, and A. Sugita

NTT Opto-electronics Laboratories, 162 Shirakata-shirane, Tokai, Naka-gun, Ibaraki 319-11, Japan

Abstract. A novel polarization beam splitter switch composed of silica waveguides on Si is demonstrated. Fiber-waveguide fiber insertion loss of 0.5dB and crosstalk attenuations of 16-25dB are attained.

1. Introduction

A polarization beam splitter (PBS) is required in constructing a variety of optical components such as polarization-diversity receivers, optical circulators and optical fiber amplifiers for optical communication systems. Conventional PBSs have been realized mostly in bulk-optic configurations using micro-lenses and prisms. These bulk-optic PBSs, however, appear to have some limitations in terms of producibility and suitability for integration. A possible approach to overcoming this problem is the introduction of waveguide technologies to form integrated-optic PBSs.

This paper describes a successful application of silica waveguide technologies to integrated-optic PBSs with switchable polarization states.

2. Principle

The configuration of the PBS switch proposed in this work is shown in Fig.1. This PBS switch has a Mach-Zehnder interferometer configuration consisting of two 3-dB couplers and two waveguide arms of the same length. One of the waveguide arms is equipped with a stress-applying film for waveguide birefringence control. The other

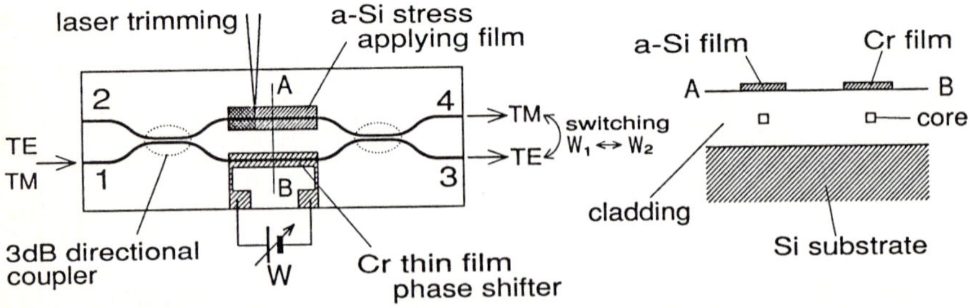

Fig. 1 Configuration of PBS switch

Springer Series in Electronics and Photonics, Vol. 29 **Photonic Switching II**
Editors: K. Tada · H.S. Hinton © Springer-Verlag Berlin, Heidelberg 1990

arm is equipped with a thin-film heater for thermo-optic phase control. Our basic idea is to adjust the optical path difference between the two couplers simultaneously to Nλ for TM polarization and to (N+0.5)λ for TE polarization by trimming the stress-applying film and tuning the thermo-optic phase shifter. This adjustment leads to the PBS operation of (1$\rightarrow$4 for TM, 1$\rightarrow$3 for TE) shown in Fig.1. This polarization state can be switched to (1$\rightarrow$4 for TE, 1$\rightarrow$3 for TM) by a further $\lambda/2$ increase in the path difference by tuning the thermo-optic phase shifter.

3. Experiment

The PBS switch shown in Fig.1 was fabricated using silica single-mode waveguide technologies. A combination of flame hydrolysis glass deposition and reactive ion etching was used to form an MZ interferometer consisting of 8µmx8µm SiO_2-TiO_2 cores and 50µm-thick SiO_2 cladding on a crystal Si substrate[1-3]. Magnetron-sputtered amorphous Si (a-Si) film was used as a stress-applying film. Thin Cr film, 10 mm long, was deposited as a thermo-optic phase shifter[4]. The chip size was 30mm x 3mm.

Figure 2(a) shows a cross-sectional cross-Nikol observation of a simple silica glass substrate sputtered with a-Si film. It is confirmed that a large stress concentration exists beneath the sputtered a-Si film. The measured waveguide birefringence B as a function of the a-Si film width is shown in Fig.2(b). The a-Si thickness of this sample was 6µm. Without the a-Si film(D=0), the waveguide exhibited a birefringence of $B_0=4\times10^{-4}$ due to compressive stress exerted by the Si substrate. As the width D was increased, birefringence B passed a peak value of 5.5×10^{-4} at D=50 µm and then decreased to 2.5×10^{-4}. This peculiar behavior, reflecting the large residual stress in sputtered a-Si films, was also supported by stress distribution analysis based on the finite element method. An a-Si film with D=50µm was used in this work. Hence, the waveguide birefringence B was changed from 5.5×10^{-4} to 4×10^{-4} by trimming the a-Si film. This trimming leads to the

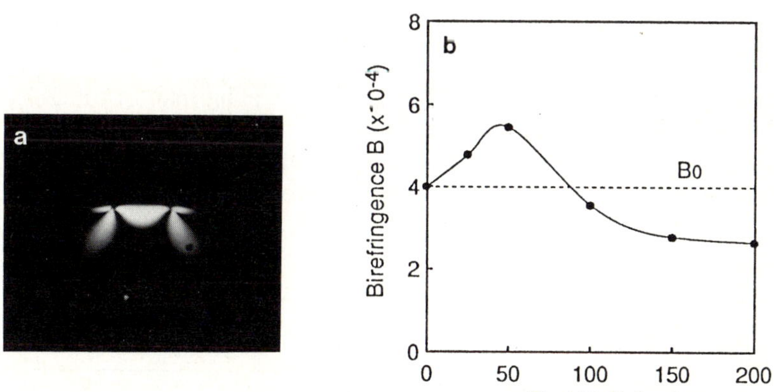

Fig.2 (a)Cross-Nikol observation in a simple silica substrate with a sputtered a-Si film, (b)waveguide birefringence as a function of the a-Si width

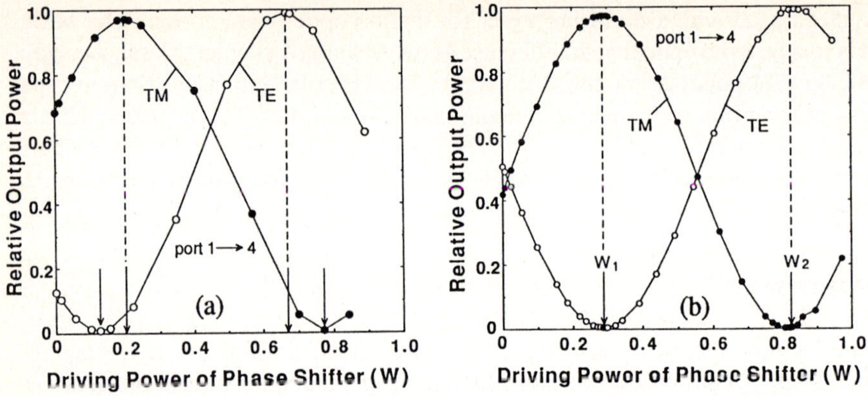

Fig.3 Thermo-optic responses of PBS switch (a) before trimming, and (b) after trimming

optical path difference between TE and TM modes. This optical path difference was precisely adjusted to $\lambda/2$ by trimming the a-Si film with a Q-switched YAG laser cutter[5].

Figures 3(a) and (b) show the thermo-optic responses of the MZ interferometer before and after laser trimming. The relative output optical powers for the TE and TM modes along the cross-path (1→4) are plotted as a function of the thermo-optic phase shifter driving power. A 1.55μm DFB-LD was used as a light source. In Fig.3(a), the phase difference between the TE and TM modes does not fit $\lambda/2$ because of uncontrolled waveguide birefringence. On the other hand, the optical path difference is successfully adjusted to $\lambda/2$ in Fig.3(b). The a-Si film length L after trimming was 4.4mm. This fact indicates that the birefringence increment ΔB by a-Si loading was 1.7×10^{-4} , since $\Delta BL=\lambda/2$. The thermo-optic phase shift required for the PBS operation shown in Fig.3(b) was $W_1=0.29$ W for the TE1→3/TM1→4 state and $W_2=0.82$ W for the TE1→4/TM1→3 state. The measured switching response time between the two states was 1 msec. The insertion loss of the PBS switch was as low as 0.5dB including input and output single-mode fiber coupling losses. The measured cross talk attenuation was 16-25 dB as listed in Table I. Based on the crosstalk measurements shown in Table I, the coupling ratio of the 3-dB couplers in the fabricated PBS switch was estimated to be 53% for the TM mode and 55% for the TE mode. The crosstalk attenuations will be improved by further adjustment of the coupling ratio to 50 %. Another problem to be solved is the

Table I Crosstalk Attenuation Characteristics

Heater Driving Power (W)		$W_1=0.29$		$W_2=0.82$	
Output Port (Input port : 1)		3	4	3	4
Relative Output Power	TE-mode	0.995	0.005	0.011	0.989
	TM-mode	0.026	0.974	0.997	0.003
Crosstalk Attenuation (dB)		15.9	23.1	19.6	24.6

reduction of the holding power of W_1 to zero. This will be readily achieved by designing a small path difference in the MZ interferometer arms, corresponding to the W_1 phase shift.

4. Conclusions

A Mach-Zehnder interferometer type polarization-beam splitter switch composed of silica based waveguides is demonstrated for the first time. The waveguide birefringence is accurately controlled by a-Si stress-applying film and laser trimming. The output polarization states can be switched by controlling the thermo-optic phase shifter. The switching response time is 1 msec. The fiber-waveguide-fiber insertion loss is 0.5 dB and polarization crosstalk attenuations are 16-25 dB. This type of waveguide PBS switch will find a wide application in the construction of a variety of integrated-optic devices with controlled polarization-sensitivity.

Acknowledgments. The authors wish to thank H. Okazaki, M. Yasu, K. Onose and K. Shimada for their help in device fabrication.

References

1. M. Kawachi, M. Yasu, and T. Edahiro: Electron. Lett. 19, 583 (1983)
2. N. Takato, M. Yasu, and M. Kawachi: Electron. Lett. 22, 321 (1986)
3. N.Takato, K.Jinguji, M. Yasu, H. Toba, and M. Kawachi: J.Lightwave Technol LT-6,1003 (1988)
4. A. Sugita, K. Jinguji, N. Takato, K. Katoh, and M. Kawachi: Trans.IEICE E 73, (1990)
5. M. Kawachi, N. Takato, K. Jinguji, and A. Sugita: in Tech. Dig. CLEO'89, TUJ17 (1989)

Part III

Optically Controlled Switches

Light-Controlled Optical Modulation by Quantum Well Structure

S. Noda, T. Uemura, T. Yamashita, and A. Sasaki

Department of Electrical Engineering, Kyoto University, Kyoto 606, Japan

Abstract. The modulation of interband-resonant light by intersubband-resonant light using quantum well is proposed, and results of theoretical analysis are reported. It is shown that a large absorption coefficient change from 100 to 3160cm^{-1} for interband-resonant light can be obtained by changing the intensity of intersubband-resonant light from 0 to 1MW/cm^2. Very fast modulation speed ∿ ps can be expected.

1. Introduction

There is an increasing interest in optical information processing and optical computing. In such fields, optical modulator is one of indispensable devices. However, the modulation has been typically made by electronic method. To achieve all-optical operation, the modulation should be made by optical method. In this paper, we report light-controlled optical modulation using quantum well structure.

There are various energy levels in quantum well structure. Recently, it has been shown that the strong optical absorption occurs due to electron transition between intersubbands or between bound to free energy levels in quantum well structure.[1-3] By utilizing this intersubband (or bound to free) transition and conventional band to band (interband) transition simultaneously, we can expect to realize light-controlled optical modulation. In the following, we propose the modulation of interband-resonant light by intersubband-resonant light and report results of theoretical analysis.

2. Modulation principle and theoretical analysis

Figure 1 shows schematic drawing to show the principle of this type of optical modulation. The energy level E_{c1} is filled with electrons, and therefore the absorption coefficient for the light with photon energy of

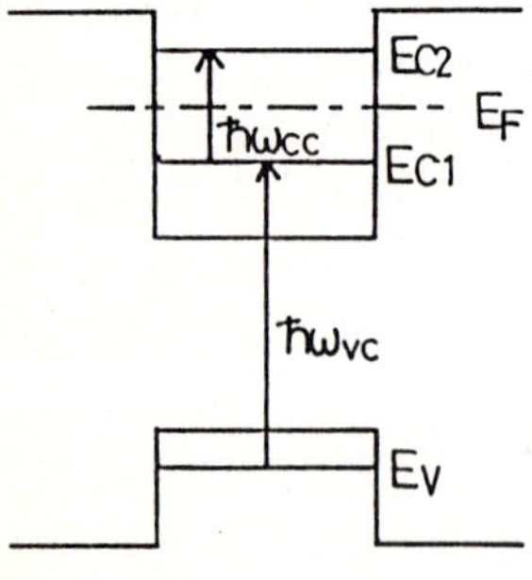

Fig.1 Schematic drawing to show the principle of light-controlled optical modulation by quantum well. The interband-resonant light is modulated by intersubband-resonant light.

Springer Series in Electronics and Photonics, Vol. 29 Photonic Switching II
Editors: K. Tada · H.S. Hinton © Springer-Verlag Berlin, Heidelberg 1990

$E_{c1}-E_v$ (interband-resonant light) is very small. However, if the light with photon energy of $E_{c2}-E_{c1}$ (intersubband-resonant light) is irradiated to the quantum well and the electrons in E_{c1} level are excited to E_{c2} level, the absorption coefficient for the interband-resonant light is increased. Therefore interband-resonant light can be modulated by the intersubband-resonant light.

Modulation characteristics were analysed by solving the density matrix equation

$$\frac{\partial \rho}{\partial t} = \frac{1}{i\hbar}[H_0 - M(F_{vc}+F_{cc}), \rho] - \frac{1}{2}(\Gamma\rho + \rho\Gamma) \tag{1}$$

where H_0 is the unperturbed Hamiltonian, M is a dipole operator, F_{vc} and F_{cc} are the electric fields of interband- and intersubband-resonant light, respectively. Γ is the phenomenological operator responsible for carrier relaxation.

The obtained absorption coefficient for the interband-resonant light is expressed as

$$\alpha = \alpha_0 - a\cdot I_{vc} + b\cdot I_{cc} \tag{2}$$

where α_0 represents the linear absorption term, $-a\cdot I_{vc}$ the absorption saturation term for interband-resonant light itself, and $b\cdot I_{cc}$ the modulation term by intersubband light, where I_{vc} and I_{cc} are the intensities of interband- and intersubband-resonant lights, respectively. α_0, a, and b are expressed as

$$\alpha_0 = \omega_{vc}\sqrt{\frac{\mu}{\varepsilon}}\frac{1}{L}\int \frac{M_{vc}{}^2[f(E_v)-f(E_{c1})]\hbar/\tau}{(E_{vc}(k)-\hbar\omega_{vc})^2+(\hbar/\tau)^2} g(k)dk \tag{3}$$

$$a = \omega_{vc}\sqrt{\frac{\mu}{\varepsilon}}\frac{4}{L\varepsilon_0 n_r c}\int \frac{M_{vc}{}^4[f(E_v)-f(E_{c1})]\hbar/\tau}{[(E_{vc}(k)-\hbar\omega_{vc})^2+(\hbar/\tau)^2]^2} g(k)dk \tag{4}$$

$$b = \omega_{vc}\sqrt{\frac{\mu}{\varepsilon}}\frac{M_{cc}{}^2}{L\varepsilon_0 n_r c}\left[\int \frac{M_{vc}{}^2 2[f(E_{c1})-f(E_{c2})]}{[(E_{vc}(k)-\hbar\omega_{vc})^2+(\hbar/\tau)^2]\hbar/\tau} g(k)dk\right.$$

$$+\int \frac{M_{vc}{}^2[(\hbar/\tau)^2-(E_{vc}(k)-\hbar\omega_{vc})^2][f(E_{c1})\ f(E_{c2})]}{[(E_{vc}(k)-\hbar\omega_{vc})^2+(\hbar/\tau)^2]^2\hbar/\tau} g(k)dk$$

$$\left.-\int \frac{M_{vc}{}^2[(\hbar/\tau)^3-3(E_{vc}(k)-\hbar\omega_{vc})^2(\hbar/\tau)][f(E_v)-f(E_{c1})]}{[(E_{vc}(k)-\hbar\omega_{vc})^2+(\hbar/\tau)^2]^6} g(k)dk\right] \tag{5}$$

where $\hbar\omega_{vc}$ and $\hbar\omega_{cc}$ are the photon energies of interband- and intersubband-resonant lights, respectively. M_{vc} and M_{cc} are the matrix elements of interband and intersubband transitions, respectively. τ is the intraband relaxation time, L is the well width, n_r is refractive index, and $E_{vc}=E_{c1}-E_v$. f is Fermi distribution function and g is density of states.

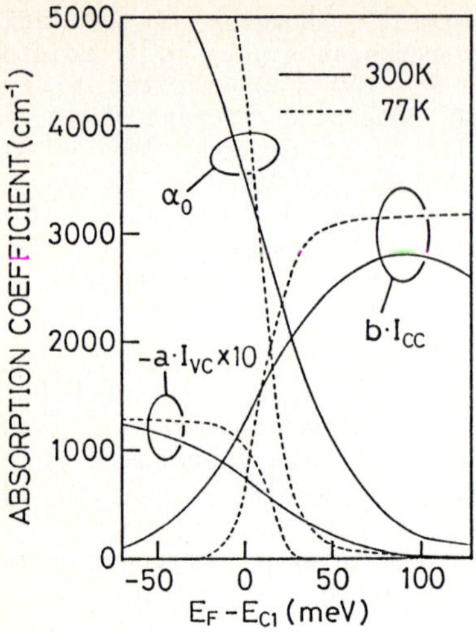

Fig.2 Calculated values of α_0, $-a \cdot I_{vc}$, and $b \cdot I_{cc}$ as a function of Fermi-energy level, where $I_{vc}=I_{cc}=1MW/cm^2$.

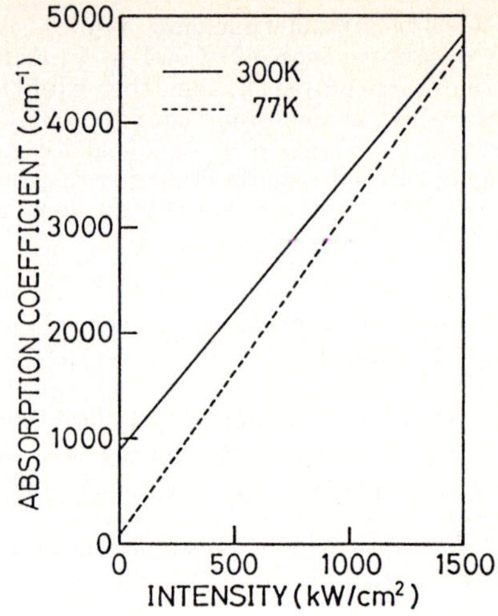

Fig.3 Absorption coefficient for interband-resonant light as a function of intensity of intersubband-resonant light.

The numerical calculation was done for the case of GaAs/AlGaAs quantum well structure, where GaAs well width is 88Å and the Al fraction of AlGaAs barrier is 0.3. Therefore, the wavelengths of interband- and intersubband-resonant light are chosen to be 836.1nm and 10.6μm, respectively, at 300K. Figure 2 shows the calculated values of each term in Eq.(2) as a function of Fermi energy level which is related to the electron concentration of quantum well. The relaxation time τ is taken to be 0.2ps and the temperature is taken to be 77 or 300K. From Fig.2, the linear absorption term (α_0) and the absorption saturation term ($-a \cdot I_{vc}$) are decreased with Fermi-energy level, while the modulation term ($b \cdot I_{cc}$) is increased with Fermi-energy level. When the Fermi energy is 57meV above the E_{c1} level (the electron concentration is $1 \times 10^{18} cm^{-3}$), the linear absorption term is $100cm^{-1}$ and absorption modulation term is $3060cm^{-1}$ at 77K. Therefore, we can obtain the change of absorption coefficient from $100cm^{-1}$ to $3160cm^{-1}$ for the interband-resonant light by changing the intensity of intersubband-resonant light from 0 to $1MW/cm^{-2}$ as shown in Fig.3. The response time of this modulation is determined by the intersubband relaxation time (∿ps) and very fast switching speed can be expected.

3. Summary

The light-controlled optical modulation using quantum well structure has been theoretically investigated. It has been shown that the interband-resonant light can be modulated by intersubband-resonant light. The absorption coefficient for the interband-resonant light can be changed from $100cm^{-1}$ to $3160cm^{-1}$ by changing the intensity of intersubband light from 0 to $1MW/cm^{-2}$. A very fast modulation speed of ∿ ps can be expected for this device.

The authors would like to thank Dr. Takeda for fruitful discussion. This work was partly supported by a Grant-in-Aid for Encouragement of Young Scientist from the Ministry of Education, Science and Culture, and also by the Inamori Foundation.

References

1. L.C.West and S.J.Eglash, Appl.Phys.Lett., **46**, 1156 (1985).
2. B.F.Levine, R.J.Malik, J.Walker, K.K.Choi, C.G.Bethea, D.A.Kleinman, and J.M.Vandenberg, Appl.Phys.Lett., **50**, 273 (1987).
3. Z.Ikonic, V.Milanovic, and D.Tjapkin, Appl.Phys.Lett., **16**, 247, (1989).

All-Optical Multiple Quantum Well Grating Switch

Y. Shibata, M. Ikeda, K. Nakashima, T. Tamamura, and T. Nishida

NTT Opto-electronics Laboratories, 3-1 Morinosato Wakamiya,
Atsugi-shi, Kanagawa-ken 243-01, Japan

A novel structure for an all-optical switch is proposed. A semiconductor multiple quantum well (MQW) waveguide with a deep grating is fabricated, and subnanosecond switching is observed in this structure using a YAG control light. Switching power is as low as 10^4 W/cm^2.

1. Introduction

An all-optical switch is indispensable in all-optical signal processing systems. Semiconductor MQW etalons, which are suitable for parallel signal processing, have been proposed as the all-optical switch because of the fast and large optical nonlinearity of the MQW, and some demonstrations have been reported[1-4]. On the other hand, a waveguide type all-optical switch is also necessary for optical fiber communication systems. In this paper, we propose a novel structure for a waveguide type all-optical switch and report the first demonstration of the switching operation.

2. Design and Fabrication of the Switch

An all-optical switch for optical fiber communication systems should have the following characteristics. First, the switching speed should be fast. Next, the extinction ratio should be large enough for system requirements. The working bandwidth should also be large for wavelength tolerance of the signal light source. Furthermore, the switch should be suitable for cascade connection. With these features in mind, the novel structure as shown in Fig.1 is proposed. The switch has a semiconductor MQW layer with a deep grating as the waveguide core. The switching principle of this device is as follows. When the control light is injected into the MQW layer, the refractive index of the MQW layer is changed by a nonlinear effect such as the ac Stark effect[2], carrier effect(band filling or plasma effect)[3] or thermal effect[4]. This leads to the shift of the Bragg wavelength of the switch as shown in Fig.2, and the transmittance at the signal light wavelength changes.

Our novel structure has the following advantage. The waveguide structure, which is suitable for cascade connection, enhances the efficiency of the nonlinear effect. The strong coupling grating enables a large extinction ratio and large operation wavelength bandwidth. For example, a 50Å bandwidth with a 15 dB extinction ratio is expected in a 50 μm long grating switch with coupling coefficient κ of 600 cm^{-1}. This structure also makes it easier to reduce the control light power because the illuminated region is very small.

The fabrication process is as follows. First, InGaAsP/InP MQWs with 20 periods of 100Å wide wells and 100Å wide barriers are grown by metal-organic molecular beam epitaxy (MOMBE). A clear exciton absorption peak is observed at 1.28 μm. Then, a first order deep grating is fabricated onto the MQW layer using electron

Springer Series in Electronics and Photonics, Vol. 29 Photonic Switching II
Editors: K. Tada · H.S. Hinton © Springer-Verlag Berlin, Heidelberg 1990

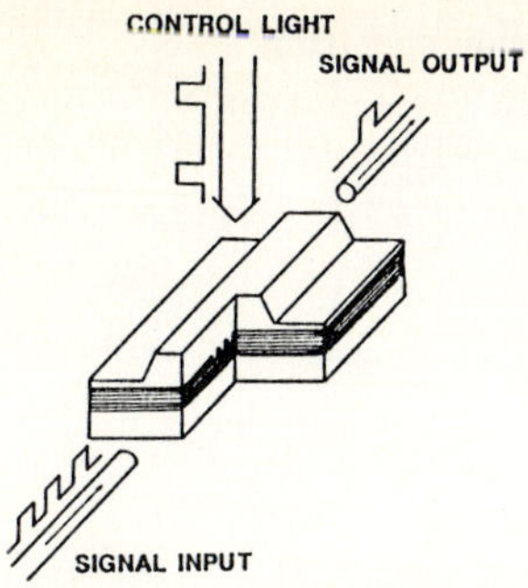

Figure 1 Schematic diagram of the grating switch.

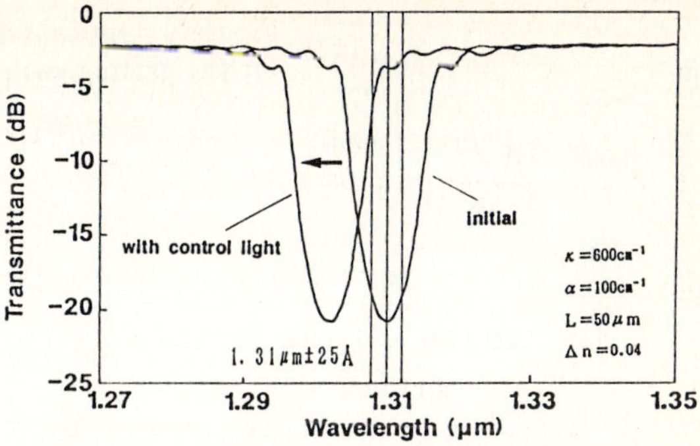

Figure 2 Calculated wavelength-transmittance characteristics of the switch.

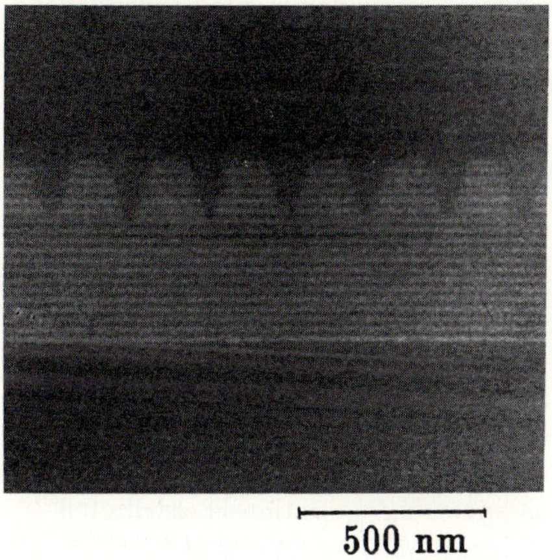

Figure 3 SEM photograph of the MQW grating.

beam lithography and argon ion beam etching (Ar-IBE). The period, depth and length of the grating are 2007Å, 1600Å and 230 μm, respectively. After the MQW grating is embedded by 1.4 μm thick InP clad using MOMBE, a ridge waveguide of 3 μm width is fabricated by Ar-IBE. Finally, the waveguide is cleaved into 300 μm length. A cross-sectional view of the MQW grating is shown in Fig.3. The deep grating is successfully embedded by InP without any deformation. No change in the photoluminescence spectrum is observed except a decrease in its intensity by about 30 percent, indicating that the MQW structure is conserved during these processes.

3. Switching Characteristics

The transmission spectrum of the grating switch is measured using a wavelength variable laser diode with an external grating mirror. The signal light is coupled into the waveguide of the grating switch using a single mode fiber with a tapered spherical end. The output optical power from the switch is coupled into a multimode fiber with a tapered spherical end and detected by an optical power meter. The measured spectrum is shown in Fig.4. A large stopband is observed at the center wavelength of 1.305 μm. The reflection bandwidth (FWHM) of the grating switch is about

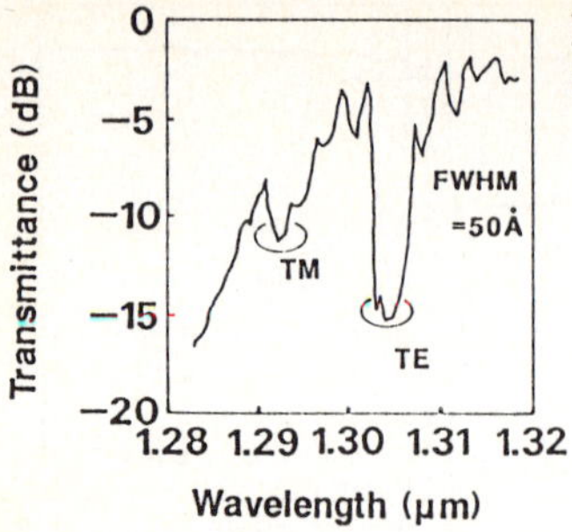

Figure 4 Optical transmission spectrum for the grating switch.

50Å. The coupling coefficient κ of the grating is estimated to be 300 cm^{-1} from the reflection bandwidth, which is in good agreement with the value expected from the coupled-mode theory[5]. The expected attenuation at the Bragg wavelength exceeds 30 dB, however, the measured value is 12 dB. This discrepancy is due to insufficient suppression of the background light level because of using multimode fiber to detect the output power. At a wavelength of 1.292 μm, another stopband is observed. This is supposed to be the TM mode stopband.

The switching response is investigated using a 1.06 μm YAG laser as the control light source. The output optical power from the switch is coupled into a multimode fiber with a tapered spherical end and detected by an Ge APD. When a 500 ns wide Q-switched control light pulse with 1 kHz repetition rate is exposed onto the switch using multimode fiber, the 1.302 μm signal light transmittance is decreased as shown in Fig.5(a) (NOR operation). At a wavelength of 1.304 μm, on the other hand, the signal light transmittance is increased by the control light as shown in Fig.5(b) (AND operation). This indicates that the Bragg wavelength of the switch is shifted to a shorter wavelength by the control light. This is caused by the negative nonlinear refractive index change, and the origin of this switching operation is attributed to the carrier effect[3]. The shift of the Bragg wavelength $\Delta\lambda_B$ is estimated to be about 50Å. The extinction ratio is estimated to exceed 10 dB from $\Delta\lambda_B$=50Å in Fig.4. The observed switching speed is limited by the control light pulse width. The switching power is as low as 10^4 W/cm^2.

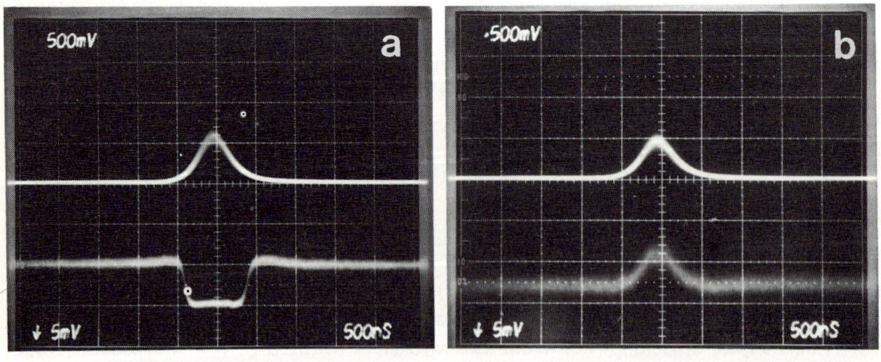

Figure 5 Switching response of the grating switch. Upper trace is the YAG control light, and lower trace is the signal light at wavelengths of (a)1.302 μm and (b)1.304 μm.

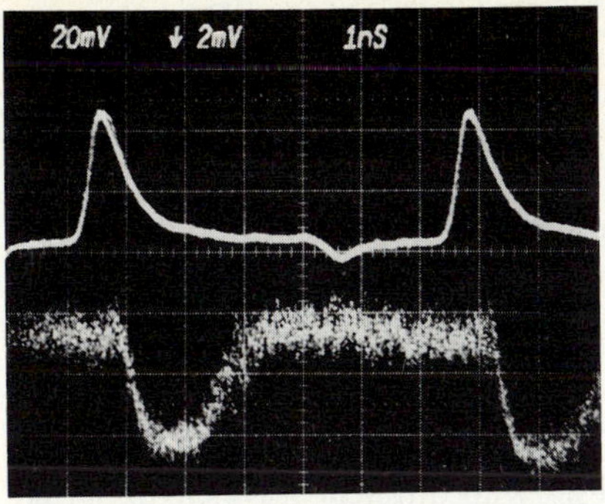

Figure 6 Switching response of the grating switch. Upper trace is the YAG control light, and lower trace is the signal light at a wavelength of 1.304 μm.

In order to estimate the switching speed, a mode locked YAG laser pulse is employed as the control light. The repetition rate and the pulse width are 160 MHz and 500 ps, respectively. Figure 6 shows the response at a wavelength of 1.304 μm. Subnanosecond switching originating from the carrier effect is observed. A thermal effect due to the high repetition rate is also observed, however, the carrier effect dominates and the switch responds to the control light pulse. In this case, the Bragg wavelength at the initial state is influenced by the thermal effect, and is shifted about 25Å to a longer wavelength.

This first demonstration of the grating switch should provide a significant step in the development of an optical wavelength switch or a matrix switch controlled by an optical gate signal.

4. Conclusion

We have proposed a novel structure for an all-optical switch and demonstrated the switching operation in this structure using a YAG control light. Subnanosecond switching is achieved with control light power of 10^4 W/cm^2. The switching operation is caused by the carrier effect.

We would like to thank Y. Kawaguchi and J. Tenmyo for their valuable support and discussions on MOMBE growth, and also K. Kurumada for many stimulating discussions.

References

[1] H. M. Gibbs et al., Appl. Phys. Lett. **35**, 451 (1979)
[2] D. Hulinet al., Appl. Phys. Lett. **49**, 749 (1986)
[3] A. Migus et al., Appl. Phys. Lett. **46**, 70 (1985)
[4] J. L. Jewell et al., Appl. Phys. Lett. **40**, 291 (1982)
[5] W. Streifer et al., IEEE J. Quantum Electron. **QE-11**, 867 (1975)

All-Optical Set–Reset Operation of a Bistable Laser Diode at a Repetition Rate of 100 MHz

P. Blixt and U. Öhlander

Department of Physics II and Institute of Optical Research,
Royal Institute of Technology, S-10044 Stockholm, Sweden

Abstract. We report on all-optical set and reset with different wavelengths of an inhomogeneously pumped bistable laser diode. The optical reset time was about 120 ps. If the excess energy for optical reset was too small, then an automatic turn-on after 0.5-1.0 ns was observed.

1. Introduction

Bistable laser diodes have been used in a system experiment performed by NEC as dynamic optical memories for exchange of information in a high-speed optical time-division system at 256 Mbit/s [1]. Fujitsu has later increased the speed to 512 Mbit/s [2]. Usually the bistable laser is set by an optical pulse and reset by a negative electrical pulse. In order to avoid many problems with high-speed electronics, such as ringing, reflections and electronic interference, it is advantageous to use photons instead of electrons for switching. To take full advantage of photonic switching it is desirable to make the switching all-optical. The fastest way to switch off a bistable laser electrically has been by extracting the charge carriers from the absorber section by applying a negative current to it [3]. However, Inoue *et al.* have demonstrated optical reset of a bistable laser utilizing the beat frequency between injected light and the lasing light of the bistable laser [4]. In addition, gain quenching by injecting a wavelength which is longer than the bistable lasing wavelength has been used by Odagawa *et al.* [5]. Here they utilized stimulated recombination to decrease the charge carrier density to a level below the lasing threshold. The reset time was about 30 ns at a light intensity of 170 μW. Both the set and the reset operation were achieved with the same wavelength. We have used the same method as Odagawa *et al.* but at higher repetition rates.

We report on optical set and reset with different wavelengths at a repetition rate of 100 MHz with transition times of less than 120 ps.

2. Experimental set-up

The experimental set-up consisted of three lasers. A bistable laser diode was optically set by a gain-switched laser diode and optically reset by an actively mode-locked laser diode. The bistable laser diode was an inhomogeneously pumped multi-section Fabry-Perot type with a main lasing wavelength at 1310 nm. The absorber section was situated at the facet where the optical switching pulses were incident. The bistable laser was biased in the middle of the hysteresis loop. It was biased 0.40 mA below electrical switch-on and 0.55 mA above electrical switch-off. Further details about this bistable laser diode can be found in Refs. [6-8]. The optical set pulses were obtained from a gain-switched laser at a wavelength of 1310 nm with a Full-Width at Half Maximum (FWHM) of 2.8 nm. The optical reset pulses were obtained from an actively mode-locked laser. The repetition rate was 100 MHz and the pulse duration was less than 50 ps as measured by a *p-i-n* diode. The wavelength of the mode-locked laser was controlled by a grating acting as the reflector of the external cavity. During switching

Springer Series in Electronics and Photonics, Vol. 29 **Photonic Switching II**
Editors: K. Tada · H.S. Hinton © Springer-Verlag Berlin, Heidelberg 1990

the wavelength was 1322 nm. By focusing on the grating, we obtained a rather broad spectrum of about 1.1 nm FWHM, thus avoiding cavity resonance effects.

Adjacent to the collimating microscope objectives of the gain-switched laser and the mode-locked laser, optical isolators were inserted to prevent feedback. The pulses from the two laser were directed co-linearly into the bistable laser diode via a beam splitter. Microscope objectives were used for collimating and focusing light from and to the laser diodes. After the bistable laser diode, a Fabry-Perot étalon with a bandwidth of 5 nm FWHM was inserted to filter out the transmitted long wavelength pulse and its amplified part. The signal was detected by a *p-i-n* diode connected to a 4 GHz broadband amplifier and then fed to a sampling oscilloscope.

3. Results

The spectrally filtered output signal can be seen in Fig. 1a (lower trace) when switching with optical input pulses at different wavelengths (upper trace). As can be seen in Fig. 1a, the set pulse from the gain-switched laser consists of several small pulses. However, the total input coupled set pulse energy was 220 fJ. The set pulse had the same wavelength as the bistable laser (1310 nm). Also seen in the figure is that the reset pulse from the mode-locked laser had a small trailing pulse. The input coupled reset pulse energy was 290 fJ at a wavelength of 1322 nm.

In Fig. 1b the fall time of the bistable laser is shown (lower trace) and the long wavelength reset pulse (upper trace). The bistable fall time was measured to be about 120 ps and the detection system could measure fall times down to about 100 ps.

We then upgraded the 100 MHz system for switching with a repetition rate of 550 MHz. The mode-locked reset pulses had less than 50 ps duration and were tuned to a wavelength of 1326 nm. The resulting output signal from the bistable laser diode is shown in Fig. 2. However, in this set-up we did not filter out the input signal, which resulted in a large transmitted input signal with a duration of approximately 50 ps. We obtained a fall time below 100 ps with an optical input energy of 75 fJ. The on:off contrast ratio was 5:1. However, when the duty cycle was changed, the bistable laser returned to the on-state at 0.5-1.0 ns after switch-off. This is similar to the automatic turn-on obtained with electric reset pulses with insufficient overdrive [9].

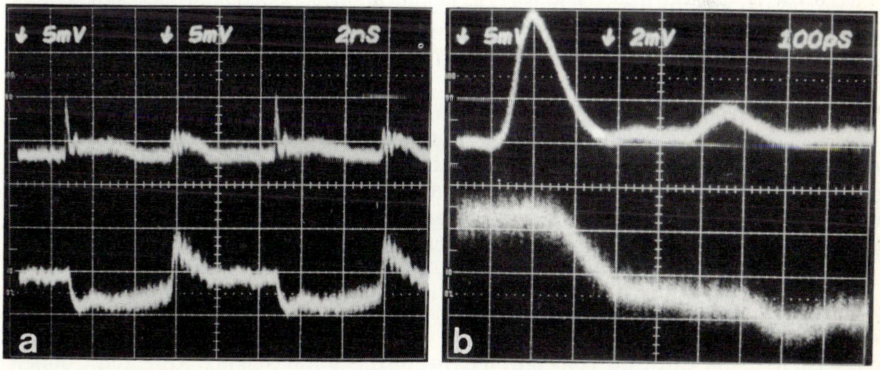

Fig. 1a. The upper trace is the optical input signal from the gain-switched laser (set pulse) and the mode-locked laser (reset pulse). The lower trace is the output signal from the bistable laser when biased in the middle of the hysteresis loop.

Fig. 1b. The bistable fall time was about 120 ps (lower trace). The upper trace is the reset pulse.

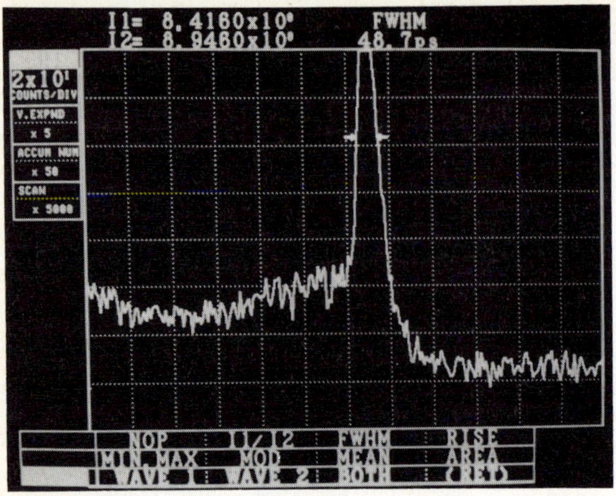

Fig. 2. Fast fall time of light from a bistable laser diode, reset with 50 ps long 75 fJ input pulses of 1326 nm wavelength. This was measured by a Hamamatsu optical-sampling oscilloscope. Time scale: 100 ps/div.

4. Conclusions

We have demonstrated all-optical switching of a bistable laser diode with two different wavelengths at a repetition rate of 100 MHz. The optical reset was performed by means of stimulated recombination which decreased the charge carrier concentration below threshold. The reset wavelength was 12 nm longer than the bistable lasing wavelength. With an input coupled reset energy of 290 fJ, the bistable fall time was measured to be about 120 ps. When performing the optical reset operation with insufficient overdrive, an automatic turn-on after 0.5-1.0 ns was observed.

The reset time could maybe be decreased further by forcing the carrier concentration to even lower values by the reset pulse. The resulting carrier concentration is determined by two factors, namely, the reset wavelength and the gain coefficient at the reset wavelength. To induce stimulated recombination for as many electron-hole pairs as possible a long reset wavelength is desirable. All electron-hole pairs with higher energy than the corresponding reset wavelength can recombine via stimulated recombination. However, the reset wavelength must be shorter than the bandgap of the bistable laser (during lasing) and longer than the bistable lasing wavelength. On the other hand, the gain coefficient decreases with increasing wavelength and consequently the stimulated recombination rate decreases. There is a trade-off between longer reset wavelengths and high gain coefficient at the reset wavelength.

5. Acknowledgments

We wish to thank Ericsson Telecom, the Swedish Natural Science Research Council (NFR), and the National Swedish Board for Technical Development (STU) for their support of this work and Ericsson Components for fabricating the lasers. We are also grateful to Hamamatsu Photonics Norden AB, for lending us the optical-sampling oscilloscope.

6. References

1. S. Suzuki, T. Terakado, K. Komatsu, K. Nagashima, A. Suzuki, and M. Kondo, J. Lightwave Technol. **LT-4**, 894 (1986).
2. S. Masuda, H. Rokugawa, K. Yamaguchi, N. Fujimoto, and S. Yamakoshi, in *OSA Proc. On Photonic Switching*, Vol. 3, p. 286 (1989).
3. A. Tomita, S. Ohkouti, and A. Suzuki, in *Photonic Switching*, edited by T.K. Gustafson and P.W. Smith (Springer, Berlin, 1988), p. 88.
4. K. Inoue and K. Oe, Electron. Lett. **24**, 512 (1988).
5. T. Odagawa and S. Yamakoshi, Electron. Lett. **25**, 1428 (1989).
6. P. Blixt and U. Öhlander, Electron. Lett. **25**, 699 (1989).
7. U. Öhlander and O. Sahlén, Appl. Phys. Lett. **54**, 1198 (1989).
8. U. Öhlander, P. Blixt, and O. Sahlén, in *Proceedings of 15th ECOC*, Vol. 1, p. 288 (1989).
9. H.-F. Liu, Y. Hashimoto, and T. Kamiya, IEEE J. Quantum Electron. **QE-24**, 43 (1988).

Electrically and Optically Controlled p-i(MQW)-n Vertical Coupler Switch with Electro-Optic Feedback

J.A. Cavaillès [1], *M. Erman* [1], *K. Woodbridge* [2], *and A. Goutelle* [1]

[1]Laboratoires d'Electronique Philips, 22 Avenue Descartes, F-94453 Limeil Brévannes, France
[2]Philips Research Laboratories, Redhill, UK

Abstract: We have designed and fabricated a vertical directional coupler switch, which uses p-i(MQW)-n waveguides. Switching can be induced electrically (driving voltage 5V) or optically via an electrical feedback loop.

Photonic switches are expected to play an important role in future communciation networks for both routing and processing types of applications [1]. Optically controlled switches avoid complex electrical interfaces, are high speed devices and open the possibility of achieving self-routing devices.

Optically controlled semiconductor optical signal processing devices can roughly be divided into two separate groups. The first group exploits intrinsic non-linear optical properties of materials, whereas the second group relies on electro-optic properties in conjunction with electrical feedback loops acting on a photogenerated current. This second group comprises mainly the SEED device [2] and its numerous derivatives, which are based on p-i(MQW)-n optical *modulators* electrically connected to more or less complex feedback circuits. The optically controlled *routing* devices proposed so far belong to the first group and derive from non-linear optical coupling phenomena [3,4] .

In this paper, we demonstrate that the concept of electro-optic feedback in Multiple Quantum Well structures can be extended to a one input-two output switch which can be controlled *either* electrically *or* optically. We report experimental results on a 1X2 electro-optic switch based on a vertically arranged GaAs/GaAlAs p-i(MQW)-n directional coupler electrically connected to a constant current source.

The vertical directional coupler, comprising two superimposed 0.6 μm thick MQW waveguides (30 periods 100 Å / 100 Å $GaAs/Ga_{0.7}Al_{0.3}As$), separated by a 0.9 μm coupling region ($Ga_{0.84}Al_{0.16}As$) has been grown using MBE. Top and bottom optical confinement is provided by 1 μm thick $Ga_{0.84}Al_{0.16}As$ layers. A n^+ GaAs substrate was used and the whole device forms a p-i-n junction, the intrinsic region of which corresponds to the upper MQW. The $Ga_{0.86}Al_{0.14}As$ cladding layer is thus p-doped and capped by a 400 Å p^+ GaAs layer. The optical measurements have been performed using an end-fire coupling scheme and an Ar-laser pumped tunable dye laser beam (Styryl 9 dye). In order to have well defined initial conditions for the optical coupling phenomenon, we have boron-implanted the sample down to the bottom of the upper waveguide on the first part of light propagation after the input facet. As a matter of fact, implantation makes the upper waveguide very absorbing

Springer Series in Electronics and Photonics, Vol. 29 **Photonic Switching II**
Editors: K. Tada · H.S. Hinton © Springer-Verlag Berlin, Heidelberg 1990

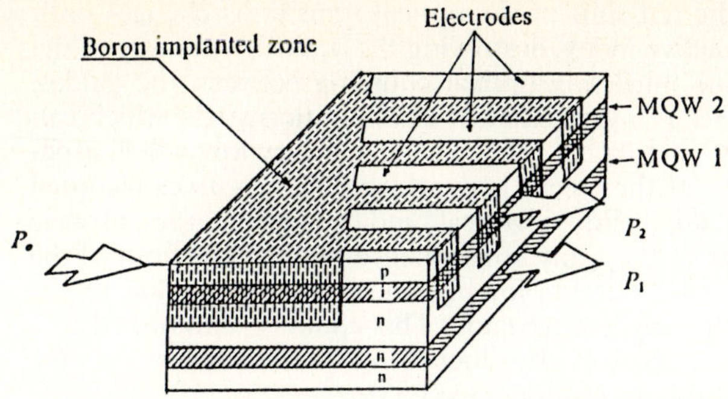

Figure 1: Schematic structure of the sample. The boron implanted region is tilted with respect to the sample edge, so that various waveguide lengths have been tested by simply translating the sample.

and after some ~ 100-300 μm propagation, all the remaining light is localized in the lower waveguide and light arrives in the unperturbed vertical coupler with a field profile corresponding to the lower waveguide guided mode. We have checked that this filtering was sufficient by observing that, when displacing the input beam around the optimum coupling position, the output facet near field pattern was unchanged except for its overall intensity.

In order to enable electrical contacts to the sample, we used an ohmic contact alloy (Ti/Pt/Au) deposited in 40 μm wide stripes on top of the sample. We then implanted boron between the stripes in order to isolate them electrically from their neighbours. Contacts were taken using a probe needle.

The coupling length of the device was determined as described in [5] : it was found to be 110 μm for a 880 nm wavelength. It was however much more practical to work with a 330 μm long coupler (3 times the coupling length). The switch has been first tested using electrical control, i.e. reverse biasing the p-i(MQW)-n structure. These measurements have been made for two very different input powers in order to check for intrinsic optical non-linear effects.

The results are shown in figure 2, where we plotted, as a function of the bias, both the fractional output from the input guide ($P_1/(P_1 + P_2)$) and the total output normalized to its zero-bias value ($(P_1 + P_2)(V)/(P_1 + P_2)(0)$). Switching action is due to the quantum confined Stark effect in the upper multiple

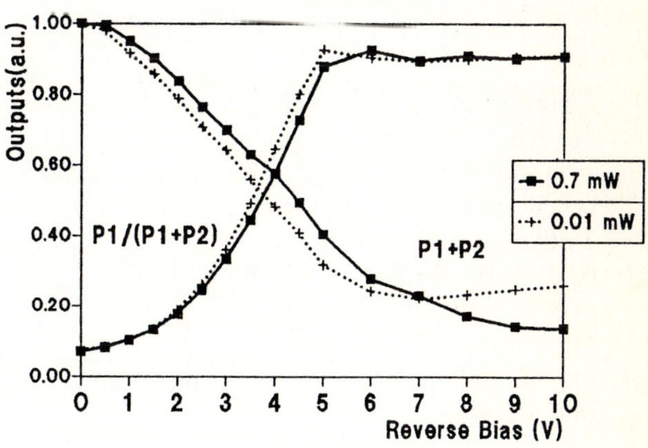

Figure 2: Fractional output power from the input guide and total output power normalized to its zero bias value as a function of the reverse bias. The results have been recorded for two different input powers: 0.01 mW (dotted curves) and 0.7 mW. ($\lambda = 880nm$, TE polarisation)

quantum wells in which the red shift of the optical transitions changes both the absorption and the refractive index, destroying the optical symmetry of the directional coupler and thus inhibiting optical coupling between the guides. It can be seen that the device is a very efficient electrooptic switch, which can change from 'cross' to 'bar' state with only 5 V bias, a very low value, especially when the short length of the device is considered. The curves recorded at different input powers are nearly identical, indicating that the intrinsic non-linearities are too weak to have any consequence on the behaviour of the coupler. The decrease in total outputs for high reverse biases is due to the increase in absorption in the upper waveguide. This could be diminished to a large extent in subsequent devices by increasing the thickness of the $Ga_{0.84}Al_{0.16}As$ layer between the waveguides (see [5]).

The previously described electro-optic switch can be used to obtain all-optical switching as follows: When the reverse bias across the p-i(MQW)-n diode increases, absorption in the upper waveguide increases as an immediate consequence of the quantum confined Stark effect and photocurrent is generated. If, for a given input power, one imposes with a current source the current flow through the device, the voltage drop across the diode adjusts itself in such a way that the photocurrent is equal to the imposed current. The equilibrium state reached thus imposes the state of the switch, but since this equilibrium depends upon the input power (through the photocurrent) the switch is optically controllable.

The device was thus connected to a constant current source. The outputs from both waveguides were monitored as a function of the imposed current for several values of the input optical power. The results are shown in figure 3. The indicated values of the power do not take coupling efficiency into account and are therefore larger than the effective optical power coupled into the coupler by at least 10dB. The results clearly demonstrate that the switching state can be controlled either by the current or by the input optical power.

When operated at constant current, switching can be induced by changing the intensity of the input beam (Figure 4). By changing the current in the circuit, the critical power needed for switching can be adjusted at will. This is a strong advantage of this structure with respect to conventional, intrinsically nonlinear directional couplers.

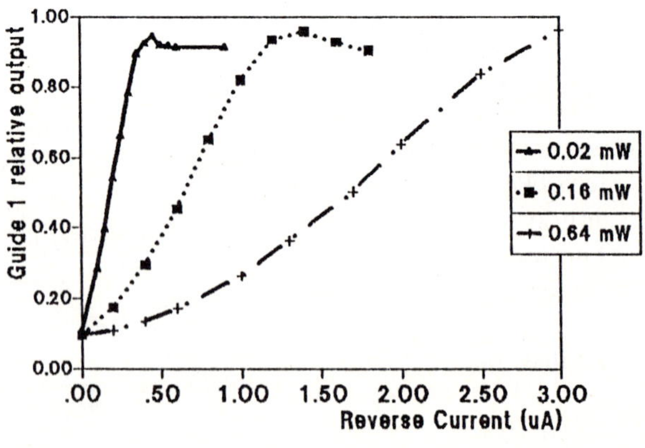

Figure 3: Fractional output from the input guide as a function of an imposed reverse current for 3 different input powers : 0.02 mW (solid), 0.16 mW (dotted), 0.64 mW (dash-dotted). The coupling efficiency is not taken into account.

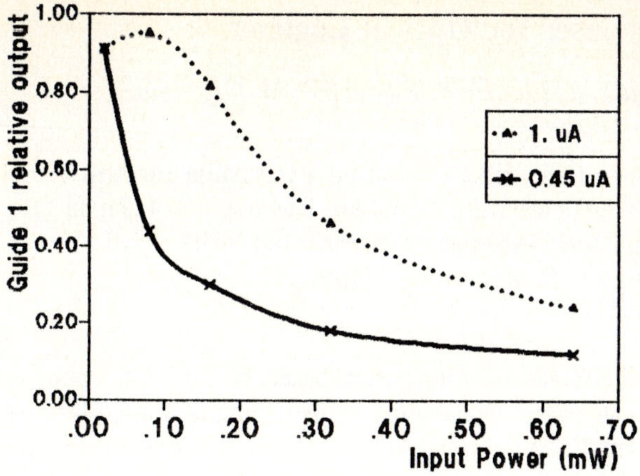

Figure 4: Fractional output power from the input guide as a function of the input power (coupling efficiency not taken into account) for two imposed values of the reverse current : 0.45 μA (solid) and 1. μA (dotted).

The above described properties give to the directional coupler the possibility of use as a non-linear self-routing element. The switch also offers interesting applications for wavelength multiplexed systems: when the input consists of two signals at different wavelengths, the state of the switch for one signal can be controlled by the second one.

This work has been partially supported by the European Community within the framework of the RACE R1033 (OSCAR) project.

References

1. N. Whitehead and N. Parsons, ECOC-89 11-14 September 1989, Paper WeA15-1
2. D.A.B Miller, D.S. Chemla, T.C. Damen, A.C. Gossard, W. Wiegmann, T.H. Wood and C.A. Burrus, Appl. Phys. Lett., vol. *45* (1), p. 13, 1984
3. P. Li Kam Wa, J.E. Sitch, N.J. Mason, J.S. Roberts and P.N. Robson, Elec. Lett., vol. *21*,p. 26, 1985
4. M. Cada, R.C. Gauthier,B.E. Paton and J. Chrostowski, Appl. Phys. Lett., vol. *49*(13), p. 755, 1986.
5. J.A. Cavailles, M. Erman and K. Woodbridge, IEEE Photonics Tech. Lett., *1 (11)* ,373

Side-Diode-Injection Laser for Optical Logic

S. Wakabayashi[1], *T. Tsuruta*, *K. Idota*[2], *Y. Takeuchi*[2], *M. Ohshima*[2], *and Y. Toyoda*[1]

[1]Optoelectronics Technology Research Corporation, Matsushita Electric Industrial Co., Ltd., 3-10-1 Higashimita, Tama-ku, Kawasaki 214, Japan
[2]Semiconductor Research Center, Matsushita Electric Industrial Co., Ltd., 3-10-1 Higashimita, Tama-ku, Kawasaki 214, Japan

Abstract. InGaAsP/InP side-diode-injection lasers based on the laser quenching effect were monolithically integrated. The degree of quenching of a main laser as a function of its injection current and the input light to a side laser has been studied. The side-diode-injection laser is found to be capable of functioning as an optical inverter at room temperature.

1. Introduction

Semiconductor laser switching devices using the quenching effect have been studied [1-3]. Switching devices that operate by laser quenching have some advantages; they have the capability of high-speed operation and do not require close wavelength matching between a laser and a quencher. Furthermore, if the direction of the output light from main laser is perpendicular to that of the input light from side laser, there is no mixing of the output light and the input light at the light received end. The optical logic operation using an AlGaAs/GaAs laser logic gate monolithically integrated was first reported by Grande and Tang in 1987 [3]. The laser logic gate was operated only by injection current to the side laser.

In this paper, we describe, for the first time, the all-optical switching using the quenching of a main laser by a side laser. The side-diode-injection laser is monolithically integrated. We measured dependence of the degree of quenching of the main laser on its injection current and the input light to the side laser, and we found that the degree of quenching is large enough to operate as an optical logic gate. We also demonstrate the optical logic gate by an input light to the side laser.

2. Experimental

We first studied the quenching effect of a main laser using a laser diode module with a single-mode fiber as a side laser. The laser diodes used in this study were 1.3 μm InGaAsP/InP buried crescent with semi-insulating layer (BCSI) lasers [4]. An incident beam was injected into the active layer of the main laser from the side using a single mode fiber. Fig. 1 shows the contrast ratio between the unquenched and quenched states versus the cavity length (L) of the main laser. The wavelengths of the main longitudinal mode of the main laser and the laser diode module were 1.291 μm and 1.297 μm. The coupling efficiency for an incident beam into the main laser was estimated to be less than 2 %. The light input power was 8.0 mW. Output powers

Springer Series in Electronics and Photonics, Vol. 29 **Photonic Switching II**
Editors: K. Tada · H.S. Hinton © Springer-Verlag Berlin, Heidelberg 1990

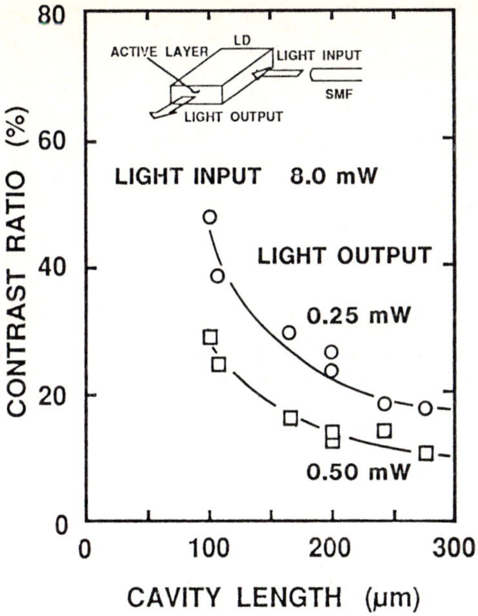

Figure 1. Contrast ratio between the unquenched and quenched states versus cavity length of the laser. Light output power of the laser and light input power of the laser module with a single mode fiber are constant.

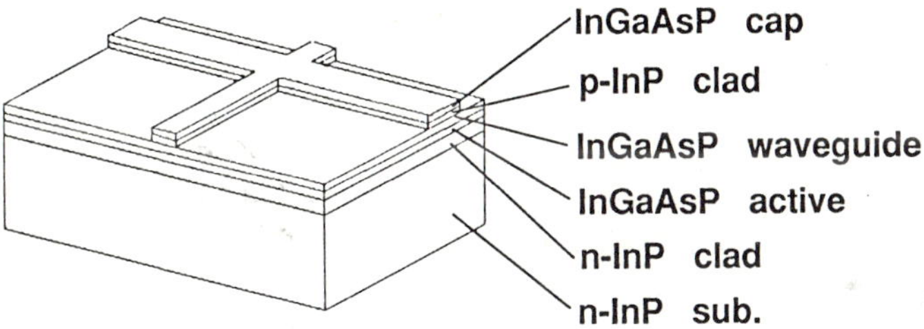

Figure 2. Schematic structure of the side-diode-injection laser monolithically integrated.

were 0.25 and 0.50 mW under no optical injection. The degree of quenching, namely the contrast ratio,decreases with increasing the cavity length (L) of the main laser. The degree of quenching is found to be proportional to 1/L. Therefore, it can be concluded that the cavity length of main laser should be as short as possible when we fabricate a side-diode-injection laser.

Fig. 2 shows the schematic structure of the side-diode-injection laser monolithically integrated. This structure was grown by liquid phase epitaxy (LPE). InGaAsP active layer with 1.3 µm bandgap was about 0.15 µm thick and InGaAsP waveguide layer above the active layer with 1.1 µm bandgap was about 0.20 µm thick. The active and waveguide layers were sandwiched between p- and n-InP cladding layers. Ridges of the main laser and the side laser were fabricated using a chemical etching method. The device consists of two ridge waveguide lasers, each of which has two cleaved facets. In this study, the main laser was 5.0 µm wide and 200 µm long. The side laser was 10 µm wide and 400 µm long. The main laser intersects the side laser, so that the main laser shares a part of its gain volume with the side laser. The resistance between the main laser and the side laser is about 100 Ω. The pulse threshold current (500 ns pulse width, 20 kHz repetition rate) of the main laser was about 90 mA, while that of the side laser was about 200 mA. The wavelength of the main longitudinal mode of the main laser and the side laser was 1.290 µm.

3. Results and Discussion

Fig. 3 shows the light output power of the main laser versus injection current of the side laser when the main laser was biased at 130 mA without injection of the light input. The light output power of the main laser begins to decrease linearly when the side laser current exceeds a threshold current. The maximum degree of quenching is achieved at about 600 mA of the side laser current. The light output power of the main laser does not go to zero because of the spontaneous emission.

Fig. 4 shows the light output power of the main laser versus the light input power injected into the active layer of the side laser, when the main laser was biased at 110 mA (500 ns pulse width, 20 kHz repetition rate) and the side laser at 200 - 400 mA (500 ns pulse width, 20 kHz repetition rate). The inverter operation was observed by the light input when the side laser was biased at 400 mA.

In this case, the wavelength of the input light to the side laser was 1.285 µm. Although the input and output light wavelengths differ by about 5 nm, this switching device can operate as an optical inverter. Therefore, this semiconductor laser device has a great advantage, when practical complex logic functions are carried out by connecting many optical logic devices. However, the contrast ratio between the unquenched and quenched was rather low. This is due to the low coupling efficiency of about 2 %. It follows that the strong light input power and the high bias current are necessary to quench the main laser enough. We believe that the smaller device with etched facets remarkably improves the contrast ratio.

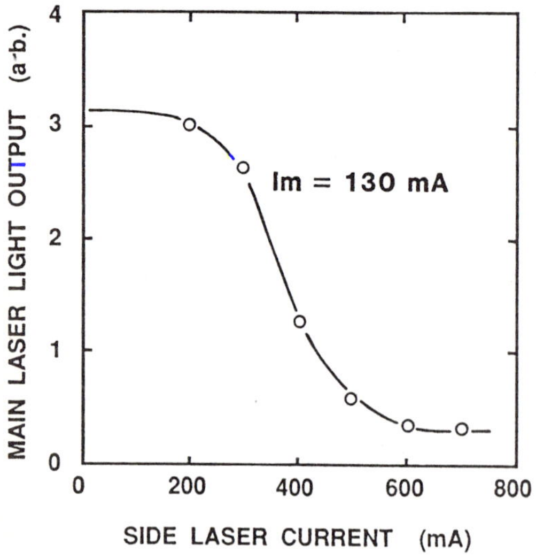

Figure 3. Light output of the main laser versus injection current to the side laser. The main laser was biased at 130 mA and no light input was injected.

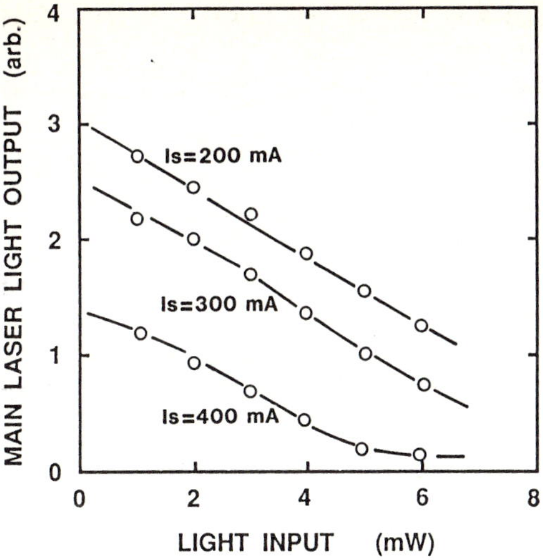

Figure 4. Light output of the main laser versus light input power to the side laser. The wavelength of the input light was 1.285 μm.

4. Summary

We have monolithically integrated the side-diode-injection laser and demonstrated for the first time the main laser switching operation using input light to the side laser without mixing the input and output light. We believe that this side-diode-injection laser is a promising device for all-optical logic operation.

Acknowledgment. The authors wish to thank I. Hayashi and K. Dazai for their continuous encouragement. The authors also would like to thank T. Kajiwara, T. Onuma and M. Mihara for their continuous encouragement and valuable advice.

References

1. A.B. Fowler, Appl. Phys. Lett. 3,1 (1963).
2. C.E. Kelly, IEEE Trans. Electron Devices **ED-12**, 1 (1965).
3. W.J. Grande and C.L. Tang, Appl. Phys. Lett. **51**, 1780 (1987).
4. M. Ohshima, Y. Toyoda, N. Hirayama, N. Takenaka, M. Matsuki, H. Kobayashi, and Y. Kino, IEEE J. Quantum. Electron. **QE-21**, 563 (1985).

Optical Flip-Flops Using Two Nonlinear Etalons

H. Tsuda, Y. Sakai, and T. Kurokawa

NTT Opto-electronics Laboratories, 3-1 Morinosato Wakamiya,
Atsugi-shi, Kanagawa 243-01, Japan

Optical flip-flop operations are demonstrated successfully with tandem and symmetric configurations using two nonlinear etalons.

1. Introduction

All-optical signal processing is necessary to realize future broadband transmission networks because it is out of the bandwidth limitation of electronic circuits. A nonlinear etalon is suitable for such an application because of its high-speed all-optical switching potential. It exhibits bistable characteristics, therefore, one nonlinear etalon itself can work as an optical flip-flop(1), (2). However, resetting it requires a reduction in incident power(3). This problem can be solved by coupling two nonlinear etalons. A tandem type optical flip-flop has been constructed with two nonlinear etalons in series by Rushford et al.(4), which was operated using two optical sources with different wavelengths. However, a single wavelength operation is desirable for cascading construction of an optical circuit. Kuszelewicz et al.(5) discussed the possibility of a symmetric type optical flip-flop with two nonlinear etalons where the reflected beams are fed back to each other. The above mentioned optical flip-flops installed on optical benches are very difficult to operate because they require a strong coupling and precise alignment. However, we fabricated an optical triode switch using a nonlinear etalon having pig-tail fibers and adjustable input-output characteristics(6). All-optical logic circuits, such as flip-flops, can be constructed easily by the coupling of optical triode switches. Moreover, this coupling enables easy and stable adjustment as well as miniaturizes the optical system. In this study, we constructed both tandem and symmetric type optical flip-flop circuits and demonstrated them successfully.

2. Tandem Flip-flop

A schematic view of a tandem type optical flip-flop is shown in Fig.1. To achieve a single wavelength operation, we used a front nonlinear etalon as the inverter gate in the reflection mode. So, reset pulses are inverted through etalon 1. A reflected hold beam and a set beam are incident on etalon 2, which is switched-on and off by set and inverted reset pulses. To attain stable operation, the hold level must be almost the same as the average value of the switch-on and off threshold, therefore, the height of the inverted pulse must exceed half the value of the difference of the switch-on and

Springer Series in Electronics and Photonics, Vol. 29 **Photonic Switching II**
Editors: K. Tada · H.S. Hinton © Springer-Verlag Berlin, Heidelberg 1990

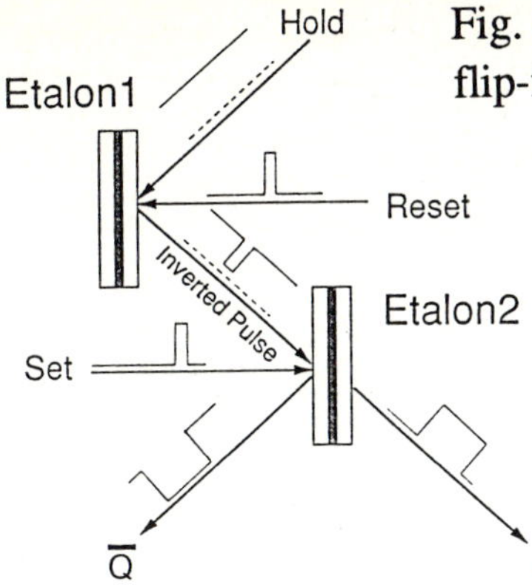

Fig. 1 Schematic view of a tandem type optical flip-flop.

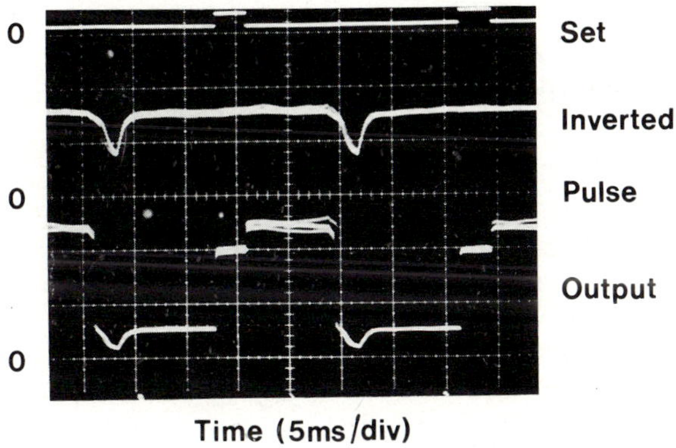

Fig. 2 Optical flip-flop operation of tandem type.

off threshold. Therefore, etalon 1 was aligned at a defocused position to enhance inverted pulses. We used ZnSe nonlinear etalons whose resonance wavelengths are about 514.5 nm. The operation of the tandem type optical flip-flop is shown in Fig.2. Upper, center and lower traces are the set beam (7.2 mW), the reflected hold beam from etalon 1 (the drop in the hold beam was about 5 mW from level), and the transmitted output beam from etalon 2 (set: 1.7 mW, reset 0.4 mW) respectively. However, getting the sufficient intensity of the inverted pulses required a very intense hold beam (46 mW) and reset pulses (80 mW), and this caused quite a different intensity of set and reset pulses.

3. Symmetric Flip-flop

Figure 3 shows the symmetric type optical flip-flop setup in which GRIN lenses and a nonlinear etalon constitute the optical triode switch. Hold 1 and a set beam are emitted to one triode switch and hold 2 and a reset beam to another triode switch. The reflected beams are again incident on each other with both nonlinear etalons working as an inverter gate in the reflection mode. Symmetric flip-flop outputs are

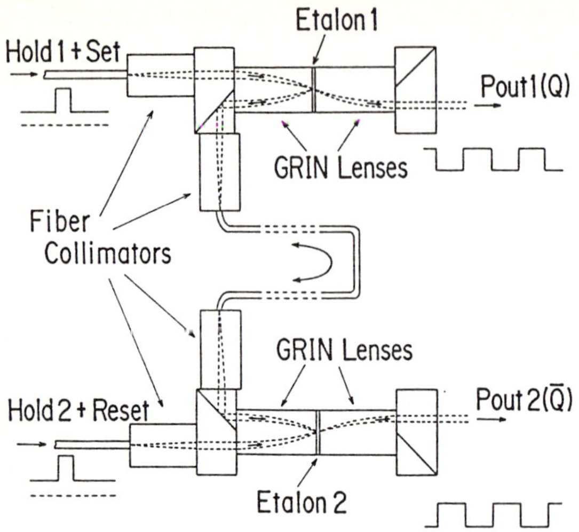

Fig. 3 Symmetric type optical flip-flop configuration.

obtained from transmitted hold beams. Input-output (P_{tra}; transmission and P_{ref}; reflection) characteristics of a nonlinear etalon is consistent with the following equations under high finesse approximation.

$$P_{in}^{(i)} = P_{tra}^{(i)}[1+3(q^{(i)}P_{tra}^{(i)}-p^{(i)})^2]/T_{max}^{(i)} \qquad (i=1,\ 2) \tag{1}$$

$$P_{ref}^{(i)} = 3P_{tra}^{(i)}(q^{(i)}P_{tra}^{(i)}-p^{(i)})^2/T_{max}^{(i)} \qquad (i = 1,\ 2), \tag{2}$$

where T_{max} is the peak transmission of an etalon, parameter q is the nonlinear coefficient and parameter p designates normalized detuning with which a nonlinear etalon shows limiter ($p<0$), differential ($0<p<1$) and bistable characteristics ($p>1$). Static characteristics of the system were obtained by equations (1),(2) and (3).

$$P_{in}^{(i)} = H^{(i)} + \eta P_{ref}^{(j)} \qquad (i,\ j) = (1,\ 2),\ (2,\ 1), \tag{3}$$

where H and η represent the holding power and coupling constant (<1). Static optical flip-flop characteristics were calculated on the assumption that the system had fully symmetric characteristics, $T_{max} = .5$, $p = 1.1$ and $\eta = .8$ as shown in Fig.4. As η becomes small, the flip-flop area decreases and finally vanishes ($\eta < 0.2$). From the view point of dynamics, feedback delay time must be much shorter than switch-on and switch-off time of the etalon.

The optical flip-flop operation based on a symmetric type configuration is shown in Fig.5. The upper two traces are hold beam 1 (3.6 mW) + set pulses (2.3 mW) and the transmitted beam from etalon 1 (H: 1.7 mW, L: 0.4 mW). The lower two traces are hold beam 2 (4.5 mW) + reset pulses (3.1 mW) and the transmitted beam from etalon 2 (H: 2.0 mW, L: 0.7 mW). As shown in Fig.5, both Pout 1 and Pout 2 waveforms complement each other.

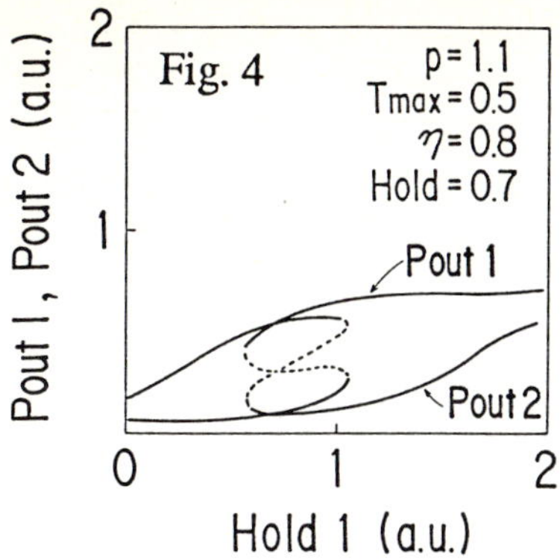

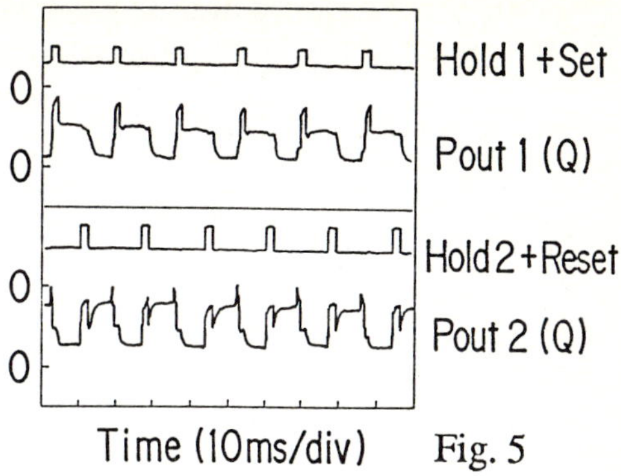

Fig. 4 Calculated static characteristics of a symmetric type optical flip-flop in the bistable region.

Fig. 5 Optical flip-flop operation of symmetric type.

4. Conclusion

We constructed tandem and symmetric type optical flip-flops using two nonlinear etalons and operated both types successfully. The tandem type has the advantage of fast switch-on and switch-off by external pulses, because most kinds of nonlinear etalons have a much shorter switch-on time than switch-off time. Contrary to the tandem type, the symmetric type has.the advantage of two complementary binary outputs suited for optical signal processing. The coupling of optical triode elements using nonlinear etalon will enable construction of combination and sequential logic circuits for all-optical signal processing.

References

[1] S. D. Smith, J. G. H. Mathew, M. R. Taghizadeh, A. C. Walker, B. S. Wherrett, and A. Hendry, *Opt. Commun.*, vol.51, 357(1984).

[2] T. Kurokawa, H. Tsuda, and S. Fukushima, *IQEC* Tul-9, (1988).

[3] S. S. Tarng, K. Tai, J. L. Jewell, and H. M. Gibbs, *Appl. Phys. Lett.*, vol. 40, 205(1982).

[4] M. C. Rushford, H. M. Gibbs, J. L. Jewell, N. Peyghambarian, D. A. Weinberger, and C. F. Li, *Optical Bistability 2* (Plenum Press, New York, (1984) p. 345.

[5] R. Kuszelewicz and J. L. Ouder, *IQEC* MP-68, (1988).

[6] H. Tsuda and T. Kurokawa, *IEEE J. Photo. Tech. Lett.*, vol.1, No12, 449(1989).

An Erbium-Doped Fiber Active Switch Controlled by Coded Pump Light

Y. Sato, K. Aida, and K. Nakagawa

NTT Transmission Systems Laboratories, 1-2356 Take, Yokosuka-shi, Kanagawa-ken 238-03, Japan

Abstract. We present an optical space-division switch which is based on Er-doped fiber amplifiers. The pumping light as the control signal contains interconnection information. The properties of pulse pumped Er-doped fiber amplifiers and the design of a large matrix switch are discussed. The performance of a 1×2 switch is demonstrated.

1. Introduction

Optical switches will be a key element of fiber optic networks. Future space-division switches are required to handle high-speed data pulses with low insertion loss, without polarization dependence. Er-doped fiber amplifiers meet these requirements and their fluorescence lifetime is about 14ms [1]-[4]. Such a lifetime makes it possible to excite a Er-doped fiber by pulsed pump light. In this paper, we propose a novel switch based on Er-doped fiber amplifiers as optical gates. It is expected to be a loss-less switch and enable the realization of a large matrix switch.

2. Switch Configuration

The proposed switch configuration is shown in Fig. 1. Input optical signals are divided using fiber couplers. Selected signals are amplified by pumped Er-doped fibers and others are blocked by the loss of Er-doped fibers. The amplified signals are combined by means of output fiber couplers and are output. By using switching amplifiers as optical gates, we can control the interconnection of the switch. The Er-doped fiber amplifiers are controlled by switching the pump power. When an amplifier is pumped by a pulsed light whose repetition period is shorter than the fluorescence lifetime, it goes active and stays active between pump pulses. This switch has an optical decoder of the pump light, it is synchronized to the pump light clock. It decodes the coded pump light and distributes pulsed pump light only to the amplifiers in the ON-state. The coded pump light contains the interconnection information. When we change the code of the pump light, interconnection of this switch is changed.

Springer Series in Electronics and Photonics, Vol. 29 **Photonic Switching II**
Editors: K. Tada · H.S. Hinton © Springer-Verlag Berlin, Heidelberg 1990

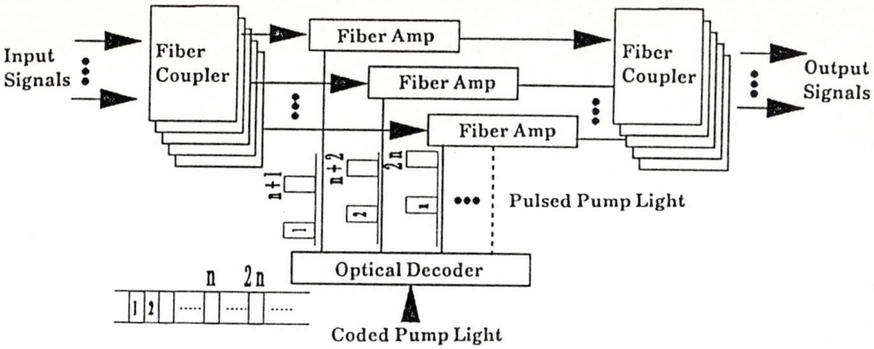

Fig. 1 Switch Configuration

3. Experimental Results

A 1×2 optical switch was composed of two Er-doped fiber amplifiers, an acousto optical deflector (AOD) , dividers, combiners, filters and a pump LD, as shown in Fig. 2. Each Er-doped fibor was 15 m long and its Er^{3+} concentration was 135 ppm. The amplifier gain was 10dB with 20mW pumping at 1.48µm wavelength and loss was 25dB without pumping. A 1.55 µm input optical signal was coupled to the two Er-doped fibers. The pump LD was driven by the code whose clock was 20kHz. The AOD was used as the optical decoder. In the figure, the coded pump light (a) was modulated to alternately switch on each amplifier.

The decoded pump light (b, c) had 100ms pulses at 200ms intervals. The rise time of the output optical signals (d, e) was about 10 ms for both amplifiers.

The bit error rate (BER) performance of the pulse pumped Er-doped fiber gate was measured as a function of the pump pulse repetition rate, including CW pumping, and is shown in Fig.3. Pump pulse duty was 50%, the repetition rate ranged from 500 Hz to 10 kHz and the optical wavelength

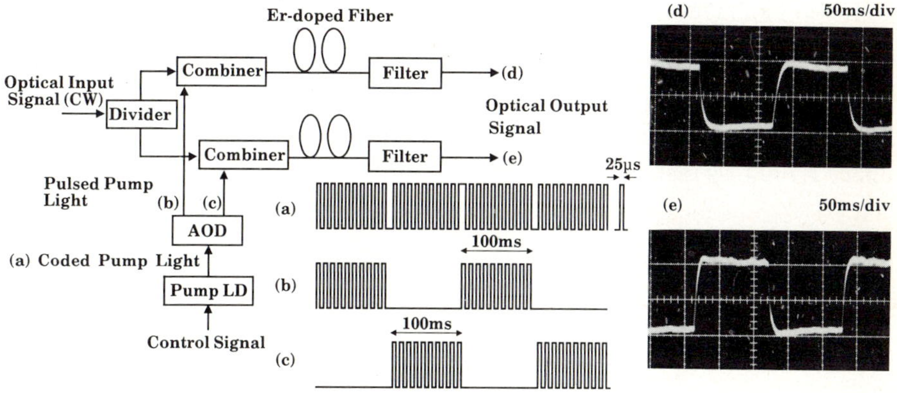

Fig. 2 A 1×2 Switch Demonstration

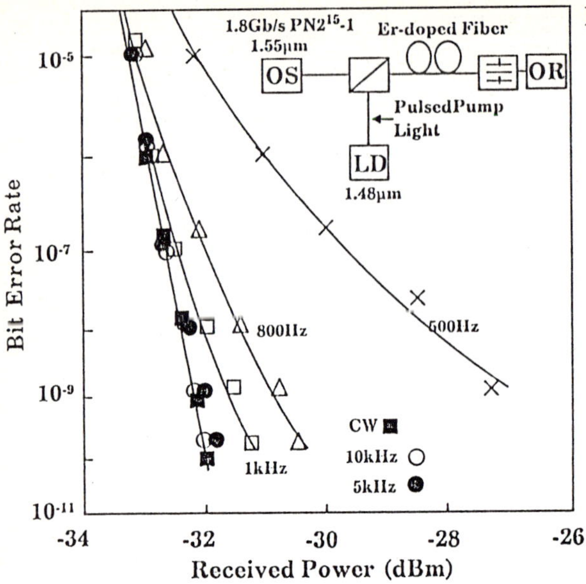

Fig. 3 Bit Error Rate Performance

was 1.48 μm. The 1.55 μm optical input is emitted from a direct-modulated DFB-LD at 1.8Gb/s with −20dBm average power. The output signal was detected with a InGaAs-APD direct detector. No degradation in BER performance was observed at pump pulse repetition rates over 5kHz.

4. Matrix Switch Design

The dimension limits for a matrix switch using Er-doped fiber amplifiers were calculated. According to insertion loss, the "Clos network", which is a three-stage matrix switch architecture, can yield larger matrix dimensions than is possible with the basic matrix switch [6] [7]. If we assume that ON-state gain and OFF-state loss of Er-doped fiber amplifiers are both 20 dB, a 200×200 "Clos" type matrix can be constructed with 18 dB isolation. The number of pump LDs required for this matrix is shown in Fig. 4. If the pump LD can supply optical power to 10 gates using the configuration proposed in this paper, a 200×200 "Clos " type optical matrix switch requires only about 1,700 LDs instead of 17,000 LDs.

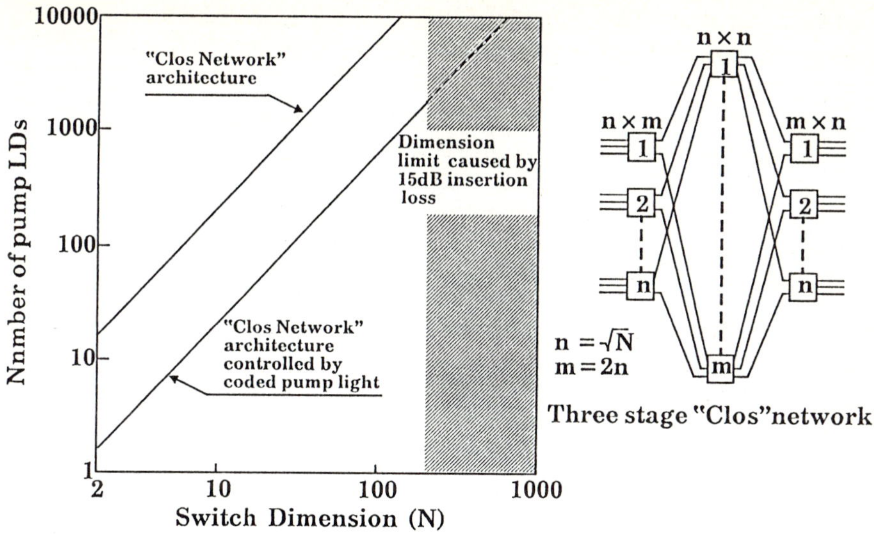

Fig. 4 The required number of pump LDs for "Clos" matrix switch

5. Conclusion

A 1×2 optically active switch controlled by coded pump light is demonstrated. No degradation of a 1.8Gb/s optical signal by an Er-doped fiber amplifier operating as an optical gate excited by pulsed pump light was observed. This switch is easy to control and reduces the number of pump LDs. It is possible to fabricate a 200×200 matrix switch with this switching configuration.

References

1. R. J. Mears, L. Reekie, I. M. Jauncey, D. N. Payne : Electron. Lett., 23, 1026-1028 (1987)
2. C. R. Giles, E. Desurvire, J. R. Simpson : CLEO'88, PD-9 (1988)
3. K. Hagimoto , K. Iwatsuki, M. Nakazawa, M. Saruwatari, K. Aida, K. Nakagawa, M. Horiguti : OFC'89, PD-15 (1989)
4. B. J. Ainslie, S. P. Craig-Ryan, S. T. Davey, J. R. Armitage, C. G. Atkins, R. Wyatt : IOOC '89, 22-23 (1989).
5. J. Yamada, K. Hagimoto, N. Ohkawa : Rev. Electr. Commun. Lab., 35, (3), 227(1987)
6. C. Clos : Bell, System, Tech. Jour., 407 (1953)
7. R. A. Spank : Jour. Quantum Electron., 22, 964, (1986)

Demonstration of 50 Gb/s All-Optical Kerr Demultiplexing Utilizing Pump–Probe Pulse Walk-off in Optical Fibers

T. Morioka[1], *M. Tsukada*[2], *and M. Saruwatari*[1]

[1]NTT Transmission Systems Laboratories, 1-2356 Take, Yokosuka-shi, Kanagawa-ken 238-03, Japan
[2]NTT Communication Switching Laboratories, 3-9-11, Midori-cho, Musashino-shi, Tokyo 180, Japan

Abstract:An all-optical 50 Gb/s Kerr demultiplexing experiment is successfully demonstrated utilizing a 20ps square Kerr shutter profile induced by a 6ps Gaussian pump pulse through pump-probe pulse walk-off in optical fibers.

1. Introduction

Ultrafast all-optical information processing schemes have been studied intensively [1]-[4] to realize the future high-bit-rate optical fiber transmission systems. All-optical multiplexing/demultiplexing of an ultrafast optical pulse train [1] is one of the basic functions and is expected to play a key role in the future multi hundred Gb/s systems. One of the problems associated with demultiplexing a probe (signal) pulse train by another short pump pulse is that the bell-shaped pump pulse, which usually has a longer duration than that of a signal pulse, tends to switch the wings of the two adjacent signal pulses as well. This results in degradation in switching crosstalk, hence limits the effective switching speeds.

To overcome this problem, we propose the use of a pump pulse that is shorter than the signal pulse, and the use of pulse walk-off due to group velocity mismatch between the pump and signal to obtain an effective square Kerr shutter profile. We then report the first demonstration of an all-optical Kerr demultiplexing experiment in the 50 Gb/s regime utilizing the pulse walk-off technique. Application of this technique to cascaded, in-line demultiplexing is also proposed.

2. Principle of Operation

The present Kerr demultiplexer utilizes the conventional Kerr switching configuration using polarization-maintaining (PM), single-mode (SM) fibers where pump and signal polarizations are both linear and crossed at 45 degrees to each other. The net induced phase change $\Delta\phi$ between the two orthogonal polarization components of the signal is given by the following equation for a Gaussian pump pulse (FWHM:Δt_0) if polarization dispersion τ_p(ps/m) is much smaller than pulse walk-off between the pump and signal τ_G(ps/m) (i.e.,$\tau_G \gg \tau_P$):

Springer Series in Electronics and Photonics, Vol. 29 **Photonic Switching II**
Editors: K. Tada · H.S. Hinton © Springer-Verlag Berlin, Heidelberg 1990

$$\Delta\phi(t)=\Delta\phi_x-\Delta\phi_y\simeq\frac{4}{3}\frac{2\pi}{\lambda_s}n_2 I_0 L\,[Erf(a)-Erf(a-\gamma)]/\gamma, \qquad (1)$$

where λ_s is the signal wavelength, I_0 is the pump intensity (W/m^2), L is the fiber length, n_2 is the nonlinear refractive index and

$$a=\left(\frac{t}{\Delta t_0}\right)\sqrt{(4\ ln\,2)},\ \gamma=\frac{\tau_G L}{\Delta t_0}\sqrt{(4\ ln\,2)}\ ,\ Erf(a)=\int_0^a e^{-s^2}\,ds.$$

Complete (100%) Kerr switching occurs when the peak value of $\Delta\phi(t)$ coincides with π. The last part of Eq. (1), $[Erf(a)-Erf(a-\gamma)]/\gamma$, corresponds to the pulse walk-off term that broadens the effective pump pulse duration. Since the rise and fall times of the induced phase profiles are determined merely by those of the pump pulse, one can obtain a square Kerr shutter profile with sharp rise and fall times if the pump pulse is short enough.

The most striking feature of Eq. (1), is that when $\gamma'=\gamma/\sqrt{(4\ ln\ 2)}=2$ or above, which means the pulse walk-off is more than twice as much as the original pump pulse width, the peak value of the induced phase profile $\Delta\phi$ saturates to a constant value regardless of the fiber length. In addition, an additional increase in γ' only gives rise to linear broadening of the phase profile. Thus, utilizing these induced phase broadening and saturation properties, one can obtain a square Kerr shutter profile $\Delta\phi(t)$ with a constant peak value, regardless of the Kerr fiber length. This leads to the idea of an in-line, cascaded demultiplexer that is capable of truncating a pulse or a set of pulses one by one with just one pump source as the pump pulse sweeps the signal pulse train through pulse walk-off in the Kerr fiber.

3.Results and Discussion

The schematic experimental set-up is shown in Fig. 1. A 6 ps, 82 MHz pump pulse at 1.06 μm is generated from a 90 ps mode-locked YAG pulse through a fiber grating-pair pulse compressor. A 15ps signal pulse is generated from the 1st Raman Stokes ($\lambda=1.117$ μm) induced by the YAG pump in a PM-SM fiber (4.6 μm core diameter, 40 m long). The signal is then multiplexed to a 50 Gb/s burst through a Fabry-Perot etalon with a free spectrum range (FSR) of 50 GHz. The Kerr medium consists of a pair of PM-SM fibers (7.5+7.5 m) that are concatenated with their fast axes at 90 degrees in order to compensate the fiber birefringence [1]. The pump and the burst signal are then coupled into the concatenated Kerr PM-SM fiber with pump polarization along one of the principal axes after temporal synchronization by an optical delay line. The switched and unswitched portions of the burst signals are observed with a synchroscan streak camera (6 ps temporal resolution) in a time-resolved configuration after phase adjustment using a Babinet-Soleil compensator and a polarizer. The pump pulse is then separated from the signal by the monochromator. The demultiplexing Kerr fiber has a 4.6 μm diameter with polarization dispersion of 1.0 ps/m at $\lambda=1.117$ μm. The pulse walk-off between the pump ($\lambda=1.06$ μm) and the

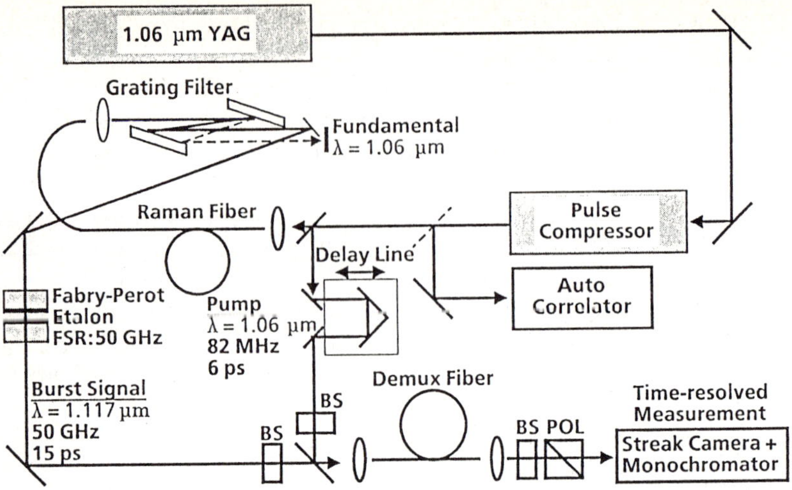

Figure 1 Experimental setup for 50 Gb/s all-optical demultiplexing

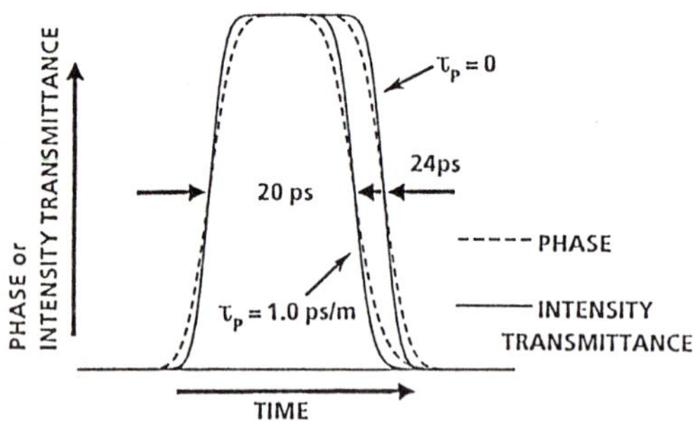

Figure 2 Calculated phase and Kerr shutter profiles for a 15 m fiber

signal (λ=1.117 μm) is 1.6 ps/m. Fig. 2 shows the calculated induced phase and Kerr shutter profiles (=intensity transmittance∝$\sin^2(\Delta\phi/2)$) under experimental conditions and in the case of no polarization dispersion as a reference considering the effects of polarization and chromatic dispersions [1]. The effective temporal duration of the Kerr shutter profile is calculated to be 20 ps. The figure clearly shows the square-shaped intensity transmission.

50 Gb/s demultiplexing results are shown in Fig. 3. The original 50 Gb/s burst signal, switched, and unswitched portions of the signal are shown in Figs. 3(a), (b), and (c), respectively. It can be seen that 100 % switching with relatively low crosstalk has been demonstrated. The required pump power is 48W for the 15 m Kerr fiber. Double-pulse and triple-pulse demultiplexing (extraction) results using 30 m (15+15 m), and 50 m (25+25 m) Kerr fibers are also shown in Figs. 4(a) (b), and (c) (d), respectively. The pump powers for the two cases are 33W and 31W, respectively. The relatively poor crosstalk probably comes from the deviation of the fiber length from the ideal one. Fig. 4 may suggest the possibility of constructing an in-line, cascaded

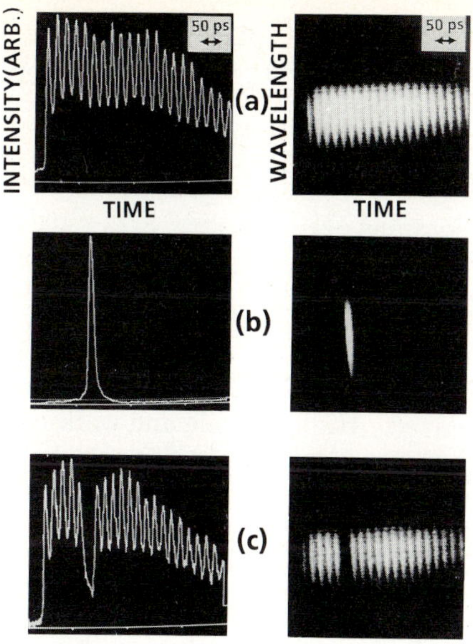

Figure 3 Experimental results of 50 Gb/s all-optical demultiplexing. (a) Original burst signal pulses, (b) a switched signal pulse, (c) unswitched signal pulses

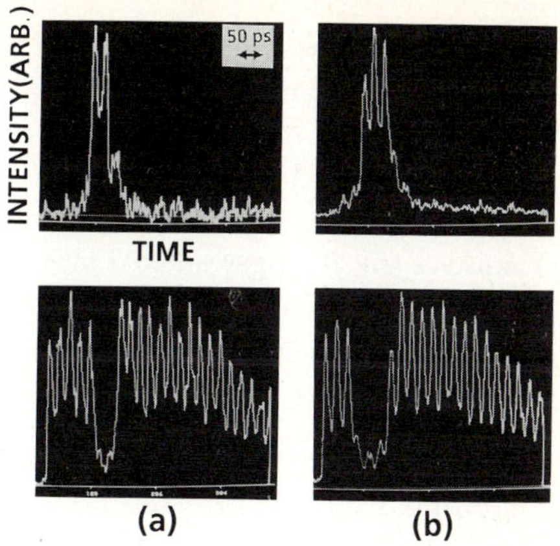

Figure 4 Experimental results of (a) (b) double-pulse and (c) (d) triple-pulse demultiplexing.

demultiplexing (i.e., in-line demultiplexing one by one at a point an integer multiple of a Kerr fiber length L_0) utilizing the pulse walk-off technique.

4. Conclusion

All-optical, stable Kerr demultiplexing of a 15 ps burst signal has been demonstrated in the 50 Gb/s regime for the first time to our knowledge utilizing the pulse walk-off between the pump and the signal in optical fibers. It has been verified that the use of pulse walk-off provides a convenient way to produce square Kerr switching profiles of arbitrary durations with fast rise and fall times.

Acknowledgement

The authors wish to thank Drs. S. Shimada, H. Ishio and, K. Nakagawa for their constant encouragement and support.

References

(1) T. Morioka and M. Saruwatari, IEEE Journal of Selected Areas in Communication. Vol. 6, 1186-1198(1988)
(2) M. N. Islam, Opt. Lett. 14(22) 1257-1259(1989)
(3) N. J. Doran and D. Wood, J. Opt. Soc. Am. B 4 1843-1846(1987)
(4) S. R. Friberg, Y. Silberberg, M. K. Oliver, M. J. Andrejco, M. A. Saifi, and P. W. Smith, Appl. Phys. Lett. 51, 1135-1137 (1987)

Ultrafast, Low Power, and Highly Stable All-Optical Switching in an All Polarization Maintaining Fiber Sagnac Interferometer

M. Jinno and T. Matsumoto

NTT Transmission Systems Laboratories, 1-2356 Take, Yokosuka-shi, Kanagawa-ken 238-03, Japan

Abstract. Ultrafast, low power, and highly stable all-optical switching in a nonlinear Sagnac interferometer is reported. Highly stable and walk-off free all-optical switching was successfully demonstrated at 5 Gb/s.

1. Introduction

Ultrafast all-optical switching devices are key elements for future optical communication and computing systems. Optical fiber Kerr switches are the most promising because of their sub-picosecond response, low switching power due to extremely long interaction length, and high intensity. Increasing the fiber length to reduce the switching power strengthens the deleterious effect of thermal and acoustic fluctuations, and increases the walk-off between the input signal and the control pulse. To overcome the fluctuation problems, the birefringence compensation technique[1] and the temporal separation technique[2] were proposed. The Sagnac interferometer configuration is another way to achieve a highly stable switch[3,4], which is essentially free from the above perturbations because it uses balanced interferometer arms.

In this paper, we report on ultrafast, low power, and highly stable all-optical switching in a nonlinear Sagnac interferometer. To achieve low power and walk-off free switching, a small core dispersion-shifted polarization maintaining fiber (200 m in length) was used. In addition, to achieve the complete polarization stability, a wavelength-sensitive polarization maintaining fiber coupler was used. All-optical time domain switching at 5 Gb/s was successfully demonstrated.

2. Principle

Figure 1 shows the schematic configuration of our nonlinear Sagnac interferometer switch. The key component is a dispersion shifted polarization maintaining fiber loop, which has equal group velocities at the wavelengths of the input signal and the control pulse. In addition, we employed a wavelength-sensitive polarization maintaining fiber coupler, whose power coupling ratio is nearly 50% at the wavelength of the input signal and nearly 0% (no coupling) at the wavelength of the control pulse.

Springer Series in Electronics and Photonics, Vol. 29 Photonic Switching II
Editors: K. Tada · H.S. Hinton © Springer-Verlag Berlin, Heidelberg 1990

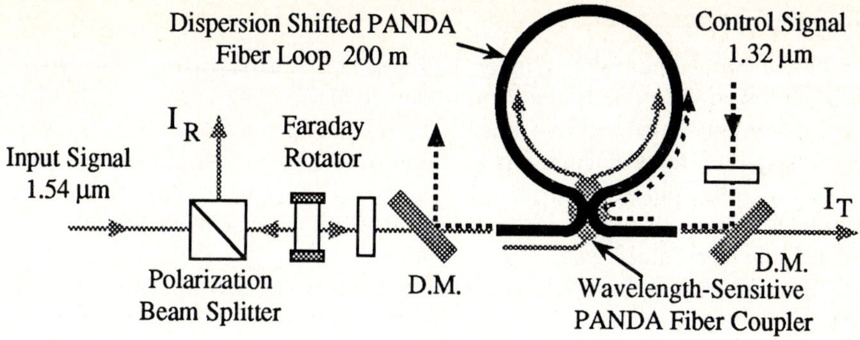

Fig. 1 Schematic configuration of the nonlinear Sagnac switch

A weak input signal, to avoid the nonlinear effect induced by the signal itself, is launched into a port of the wavelength-sensitive polarization maintaining fiber coupler through a 45° Faraday rotator, and is split into two counterpropagating signals in the polarization maintaining fiber loop. A high intensity control pulse train is launched into the other port of the coupler, and propagates in the counterclockwise direction.

In the absence of the control signal, the two counterpropagating signals return in phase through the exact same path length to recombine at the coupler and are fully reflected. Passing through the 45° Faraday rotator twice, this reflected signal is orthogonally polarized with respect to the input signal, and is extracted by the polarization beam splitter.

In the presence of the control signal, the symmetry between the counterpropagating paths is broken due to the induced cross phase modulations, and some part of the input is transmitted. The reflected and transmitted signals, I_R and I_T, can be written by $2k_s(1-k_s)\{1+\cos\Delta\phi(t)\}$ and $1-I_R$, respectively, in terms of the phase shift difference in the counterpropagation signals, $\Delta\phi(t)$ as

$$\Delta\phi(t) = \frac{4\pi n_2 (1-2k_c)}{\lambda_s A_{eff}} \left\{ \int_0^L I_p(t-\tau l)\,dl - \overline{I_p} L \right\}$$

where k_s and k_c are the power coupling ratio of the coupler at the wavelengths of signal and control pulse, respectively. n_2 is the nonlinear coefficient, λ_s is the signal wavelength, A_{eff} is the effective core area, L is the fiber loop length, t is the time, $I_p(t)$ is the power profile of the control pulse, τ is the group velocity difference between the input signal and the control pulse, and I_P is the average power of the control pulse train. It was assumed that the period of control pulse train T is much smaller than twice the loop transit time.

3. Experiments

A 200 m long dispersion shifted polarization maintaining fiber with a SiO_2-GeO_2 core and SiO_2 cladding was used for the Sagnac loop. This fiber was designed so as to have nearly equal group delay time at 1.32 μm (control pulse, mode-locked Nd:YAG) and at 1.54 μm (input signal, DFB-LD) and the group delay difference was 5 ps / 200 m. This fiber has a small effective core area of 28 μm^2 and thus a large Kerr effect. The polarization crosstalk was −31 dB / 200 m.

The wavelength-sensitive polarization maintaining fiber coupler, which has power coupling ratios of 1% at 1.32 μm and 45% at 1.54 μm, was fabricated by the fused elongation method. The deviation of the coupling ratio from the ideal 50% at the signal wavelength decreases the switching contrast to 20 dB. The excess loss and the crosstalk were <0.5 dB and <−20 dB, respectively.

At first the temporal Kerr modulation profiles were investigated with a cw input signal and short control pulses (pulsewidth 110 ps, repetition rate 100 MHz). Figures 2 (a) and (b) show the control pulse and the Kerr modulation profiles observed with an InGaAs pin photo diode (bandwidth >10 GHz) when the control pulse has the same polarization as the input signal. It corresponds to the peak phase shift differences of π and the required peak power 1.8 W shows good agreement with the theoretical value 1.7 W. No significant effect of walk-off was observed. This pump level is available with a gain-switched laser diode coupled with an Er^{3+} doped fiber amplifier. Figure 2 (c) shows the Kerr modulation profile when the control pulse was orthogonally polarized with respect to the input signal. A large walk-off of 520 ps due to the large birefringence of the fiber loop (6.5×10^{-4}) was observed and 100% Kerr modulation was not obtained even when the peak power of the control pulse was 11.3 W. From these experiments, the validity of group velocity matching between the input signal and the control pulse using the dispersion shifted fiber was confirmed

Using this switch, all-optical time domain switching was successfully demonstrated at 5 Gb/s. Figure 3 (a) shows the input fixed "11····1" signal

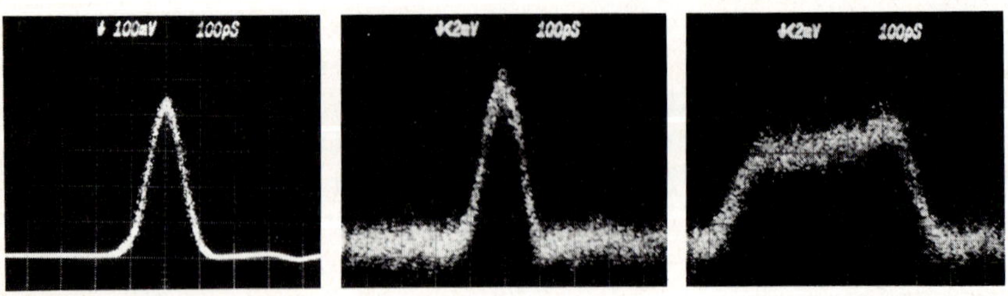

(a) Control Pulse (b) ∥ , $\Delta\phi=\pi$, P=1.8 W (c) ⊥, $\Delta\phi=0.6\pi$, P=11.3 W

Fig. 2 Temporal Kerr modulation profiles

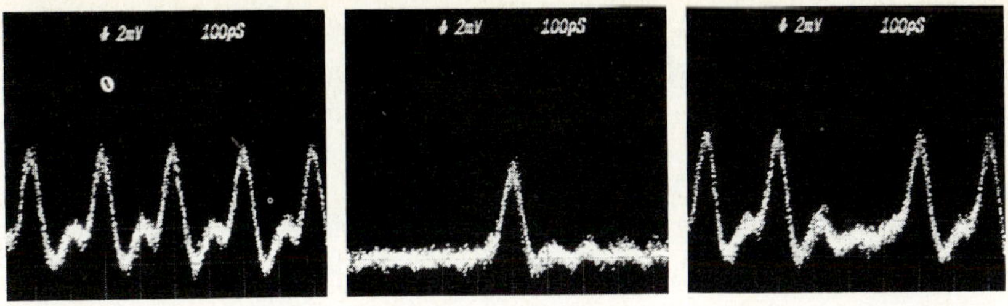

(a) Input "11···1" signal (b) Transmitted signal (c) Reflected signal

Fig. 3 All-optical switching at 5 Gb/s

at 5 Gb/s obtained from a gain-switched DFB-LD (pulse width 43 ps). Figures 3 (b), and (c) show the transmitted and the reflected signal, respectively. The required switching power was the same value for cw input and no waveform degradation was observed.

The switch demonstrated here was highly stable because of its all polarization maintaining fiber Sagnac interferometer structure, and no polarization or operating point readjustment was required.

4. Conclusion

We have reported on ultrafast, low power (1.8 W_{peak} / 200m), and highly stable all-optical switching in a nonlinear Sagnac interferometer switch. To achieve low power and walk-off free switching, a small core area (28 μm^2) dispersion-shifted polarization maintaining fiber which has a small group delay difference (5 ps / 200 m) between the input signal (1.55 μm) and control pulse (1.32 μm), was used. In addition, to avoid polarization fluctuations, a wavelength-sensitive polarization maintaining fiber coupler was employed All-optical time domain switching was successfully demonstrated at 5 Gb/s.

Acknowledgment The authors would like to express their appreciation to K. Kitayama and F. Ito for their helpful discussions. They also wish to thank A. Takada for his valuable technical advice.

References

1 Morioka and M. Saruwatari: IEEE J. on Sel. Areas in Com., **6**, 1186 (1988)
2 M. J. LaGasse, D. Liu-Wong, J. G. Fujimoto, and H. A. Haus: Opt. Lett., **14**, 311 (1989)
3 K. J. Blow, N. J. Doran, and B. K. Nayar: Opt. Lett., **14**, 754 (1989)
4 M. C. Farries and D. N. Payne: Appl. Phys. Lett. **55**, 25 (1989)

Optical Bistability in a Tapered Multimode Fiber with a Nonlinear Overlay

P.C. Subramaniam

Optical Waveguide Group, Department of Physics,
Indian Institute of Technology, New Delhi-110016, India

Abstract. Optical bistability is achieved in a tapered multimode fiber which has a nonlinear medium at the taper region.

1. Theory

The proposed configuration is shown in fig.(1). For an adiabatic taper, assuming that all the modes are equally excited and that the nonlinear medium does not change the modal pattern and distribution appreciably (which is the case with almost all the known nonlinear media), the output power P_o and the input power P_i are related through the expression

$$P_o = P_i \frac{[n^2_{co} - n^2_L]}{(\alpha/\beta)^2 [n^2_{co} - n^2_{cl}]} \qquad (1)$$

where, n_{co}, n_{cl} and n_L are the refractive indices of the core, cladding and the nonlinear overlay respectively, α and β are the maximum and minimum core dimensions of the taper.

Eq.(1) has been derived by considering the fact that the number of modes at a particular cross section and, hence the power, is related to the square of the local V-value. The output depends on the minimum V-value in the whole length of the device.

The refractive index n_L is modelled as a saturating nonlinearity through the expression

$$n_L = n_l + \delta n \frac{(P_i - P_o)}{(P_s + P_i - P_o)} \qquad (2)$$

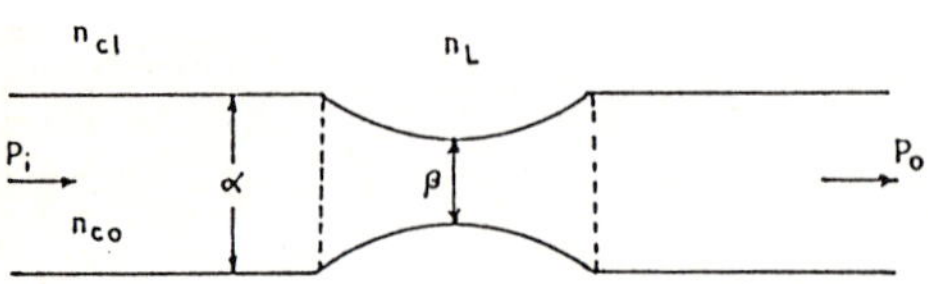

Fig.1. The proposed configuration.

Springer Series in Electronics and Photonics, Vol. 29 **Photonic Switching II**
Editors: K. Tada · H.S. Hinton © Springer-Verlag Berlin, Heidelberg 1990

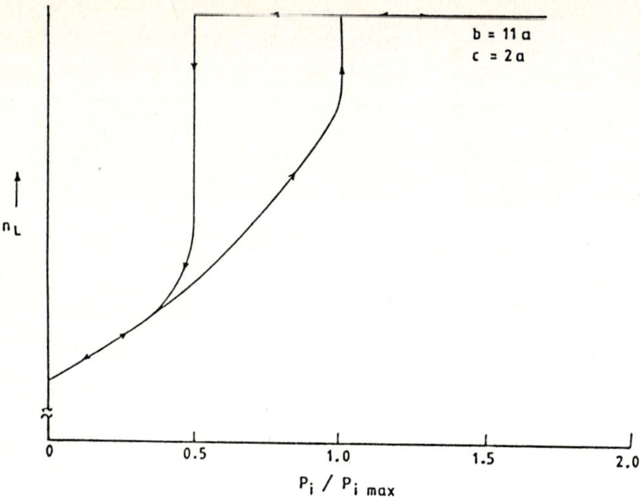

Fig.2. Variation of n_L with P_i for the saturating nonlinearity

where P_s is a saturation power characteristic of the medium. For large powers, the refractive index saturates to the value $n_l + \delta n$. Here, we have assumed that only the real part of the complex refractive index changes with the power and not the imaginary part.

With $n_l \gg \delta n$, $[n^2_{co} - n^2_{l}] = a$, $2n_l . \delta n = b$, $(\alpha/\beta)^2(n^2_{co} - n_{cl}) = c$, $P_i/P_s = y$ and $P_o/P_s = x$, we get from equations (1) and (2) the relation

$$c\,x = y\,[a - b\,(y - x)/(1 + y - x)\,]. \tag{3}$$

Eq.(2) is valid till $P_i - P_o = P_A$, at which point it is assumed that the material changes its state - its refractive index n_L increases sharply as shown in fig.(2). In the higher refractive index state (refractive index > core index in the high state), the material is assumed to be isotropic, resulting in reduced scattering, as opposed to the lower index state which is characterized by strong scattering. Hence on reaching the higher index state, even on reducing the input power, the refractive index remains in the high state, falling to the lower state only at a lower power level as compared to the power for up-switching. This results in optical bistability. Liquid crystals are expected to yield this type of nonlinearity, particularly the nematic type which change their structure from a highly scattering to a low scattering state above a critical temperature.

Fig.(3a) is drawn assuming that the nonlinear medium switches from the lower to the higher index state at the

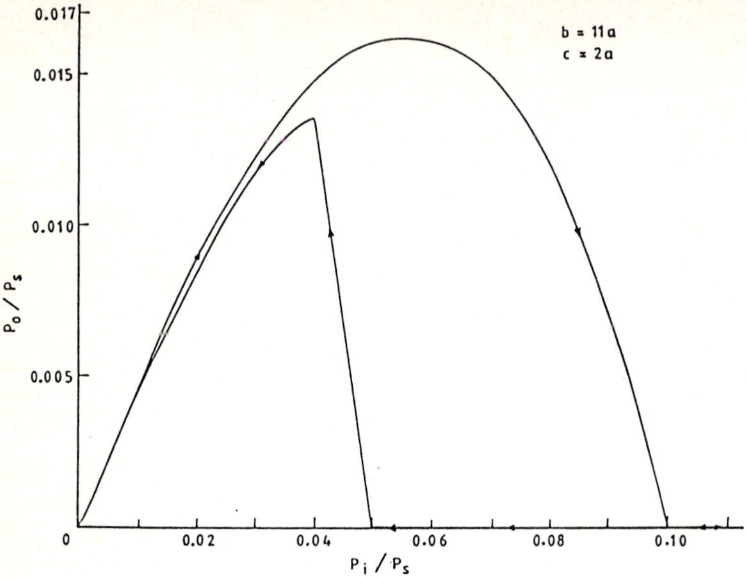

Fig.3a. Bistability curve for b = 11a and c = 2a

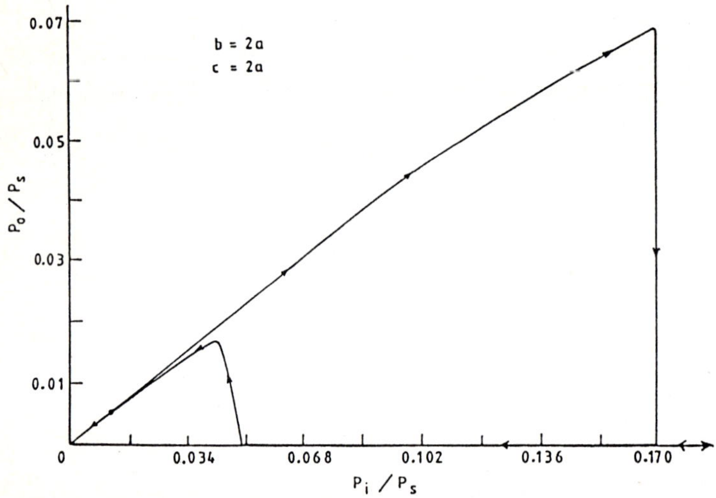

Fig.3b. Bistability curve for b = c = 2a

input power required to make the output zero. This corresponds to a value of y = 0.1. The material is assumed to switch back to the lower state at a y value of 0.05. With these same parameters, fig.(3b) is drawn for the case b = c = 2a. Here, the output drops to zero abruptly because of the fact that even before the output goes to zero due to the saturating nonlinearity, the material switches from its low to the high state.

2.Conclusion

A low power optical bistable device (due to the taper) is proposed. For certain combinations of the variables a,b and c, this device can be used as a constant output source when illuminated by a laser which is not intensity stabilized.

Ultrafast Switching Using Two-Color Soliton Collisions

S.R. Friberg and W. Jiang

NTT Basic Research Laboratories, 3-9-11 Midori-Cho,
Musashino-shi, Tokyo 180, Japan

Abstract: We introduce an ultrafast all-optical switch based on two-color soliton collisions. We describe a Mach-Zehnder interferometer that responds to but does not alter an input signal soliton.

1. Introduction

The availability of ultrashort-pulse lasers that can produce pulses of light shorter than 10 fsec in concert with the 100 THz bandwidths available in optical fibers has prompted a search for ultrafast all-optical switching devices. One approach being pursued is to make use of the small, fsec-responding Kerr optical nonlinearity of glasses in optical fibers. For example, 100 fsec optical pulses have been switched between the cores of a dual-core optical fiber [1]. Another approach uses the larger nonlinearities associated with band-gaps and other resonances in semiconductor materials. Recently, ultrafast switches based on solitons have been introduced, including the nonlinear loop mirror [2] and the soliton-trapping switch [3].

We introduce an ultrafast switch based on the phase-shift acquired by a soliton after it collides with another soliton. Two solitons with different center-frequencies propagate with different group velocities, and can be made to collide, as illustrated by computer simulation in Fig. 1. Such collisions do not affect the shape and energy of the solitons involved, rather they produce shifts in the phase (and position) of the solitons. This is the basis for two-color soliton-collision switching. Here, we describe one implementation of a two-color soliton-collision switch.

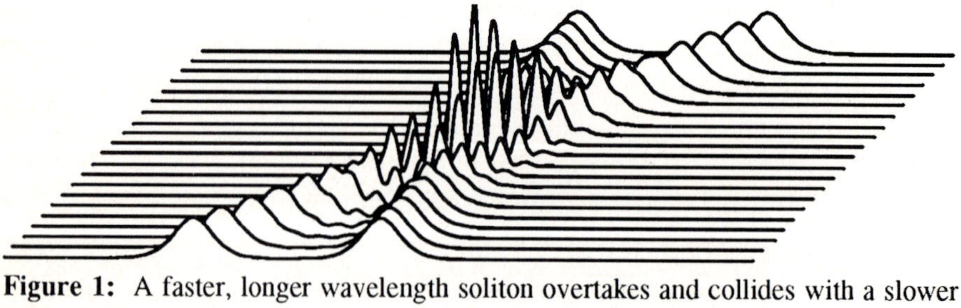

Figure 1: A faster, longer wavelength soliton overtakes and collides with a slower soliton. Note that the solitons resume their shape after colliding.

Springer Series in Electronics and Photonics, Vol. 29 **Photonic Switching II**
Editors: K. Tada · H.S. Hinton © Springer-Verlag Berlin, Heidelberg 1990

2. Solitons and two-color soliton collisions

Solitons and soliton collisions were described by Zahkarov and Shabat [4] in 1972 and were shown to be possible in optical fibers by Hasegawa [5] in 1973. One of the more striking properties of solitons is that they can propagate and collide without changing shape. For nonlinear pulse interactions, this eliminates dispersive broadening, thus permitting long interaction lengths. They also exhibit an absence of frequency chirp - the phase of a soliton, even after a collision, is uniform. Accordingly, solitons interfere much like monochromatic plane waves, producing large contrast ratios and intensity modulations. These qualities make it possible to construct efficient nonlinear switches, and can also be used for quantum non-demolition (QND) measurements of photon number [6],[7].

Analytic expressions for the nonlinear phase shifts produced by soliton collisions were obtained by Zahkarov and Shabat [4]. In Fig. 2a, we plot the phase shift acquired by a probe soliton that collides with a signal soliton as a function of the amplitude of the signal soliton. We show values for different frequency separations of the two solitons. The frequency separations are equal to $f \times f_o$, where f_o is the bandwidth of the probe soliton.[7] A phase shift of π radians or greater is easily attainable, but the frequencies of the solitons then overlap. Corresponding timing shifts are shown in Fig. 2b.

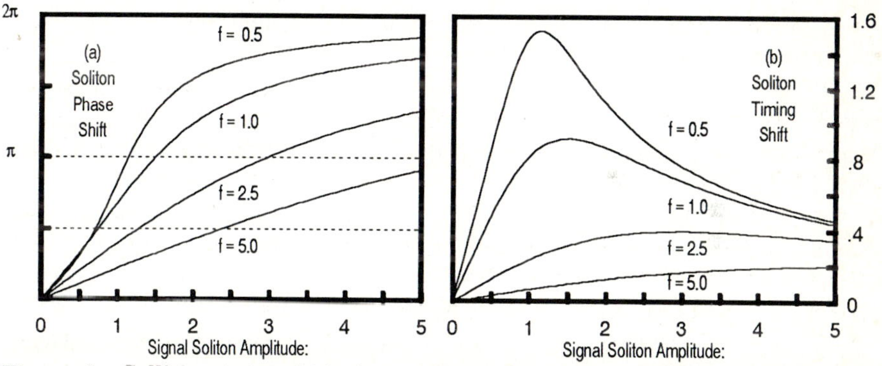

Figure 2: Collision-induced (a) phase shifts and (b) timing shifts of a probe soliton. Phase shifts (radians) and timing shifts (relative to pulse width) are plotted against the amplitude of the signal soliton for various frequency separations.

3. Two-color soliton collision switch

A nonlinear Mach-Zehnder interferometer switching device that utilizes the soliton-collision phase shifts is diagrammed in Fig. 3a. It uses three optical solitons, a probe, reference and signal. The probe and reference solitons are identical pulses emerging from the two ports of an input beamsplitter, and propagate down the two equal-length arms of the interferometer. A signal soliton, offset in frequency from the other two, co-propagates and collides with the probe soliton in one arm, and then

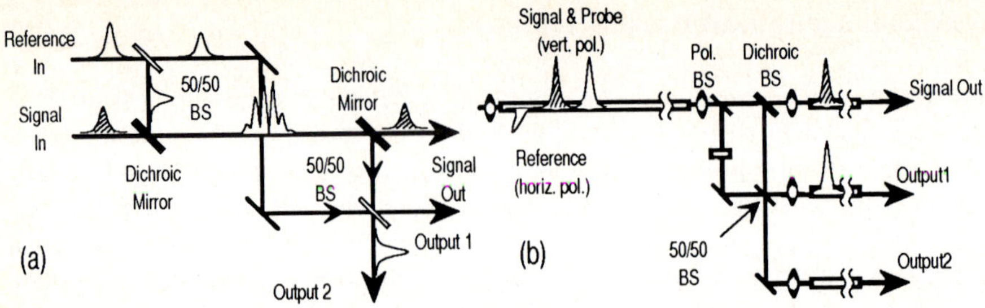

Figure 3: Two-color soliton collision Mach-Zehnder switch in (a) schematic form, and (b) fiber optic implementation. A probe and signal soliton collide in an arm of the interferometer, and interfere with a reference soliton.

exits through a dichroic beamsplitter. The probe and reference solitons interfere to produce switching at a 50/50 beamsplitter. When a signal soliton is not present, the probe and solitons emerge together at Output 1. When a signal soliton is present and collides with the probe, the probe acquires a phase shift and interferes with the reference to shift the output to Output 2. Two points are worth noting: (1) the switching is independent of the relative phase of the probe and signal solitons, and (2) the switching is insensitive to timing jitter. The second point follows because the exact position of the collision in a long interferometer arm is unimportant.

For some combinations of frequency offset and signal soliton amplitude, the probe can acquire a π phase shift. In the absence of other effects, the probe and reference would join together to emerge completely from Output 2. However, the signal soliton cannot then be separated from the probe, as the two overlap in frequency. Phase shifts smaller than or equal to $\pi/2$ allow the signal to be separated from the probe by the dichroic mirror. The energy transferral from Output 1 to Output 2 is no longer complete, although more than adequate for effective switching. For example, if the phase shift is $\pi/2$, 50% of the combined energy of the probe and reference solitons are transmitted. This is exactly the energy of the probe or reference separately, so that a soliton is put out from Output 2.

An optical fiber implementation of a two-color soliton-collision switch is shown in Fig. 3b. To maintain equal path lengths, the probe, reference and signal solitons propagate down the same polarization-maintaining fiber, with the reference soliton polarized perpendicularly to the other two. A polarizing beamsplitter and a dichroic beamsplitter separate the reference and probe solitons from the signal soliton after the collision, and they interfere at the 50/50 beamsplitter to produce switching. The resulting $\pi/2$ phase-shift of the probe soliton by the collision allows 50% of the combined probe and reference energy to be shifted to Output 2.

The energies needed for switching are the soliton energies W_E:

$$W_E = 0.88\,\frac{A_{eff}|D|\lambda^3}{c n_2 \pi^2 \tau_0}.$$

Here, A_{eff} is the effective core area, D is the group velocity dispersion, n_2 is the fiber nonlinearity and τ_o is the soliton width. For $A_{eff} = 60\ \mu m^2$, $D = 12$ psec/(km-nm),

$\lambda = 1.465$ μm, and $\tau_o = 2.0$ psec, soliton energies are ~ 11 pJ, and the signal-probe wavelength separation is ~ 2.3 nm. Switching lengths are ~ 90 m for $\pi/2$ switching. If $A_{eff} = 10$ μm^2 and D = 3 psec/km-nm, then energies are ~ 400 fJ for the same pulse durations.

The two-color soliton collision switch operates in several different modes. The signal soliton can act as a control signal, switching energy to Output 2. Or the switch can be made to act as an AND gate. Then Output 2 ANDs the signal soliton and the probe/reference soliton. It can also make a QND measurement, accurately determining the energy of the signal soliton with perturbing it [7]. Provided the signal and probe frequencies are well-separated, the signal soliton can in principle be transmitted through the device with no change in energy or shape. Indeed, all energy introduced into this type of switch emerges from one output port or another.

4. Discussion and Conclusions

Difficulties for this switching device result from Raman effects. As pulse widths decrease below ~ 1 psec, Raman-induced soliton-self-frequency shifts become important. For pulse widths below several hundred fsecs, Raman effects make the phase-shift depend on the relative phase of the two solitons. Nevertheless, the switches should be able to operate at greater than THz repetition rates, and the Raman effects give rise to other possible switching geometries, to be discussed later.

A number of different modes of operation should be possible for the soliton-collision switch. It shows many of the advantages of optics for switching: ultrafast operation, QND measurement capabilities, and energy-conserving operation.

References

1. S. R. Friberg, A. M. Weiner, Y. Silberberg, B. G. Sfez, and P. W. Smith, Opt. Lett. **13**, 904 (1988).
2. K. J. Blow, N. J. Doran, and B. K. Nayar, Opt. Lett. **14**, 754 (1989).
3. M. E. Islam, Opt. Lett. **14**, 1257 (1989).
4. V. E. Zahkarov and A. B. Shabat, Sov. Phys. JETP **34**, 62 (1972).
5. A. Hasegawa and F. Tappert, Appl. Phys. Lett. **23**, 142 (1973).
6. H. A. Haus, K. Watanabe, and Y. Yamamoto, J. Opt. Soc. Am. B **6**, 1138 (1989).
7. Y. Sakai, R. J. Hawkins and S. R. Friberg, Opt. Lett. **15,** 239 (1990).

Part IV

Free-Space Devices

Two-Dimensional Array Microlasers for Photonic Switching

J.L. Jewell[1], *Y.H. Lee*[1], *S.L. McCall*[2], *A. Scherer*[3], *J.P. Harbison*[3], *L.T. Florez*[3], *N.A. Olsson*[2], *R.S. Tucker*[1], *C.A. Burrus*[1], *C.J. Sandroff*[3], *A.C. Gossard*[4], *and J.H. English*[4]

[1]AT&T Bell Laboratories, Room 4G-520, Holmdel, NJ 07733, USA
[2]AT&T Bell Laboratories, 600 Mountain Avenue, Murray Hill, NJ 07974, USA
[3]Bellcore, 331 Newman Springs Rd., Red Bank, NJ 07701, USA
[4]Present address: Engineering Materials Department, University of California, Santa Barbara, CA 93106, USA

Abstract. Electrically-driven microlasers are shown to be attractive devices for photonic chip-to-chip communication or as elements of a photonic switch. Picosecond optical techniques allow us to study properties of microresonators with submicron diameters.

1. Introduction

Vertical-cavity surface-emitting lasers [1-4] are generating much interest due to their geometric suitability for two-dimensional array fabrication and their potential for achieving ultra-low thresholds. Here, we report on optically- and electrically-pumped microresonator devices. Electrically-driven vertical-cavity microlasers (μlasers) have transverse dimensions of a few microns and active material lengths of a few hundred Å. The very small volumes are a key factor in achieving low thresholds. So far however, surface recombination has prevented us from achieving thresholds much below 1 mA. Optical pumping has allowed us to study the characteristics of microresonators down to <0.5 μm diameter. In the ≤ 1 μm regime waveguide dispersion is very large and transverse modes are clearly seen. Also, surface recombination results in carrier lifetimes ~ 10 ps for the 0.5 μm devices.

2. Microlasers

We have constructed more than one million electrically-pumped vertical-cavity surface-emitting μlasers with dimensions of a few μm on a single GaAs chip [5]. Cylindrical μlasers have diameters 1, 1.5, 2, 3, 4, and 5 μm with heights about 5.5 μm (Fig. 1). Device density is greater than two million per square cm with a typical chip size about 7×8 mm^2.

Springer Series in Electronics and Photonics, Vol. 29 **Photonic Switching II**
Editors: K. Tada · H.S. Hinton © Springer-Verlag Berlin, Heidelberg 1990

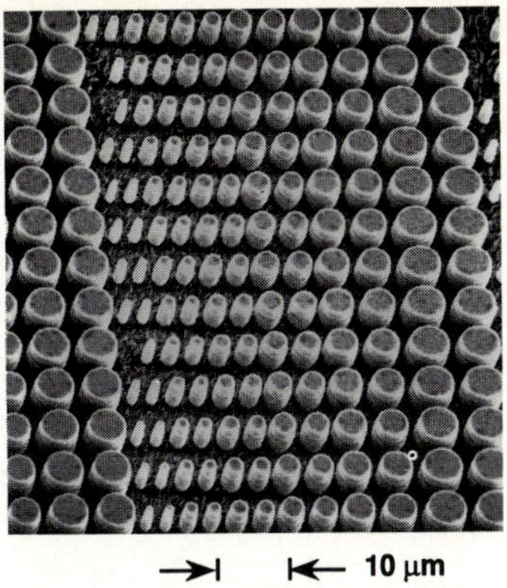

Fig.1 - Small portion of an array of μlasers.

Cylindrical and rectangular-shaped devices of various sizes up to 100 μm across were also tested. The lowest thresholds were achieved in a single quantum well (SQW) structure and were 1.1 mA pulsed in a 4-μm diameter μ-laser, and 1.5 mA CW in a 5-μm square device [6]. Additionally, a 5-μm diameter three-quantum-well (3QW) device which was treated to reduce surface recombination had a 0.8 mA pulsed threshold. In most of the chips tested the 5-μm diameter devices had yields of 95-100%. Lasing was observed in 3QW μlasers as small as 1.5-μm in diameter with an active volume $<0.05\ \mu m^3$. In SQW μlasers $10\times 10\ \mu m$ square, modulation of pseudorandom data streams at rates of 1 Gb/sec were demonstrated with $<10^{-10}$ bit error rates measured by the receiver. Sinusoidal modulation of 3QW lasers of the same size showed a -3 dB cutoff at >8 GHz.

Molecular beam epitaxy (MBE) was used to grow the two samples. Both substrates were GaAs Si-doped at $3\times 10^{18}\ cm^{-3}$ on each of which alternate layers of AlAs and GaAs were grown, forming an interference mirror. These layers, each nominally a 1/4 wave optical thickness at 980 nm, were Si-doped, also at $3\times 10^{18}\ cm^{-3}$. In the three-quantum-well (3QW) sample the undoped active region consists of three 80 Å $In_{0.2}Ga_{0.8}As$ quantum wells interleaved with four 100 Å GaAs barriers. The SQW sample had an undoped 100 Å $In_{0.2}Ga_{0.8}As$ active layer. Surrounding each active region are 200 Å compositionally graded undoped $Al_xGa_{1-x}As$ layers. Above and below the graded layer active region complex are spacer regions of doped $Al_{0.5}Ga_{0.5}As$ of thicknesses so that a full wave exists between mirrors. The second mirror was constructed similarly to the first except that Be was used as the dopant at $5\times 10^{18}\ cm^{-3}$ density. In the 3QW (SQW) sample the Si-doped mirror had 20 1/2 (23 1/2) half-wave periods while the Be-doped mirror had 12

(15). The well/barrier nature of the heterostructure of the mirrors introduces high electrical resistivity. For this reason both mirrors, the Be-mirror in particular, had superlattices at the interfaces where the carriers had to climb up the potential barriers. The non-abrupt interfaces help reduce the electrical resistance without substantially degrading the reflectivity. Above the Be-doped mirror were grown a $\sim$0.2 wave optical thickness of AlAs/superlattice and a 30-Å GaAs cap, both Be-doped at 5×10^{18} cm^{-3}. For low contact resistance, the cap was also delta-doped with 10^{13} cm^{-2} of Be. The cap layer was covered with unalloyed Au 1500 Å thick.

Standard contact photolithography with liftoff was used to create spots of Ni 1 to 5 μm diameter and 5-200 μm square on the Au. Chemically-assisted ion beam etching (CAIBE) was used to etch through the Au and the 5-6 μm heterostructure, leaving the Ni covered areas [7]. The Ni (etch mask) thickness was about 1500 Å and it barely eroded away. Residual Ni should have no significant effect on laser operation. The other side of the substrate was polished and a wire serving as ground was spot welded to it. Positive voltage pulses were applied to the Au caps using a small probe tip of radius $\sim$5 μm. Laser outputs went through the 500-μm-thick substrate and out the polished side. The configuration is shown in Fig. 2.

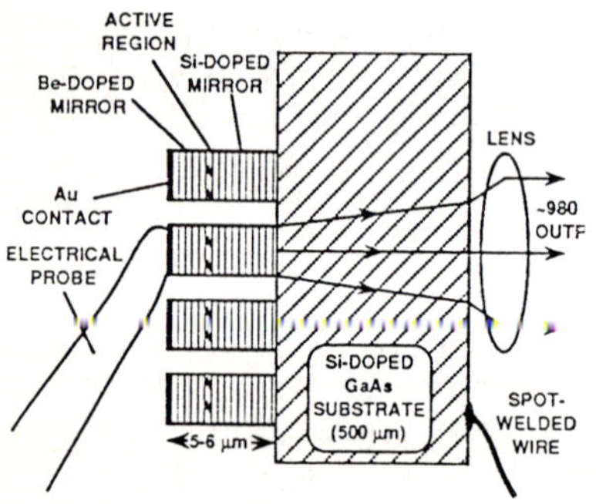

Fig. 2 - Schematic of the electrical pumping and optical output. Devices are enlarged to show detail.

All experiments were performed at room temperature. For all lasers below threshold the luminescence was unpolarized, while above threshold the laser light was linearly polarized. Slight asymmetry in the structures, rather than crystal orientation, appears to cause the polarization as with previous optically-pumped microlasers [8]. Figure 3 displays output light at 958 nm vs driving current for a 3-μm 3QW μlaser on the first chip tested. Voltage pulses were 50 ns long at low duty cycle. The measured

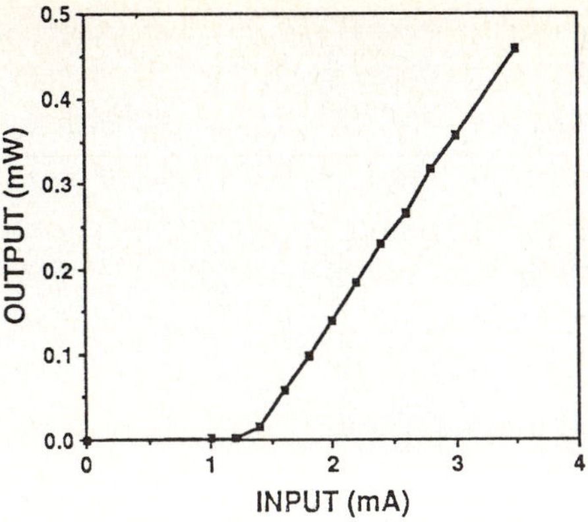

Fig. 3 - Pulsed output light vs current in a 3-μm diameter 3QW μlaser. Differential quantum efficiency is ~16 %.

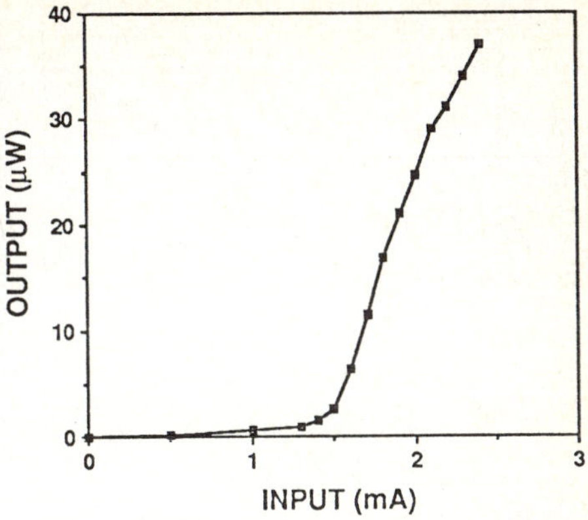

Fig. 4 - Room-temperature CW input/output for a 5-μm diameter SQW μlaser with no applied heatsinking.

single-facet differential quantum efficiency was about 16% despite some absorption of the laser output in the doped substrate. At threshold the voltage was about 15 V rising to 20 V at 3.5 mA. The 5-μm 3QW μlasers required about 8 V and 2 mA for threshold. 4-μm diameter SQW μlasers had typical pulsed thresholds of 1.1 mA with ~7 % differential quantum efficiency. Other devices were etched to a level slightly below the active layer, resulting in lower electrical resistance and higher thermal conductivity [6]. Diffractive losses in the non-waveguiding part of the bottom mirror were sufficiently small in ≥ 5 μm devices to be negligible. Room-temperature CW operation was achieved in these devices with no heat-sinking applied. The main heat flow was conduction through the bottom mirror into the substrate. Fig. 4 shows the input/output of a 5-μm square device having a 1.5 mA threshold and 5 volts applied. The wavelength was 983 nm with a measured width of 3.5 Å limited by the spectrometer resolution. A 7-μm diameter 3QW laser had a CW threshold 2.5 mA at 4 V and output over 0.3 mW with 6 mA current. The same device, at 1% duty cycle, had 18 mW peak output power and >28 % differential quantum efficiency (Fig. 5).

Modulation speeds of the CW-operating SQW lasers were tested using a 3 GHz coaxial probe tip and a pseudorandom pulse generator. A 50 Ω resistor and 0.1 μf chip capacitor wired onto the probe tip suppressed large reflections from the highly resistive devices. The data was continuously incident on the μlaser with no gating. Despite

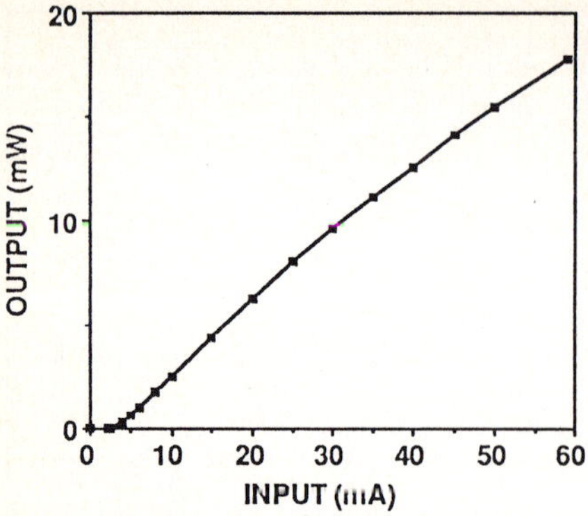

Fig 5 - Peak input/output for a 7-μm diameter 3QW μlaser run at 1 % duty cycle. Differential quantum efficiency is ~28 %.

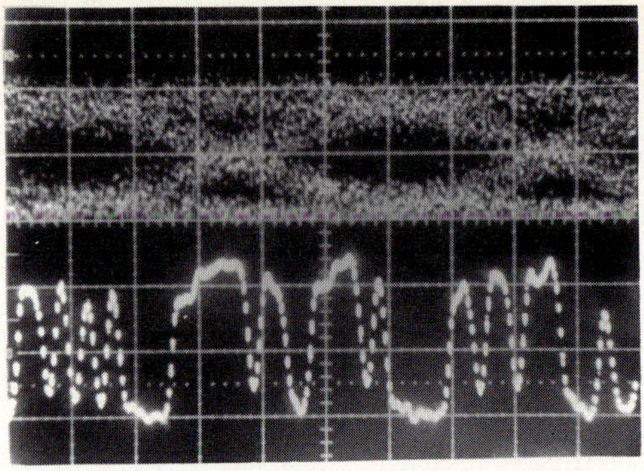

Fig. 6 - (Upper trace) Eye diagram of 1.0 Gb/sec modulation of a 10-μm$\times$10-μm SQW μlaser. Time scale is 200 ps/div. (Lower trace) Direct laser output of the continuous stream of pseudorandom bit-generated data. (5 ns/div)

electrical limitations due to detector speed and long wire lengths we obtained bit error rates $<10^{-10}$ at a modulation speed of 1 Gb/sec from a 10 μm square laser. Fig. 6 shows the wide open eye diagram and the direct laser output. We saw a clear trend of smaller μlasers showing faster modulation capability presumably due to their lower capacitance. Due to the probe tip size and spacing of the devices we were unable to contact the 5-μm lasers without also touching the larger ones. More recently, sinusoidal modulation experiments were performed on 10 μm square 3QW lasers with very low inductance mountings. To a CW bias current was added a small sinusoidal modulation. Modulation of the laser output was then monitored as a function of the current modulation frequency. A typical result is shown in Fig. 7 with the output modulation falling to -3 dB at >8 GHz. The same devices showed gain switching times of about 25 ps produced by a 2 GHz sinusoidal current pump and observed with a streak camera. Micron-diameter μlasers packaged in an integrated circuit driver should be capable of many Gb/sec operation.

We saw lasing in μlasers as small as 1.5 μm diameter with an active material volume less than 0.05μm^3 in the 3QW sample. They typically required 2.3 mA and 27 volts for threshold. The current is higher than that required for the 3- and 5-μm devices because surface

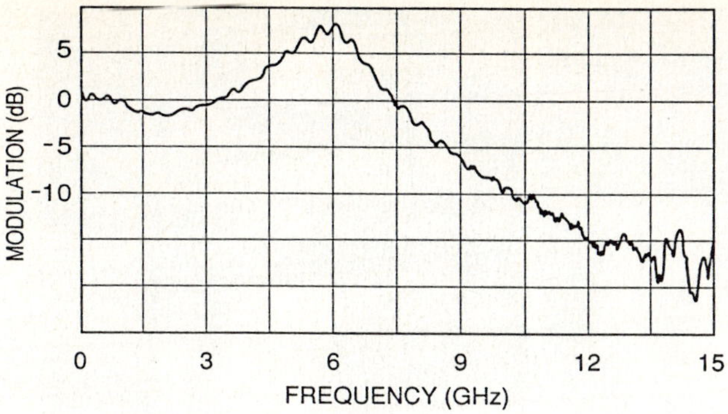

Fig. 7 - Modulation vs frequency for a continuously-driven 10-μm square 3QW laser. The -3 dB cutoff is $\sim$3 GHz.

recombination [10] on the sidewalls greatly reduces the carrier lifetimes. The expected 1/e lifetime in 1.5 μm diameter devices is $<$100 ps if the recombination velocity in $In_{0.2}Ga_{0.8}As$ is as large as that of GaAs. The high current density coupled with the high resistance of this small diameter causes the 27 V requirement.

Both the current thresholds and voltages can be greatly reduced in μlasers. Techniques are developing which can suppress surface recombination [9]. We have tried a Se-based chemical passivation on some deeply etched 3QW lasers. Figure 8 shows the characteristic for pulsed operation with submilliamp threshold in a 5-μm diameter μlaser. The 0.8 mA threshold is only about a factor of two less than the threshold before the chemical treatment, a much smaller improvement

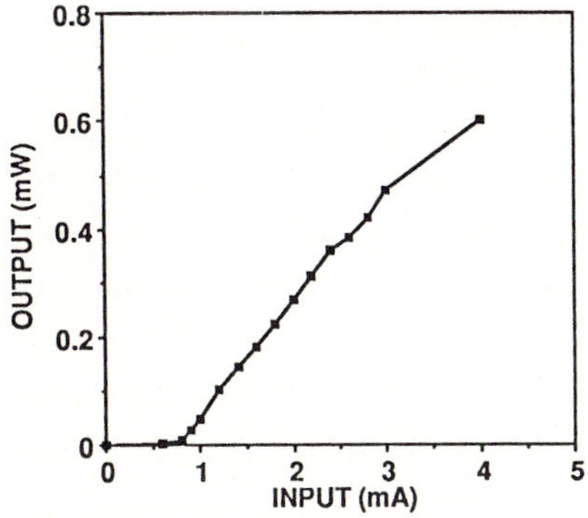

Fig. 8 - Pulsed light/current response with 0.8 mA threshold for a 5-μm diameter 3QW μlaser with some damaged sidewall material removed by chemical etching. Threshold before treatment was 1.5 mA.

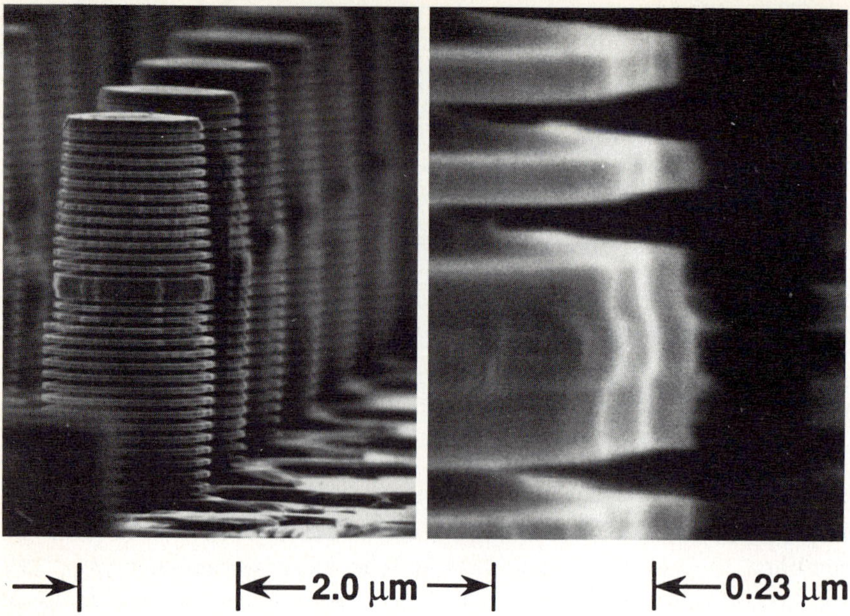

Fig. 9 - Scanning electron micrographs of a 2-μm diameter 3QW μlaser treated with the chemical etching. Such structures of this size lase with a 1.0 mA threshold (pulsed).

than hoped for. The improvement appears to be fairly permanent since it has not degraded for more than 6 months. True passivation of the sidewalls should have produced a larger improvement; also the solution used was not expected to last more than a few days. Most likely the chemical simply removed part of the outer sidewall material which had been damaged by the CAIBE processing, and any true passivation did not last long enough for us to observe. A side effect (literally) of this treatment results from the solution etching AlAs at a fast rate. A 2-μm diameter μlaser which underwent such treatment is shown in Fig. 9. The etch is seen to cut about 2000 Å into the AlAs layers, showing the laser structure very clearly. The 2-μm diameter devices do still lase with 1.0 mA pulsed threshold.

The ability to pass the laser output through the substrate allows us to use the geometry of Fig. 2, which is well-suited for micro-optic integration [11]. The output collecting lens can be replaced by lenslets with diameters on the order of 100 μm formed on the substrate back side by etching processes. These processes are non-labor-intensive and will help minimize packaging costs. Furthermore the high numerical apertures, made possible by the very high refractive index of semiconductors, are ideal for coupling in or out of ultra-small devices. Extremely compact optical systems to image data from one array of devices to another with transit times ~ 10 ps or less [11] can be built.

3. Microresonators

Semiconductor Fabry-Perot etalons have been intensively studied since the late 1970's as structures for performing all-optical switching and logic operations. Growth of the entire etalon structure by molecular beam epitaxy (MBE) greatly increased device manufacturability and structural stability. The use of chemically assisted ion-beam etching [7] to form ~ 1.5-μm diameter waveguiding "posts", or microresonators, in these structures has reduced the device cross-sectional areas, energy requirements and recovery times all by more than an order of magnitude [10]. Prior to etching the minimum controlling energy required for an optical logic etalon operation was 20 pJ. The device diameter, defined as the illuminated region, was about 10 μm and recovery time was estimated to be several ns, both deduced from measurements on comparable devices. For 1.5-μm microresonators the energy is 0.6 pJ and recovery time, ~ 200 ps. Thus, at least over this size range, the energy and time requirements scale directly with cross-sectional area. It is possible that this scaling will continue down to $\sim \lambda/n$ diameters, i.e. to $\sim 1/4$ μm for GaAs, or another factor 36 in area. This would reduce controlling energies to ~ 17 fJ and recovery times to ~ 6 ps for the same Fabry-Perot design. Gating experiments with a ~ 0.5 μm diameter device yielded a 30 ps recovery, in agreement with expectations [12]. The energy requirements could not be fairly tested since the transmission peak was at 8575 Å and the pump beam was not focused as well as the probe. Still a small fraction of the 3 pJ pump pulse produced 75% modulation by saturating the bulk GaAs absorption well into its band. Such small waveguiding etalons also exhibit waveguide dispersion which causes the transmission peak to vary with the microresonator *diameter* [12]. Figure 10 shows data taken from microresonators all lying within a ~ 50 μm area. A nearby region of >10 μm dimensions had a transmission resonance at 9010 Å, while all the microresonators had shorter wavelength transmission peaks. The measured values compare quite favorably with the calculated ones and show a shift of more than 500 Å in <0.5 μm diameter microresonators. Successful design of devices this small will therefore need to take waveguide dispersion into account.

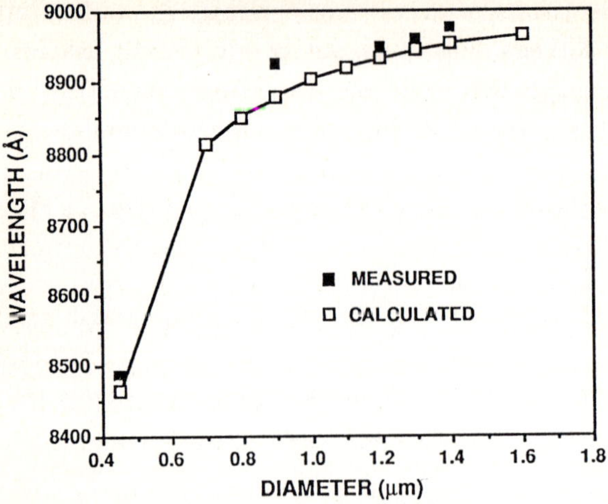

Fig. 10 - Calculated and measured resonant wavelengths for various size microresonators. The infinite-diameter wavelength is 9010 Å. Resonances near the bandgap wavelength of ~8730 Å are subject to large uncertainty in the calculation due to material dispersion and do not converge very well, hence their omission from the graph.

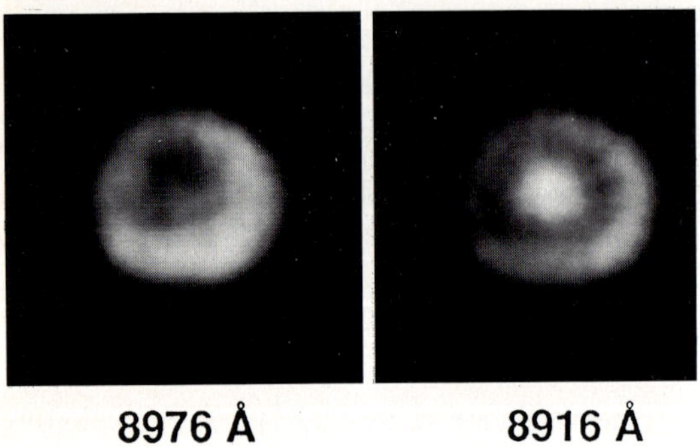

Fig. 11 - Fundamental (8976 Å) and "doughnut" (8916 Å) modes of the 1.4 μm microresonator.

Multiple transverse modes are also seen in small microresonators. Figure 11 shows the lowest order mode and also a higher order "doughnut" mode of a 1.4 μm diameter device. Since the images were recorded in reflected light the transmission resonances are indicated by *darker* images.

The energies (but not recovery times) should be much further reduced by use of thinner, higher-finesse etalons. Optically- and electrically-pumped lasing has been achieved in very-high-finesse vertical cavities containing only a *single* quantum well as the active material [6,13]. With optical pumping, the absorbed energy at threshold in our room-temperature experiments was ~ 12 fJ/μm^2 for pulsed operation and ~ 7 μW/μm^2 for cw pumping. Since optical gating typically requires 10-15 times lower carrier density than optically-pumped lasing, these results indicate that gating in a 0.5 μm device could be achieved with less than 1000 photons if scattering losses do not degrade the finesse too much. Alternatively, a system can be envisioned which would have electronic detectors whose signals, when amplified, could drive electrically driven microlasers. Although power requirements would probably be larger than for the all-optical case, this kind of system could be much more robust to imperfections in device uniformity and optical alignment of the beams. Furthermore it would be simpler and more compact since there would no longer be the need for large external lasers, array generators, and many of the beamsplitter/combiners.

4. Conclusion

We are currently investigating microresonators both as all-optical gates and as electrically-pumped lasers. Although electrical pumping appears to be the desired form of operation, optical pumping allows straightforward picosecond studies of the optical characteristics of ultra-small microresonators.

References

[1] F. Koyama, S. Kinoshita, and K. Iga, Appl. Phys. Lett. **55**, 221 (1989).

[2] A. Ibaraki, K. Kawashima, K. Furusawa, T. Ishikawa, T. Yamaguchi, and T. Niina, Jpn. J. Appl. Phys. **28**, L667 (1989).

[3] D. Botez, L.M. Zinkiewicz, T.J. Roth, L.J. Mawst, and G. Peterson, IEEE Photon. Technol. Lett. **24**, 52 (1989).

[4] K. Tai, R.J. Fischer, C.W. Seabury, N.A. Olsson, T-C.D Huo, Y. Ota, and A.Y. Cho, Appl. Phys. Lett. **55**, 2473 (1989).

[5] J.L. Jewell, A. Scherer, S.L. McCall, Y.H. Lee, S.J. Walker, J.P. Harbison, and L.T. Florez, Electron. Lett. **25**, 1123 (1989).

[6] Y.H. Lee, J.L. Jewell, A. Scherer, S.L. McCall, J.P. Harbison, and L.T. Florez, Electron. Lett. **25**, 1377 (1989)

[7] A. Scherer, J.L. Jewell, Y.H. Lee, J.P. Harbison, and L.T. Florez, Appl. Phys. Lett. **55**, (1989).

[8] J.L. Jewell, S.L. McCall, Y.H. Lee, A. Scherer, A.C. Gossard, and J.H. English, Appl. Phys. Lett. **54**, 1400 (1989).

[9] C.J. Sandroff, M.S. Hegde, L.A. Farrow, C.C. Chang, and J.P. Harbison, Appl. Phys. Lett. **54**, 362 (1989).

[10] J. L. Jewell, A. Scherer, S. L. McCall, A. C. Gossard, and J. H. English, Appl. Phys. Lett. **51**, 94 (1987).

[11] J.L. Jewell, S.L. McCall, Y.H. Lee, A. Scherer, A.C. Gossard, and J.H. English, Tech. Dig., *MOC/GRIN '89* (The Japan Society of Applied Physics, Tokyo, Japan, July 24-26 1989) paper J1.

[12] J.L. Jewell, S.L. McCall, A. Scherer, H.H. Houh, N.A. Whitaker, A.C. Gossard, and J.H. English, Appl. Phys. Lett. **55**, 22 (1989).

[13] J.L. Jewell, K.F. Huang, K. Tai, Y.H. Lee, R.J. Fischer, S.L. McCall, and A.Y. Cho, Appl. Phys. Lett. **55**, 424 (1989).

Top-Surface-Emitting GaAs Four-Quantum-Well Lasers Emitting at 0.85 μm

Y.H. Lee[1], *B. Tell*[1], *K. Brown-Goebeler*[1], *J.L. Jewell*[1], *and J.M.V. Hove*[2]

[1]AT&T Bell Laboratories, Room 4G-538, Holmdel, NJ 07733, USA
[2]APA Optics, Inc., 2950 NE 84th Lane, Blaine, MN 55434, USA

Room-temperature CW and pulsed lasing of top-surface-emitting, vertical-cavity, self-aligned, GaAs quantum-well lasers is achieved at ~845 nm. The active gain medium is four 100-Å-thick GaAs quantum wells. The whole structure is grown by molecular beam epitaxy. Deep H^+-ion implantation followed by annealing is used to control a vertical profile of resistivity for an efficient current injection at the active region. The threshold current is 2.2 mA for CW and pulsed operation using 10-μm diameter lasers. Differential quantum efficiency is about 20 %. Minimum threshold current density per quantum well of 360 A/cm^2 is obtained. Maximum CW output power better than 1.5 mW is obtained.

Vertical-cavity, surface-emitting lasers [1-9] have great potential owing to their inherent two-dimensional nature and very small gain medium volumes which are essential to low threshold currents. The possible applications are optical switching/computing, photonic interconnection, high/low power laser sources, image processing, neural networks, etc. Driven by these high promises, there have been numerous reports on vertical-cavity surface emitting lasers using InGaAs/GaAs/AlAs or GaAs/AlGaAs structures. So far most of these surface emitting lasers have emitted light through bottom mirrors, because top hybrid (or Au) mirrors were opaque and used as electrical contacts. In this paper, we report new top-surface emitting lasers in which top mirrors are used as output mirrors and deep ion implantation is used for current funnelling. Since laser light need not pass through GaAs substrates, various top-surface emitting lasers could be engineered at wavelengths opaque to bulk GaAs (< 880 nm).

The top-surface emitting laser structure is a vertical *pin* junction where electrical current is injected through the bottom and top mirrors. The whole structure is grown on Si-doped n^+ GaAs substrate by molecular beam epitaxy. The Si-doped bottom mirror has 26.5 pairs of AlAs/$Al_{.15}Ga_{.85}As$ quarter-wave stack designed at 850 nm. The undoped active region consists of four 100-Å-thick GaAs quantum wells separated by three 70-Å-thick $Al_{.3}Ga_{.7}As$ barriers. A test sample with four 100-Å GaAs quantum wells grown under the same conditions as those of the laser sample shows photoluminescence at 850 nm. The Be-doped top mirror consists of 19 periods of AlAs/$Al_{.7}Ga_{.3}As$/$Al_{.4}Ga_{.6}As$/$Al_{.15}Ga_{.85}As$. To reduce resistivity through heterostructural AlAs/$Al_{.15}Ga_{.85}As$ interfaces, 101-Å-thick $Al_{.4}Ga_{.6}As$ and 95.5-Å thick $Al_{.7}Ga_{.3}As$ layers are introduced between AlAs and $Al_{.15}Ga_{.85}As$. These two AlGaAs intermediate layers are used as two steps to lower the potential energy stairs at the heterostructural interfaces for holes. The combined optical thickness of a period

 © Springer-Verlag Berlin, Heidelberg 1990

is designed to be a half-wave at 850 nm. The introduction of intermediate AlGaAs layers slightly reduces peak reflectivity of the dielectric stack.

Following a standard photolithography, 1000-Å-thick Au/Ge electrodes are thermally evaporated. The energy of the H^+ ions is chosen to generate a high resistivity region around the active quantum well depth 2.5 μm from the top. According to our calculation, 300-keV H^+-ions introduce damage mostly around a depth of 2.3 μm from the surface, while leaving the topmost 1.5 μm relatively conductive (but still found to be highly resistive before annealing). The dose of implantation is 5×10^{14} cm^{-2}. Thermal annealing at 450 °C for 30 seconds significantly reduces the resistance for laser operation, but still it is rather high. This high resistance is responsible for the high threshold voltage of 7-8 V for top-surface emitting lasers. We believe the resistance can be reduced much more and various implantation and annealing schemes are currently under way to find an optimum condition.

The top rectangular Au-electrical contact (40 x 60 μm^2) with a circular laser output window is shown in Fig.1. The area between the rectangular contacts is chemically etched (2.5-μm deep) using H_3PO_4 to electrically isolate individual lasers. For applications where completely planar processes are essential, ion implant could replace this wet chemical etching for isolation. The window diameters are 10, 15, 20, and 30 μm. Smaller windows were not tried, since the diffraction losses for the gain-guided (optically unguided) surface emitting lasers increase rapidly with decreasing radius. Only the areas under the output windows are not damaged and current is forced to flow only through the active region as shown in Fig. 1(a). In a three-dimensional picture, the conductivity has a shape like a funnel or a wine glass. The bottom side of Si-doped GaAs substrate is Au-plated for an electrical contact. The finished samples were mounted on copper plates and individual lasers were tested using a micro-probe tip.

Fig. 2 shows a CW laser output characteristic for a 10-μm diameter top-surface-emitting laser with 2.2 mA threshold at room temperature. Up to 3.6 mA, the laser light is mostly horizontally polarised with a fundamental Gaussian transverse mode profile. Over 3.6 mA, the characteristic TEM 01* mode also begins to show up with both horizontal and vertical polarizations. The kink at 3.6 mA in Fig. 2 shows

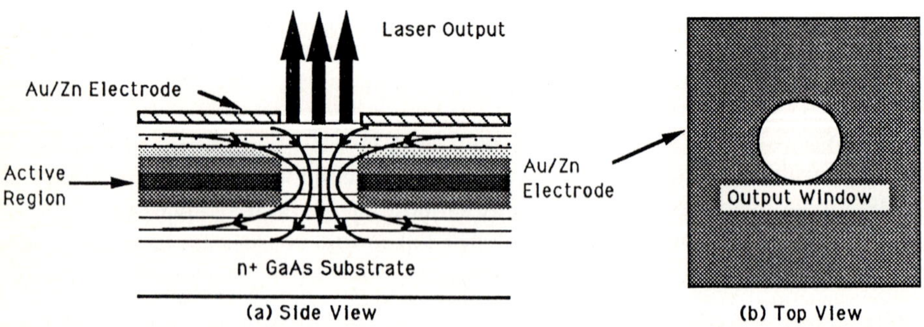

Fig. 1 Top and side views of a top-surface emitting laser. Thin arrows in (a) represent current flows. Graded shades are used to describe a vertical ion damage profile.

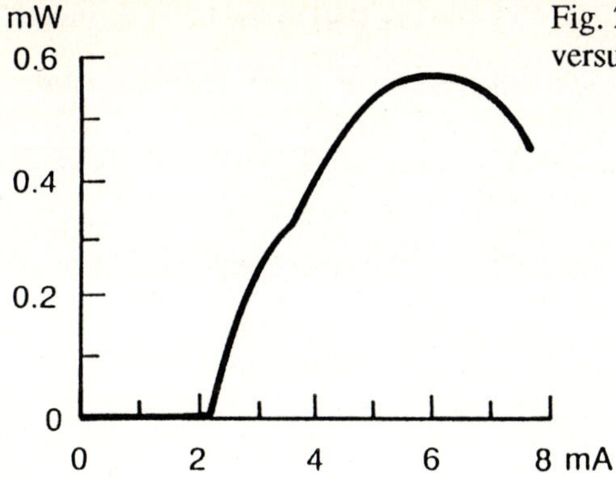

Fig. 2 Continuous laser output power versus current, at room temperature.

the onset of a new transverse mode on top of the fundamental mode. Above 5 mA, circular doughnut-shaped TEM 01* modes dominate. Lasing wavelengths red-shift with currents, from 845 nm (at threshold) to 847 nm (6 mA). A continuous output power of 0.6 mW is measured with 5.5 mA, without heatsink at room temperature. At higher currents, the output power decreases due to thermal effects as shown in Fig. 2. Differential quantum efficiency is about 20 % for currents less than 3.6 mA and decreases for higher currents. CW output power up to 1.5 mW is measured with 15-μm diameter top-surface-emitting lasers.

Using 150-ns pulses at 1 % duty cycle, typical thresholds are 2.2, 3.5, 5.6, and 10.2 mA for 10, 15, 20, and 30 μm diameters, respectively. The corresponding threshold current densities per quantum well are 700, 500, 450, and 360 A/cm^2, respectively. Larger diameter lasers experience less diffraction and diffusion losses, which results in lower threshold current densities. These threshold current densities per quantum well are only slightly higher than that of the lowest threshold current single quantum well edge-emitting laser [10]. Differential quantum efficiencies are about 20 % for all sizes.

In summary, a new style of top-surface emitting, self-aligned lasers with low thresholds is demonstrated at 845-848 nm. Deep H$^+$-ion implantation followed by annealing was successfully implemented for current funnelling only through the active region. The threshold current densities approach the best result of an edge-emitting laser. Top-surface emitting lasers are expected to open up a wide spectrum of laser wavelengths for various applications.

References

1. K. Iga, S. Kinoshita, and F. Koyama, *Electron. Lett.* **23**, 134 (1987).
2. A. Ibaraki, K. Kawashima, K. Furusawa, T. Ishikawa, T. Yamaguchi, and T. Niina, *Jpn. J. Appl. Phys.* **28**, L667 (1989) .
3. J. L. Jewell, A. Scherer, S. L. McCall, Y. H. Lee, S. Walker, J. P. Harbison, and L. T. Florez, *Electron. Lett* .**25**, 1123 (1989)
4. D. Botez, L. M. Zinkiewicz, T. J. Roth, L. J. Mawst, and G. Perterson, *IEEE Photon. Technol. Lett.* **1**, 205 (1989).

5. Y.H. Lee, J.L. Jewell, A. Scherer, S. L. McCall, J.P. Harbison, and L. T. Florez, *Electron. Lett.* **25**, 1377 (1989).
6. K. Tai, R. J. Fisher, K. W. Wang, S.N.G. Chu, and A.Y. Cho, *Electron. Lett.* **24**, 1644 (1989).
7. J. L. Jewell, Y. H. Lee, A. Scherer, S. L. McCall, N. A. Olsson, J. P. Harbison, and L. T. Florez, *Optic. Engin.* **29**, 210 (1990).
8. M. Orenstein, A. C. Von Lehman, C. Chang-Hasnain, N. Stoeffel, L. T. Florez, J. Harbison, and E. M. Clausen, Technical Digest of Optical Society of America, Orlando, Florida, Oct. 1989, paper PD-22.
9. Y.H. Lee, J.L. Jewell, B. Tell, K. F. Brown-Goebeler, J.P. Harbison, and L. T. Florez, *Electron. Lett.* **26**, 225 (1990).
10. K. Y. Lau, P.L. Derry, A. Yariv, *Appl. Phys. Lett.* **52**, 88 (1988).

Photonic Parallel Memory

K. Matsuda, J. Shibata, and T. Kajiwara

Matsushita Electric Industrial Co., Ltd., Semiconductor Research Center, 3-15 Yagumo-Nakamachi, Moriguchi, Osaka 570, Japan

Abstract. We have proposed and fabricated a photonic parallel memory (PPM) which is an array of 32 x 32 optoelectronic bistable switches. The switch consists of a heterojunction phototransistor (HPT) and a light emitting diode (LED), and optical positive feedback from the LED to the HPT is the cause of bistability. The PPM can be written in and read out with 1 kbit parallel optical signals. The PPM has been fabricated successfully and memory operation has been demonstrated as expected. A light pulse with a width of 5 ns was able to turn on the switch and the product of turn-on power and pulse width for the short pulse region was 1.5 pJ. To turn off the switch, the bias voltage is set to zero for a short period. The length of period necessary to turn off the switch was 5 - 30 ns depending on bias voltage.

1. Introduction

There has been much interest in parallel processing architecture, especially in optical parallel processing, since light has the inherent advantage of parallelism and interconnectivity. A large optical array processor with modest switching speeds per device may be able to handle a much larger number of bits per second than the fastest serial processor available today. With regard to a single switch, several kinds of optical or optoelectronic bistable switches have already been proposed and demonstrated, such as a Fabry-Perot etalon with a nonlinear optical material[1], a laser amplifier[2], a laser diode with saturable absorber[3], a self electro-optic effect device (SEED)[4], a light emitting device with pnpn or pnn structure[5,6], and an integrated device of a heterojunction phototransistor (HPT) and a light emitting diode (LED)[7-9].

In this paper we demonstrate an array of 32 x 32 optoelectronic bistable switches consisting of a HPT and an LED. In the switch the HPT and LED are connected electrically in series and light emitted from the LED is fed back to the HPT. This optical positive feedback is the cause of bistability[10]. The switch has the advantages of inherent gain and unidirectionality for optical signals[8] as well as a simple structure suitable for large scale integration[9]. The array can operate as 1 kbit memory with a function of optical parallel processing. This photonic parallel memory (PPM) can be written in and read out with 1 kbit parallel optical signals. It can operate also as a 1 kbit parallel OR gate when different signals are written in either sequentially or simultaneously.

 © Springer-Verlag Berlin, Heidelberg 1990

2. Structure and Fabrication

The structure of the PPM is shown in Fig. 1. It contains 32 x 32 mesas isolated by etched grooves. Each groove has a width of 4 μm and a depth of 5 μm, and the remaining mesa has an area of 16 x 16 μm^2. Each mesa is a bistable switch consisting of seven layers, which are an n-InP emitter, a p-InGaAsP base, an n-InP collector, an n-InP clad, an InGaAsP active layer, a p-InP clad, and a p-InGaAsP cap. The first three and the last four layers constitute the HPT and LED, respectively. All the switches are wired in parallel with a Ti/Au interconnection metal. The Ti/Au plays also the role of optical isolation for the switches since it surrounds all the side walls of the mesas. As shown in Fig. 1, input light is incident upon the HPT through the substrate and output light is emitted upward from the LED. This "passing-through" configuration can be attained by the fact that the substrate and the epitaxial layers other than the base of HPT and the active layer of LED are transparent for light with a wavelength of 1.3 μm which is used as input and output signals.

To fabricate the PPM, the seven layers mentioned above were first grown on an n^+-InP substrate by liquid phase epitaxy (LPE). Then SiN (2000 Å) and Ti (4000 Å) were deposited and patterned with reactive ion etching (RIE) using CF_4 and CCl_4, respectively. They were etched using photoresist as a mask, and they were used as a mask for etching epitaxial layers to form the grooves. The epitaxial layers were etched with RIE using a mixture of Br_2 and Ar gases[11]. After mesa formation Au-Zn ring anodes were evaporated on the tops of the mesas and Au-Sn cathodes were outside of the mesas where the emitter was exposed. After contact evaporation a Ti/Au interconnection metal was evaporated on a SiN passivation film to connect all of the anodes. Finally, the back surface of the wafer was polished so that input light can be incident upon the bottom of the switch. The photomicrograph of a fabricated chip is shown in Fig. 2. The chip has an area of 1.0 x 1.0 mm^2 including bonding pads and single switches for testing, while the active area for integration is 640 x 640 μm^2.

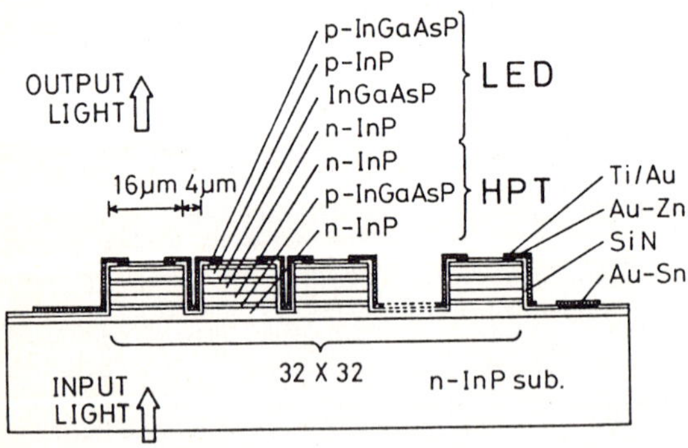

Fig. 1 Schematic cross section of a photonic parallel memory. Each mesa consisting of a heterojunction phototransistor (HPT) and a light emitting diode (LED) is an optoelectronic bistable switch.

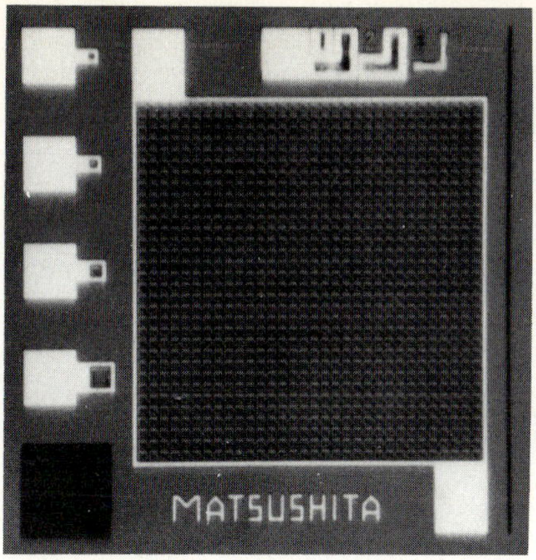

ig. 2 Photomicrograph of a fabricated PPM chip. The chip has an area of 1.0 x 1.0 mm^2 including bonding pads and single switches for testing. The active area for integration is 640 x 640 μm^2.

3. Static Characteristics

Characteristics of a single switch were first measured. As shown in Fig. 3(a), a thyristor-like voltage-current curve was observed unless input light was incident. In this curve "on-state" and "off-state" are clearly distinguished. Output light was emitted from the top of the switch in the on-state which could be observed through an infrared video camera. Then the light emitted from a 1.3 μum laser diode was incident upon the bottom of the switch through a single mode fiber with a focusing lens. As input optical power was increased, the breakdown voltage was decreased and a diode-like curve shown in Fig. 3(b) was observed when the input power measured at the fiber end was 175 μW. When operated with a constant bias voltage, the switch is turned on before this condition. In other words, the input optical power necessary to turn on the switch (turn-on power) depends on the bias voltage and the maximum value is equal to the input optical power necessary to obtain a diode-like curve.

The experimental setup for operating the PPM is shown in Fig. 4 together with a schematic diagram of memory operation. The power supply voltage ranging from 1.6 V to 2.2 V is first

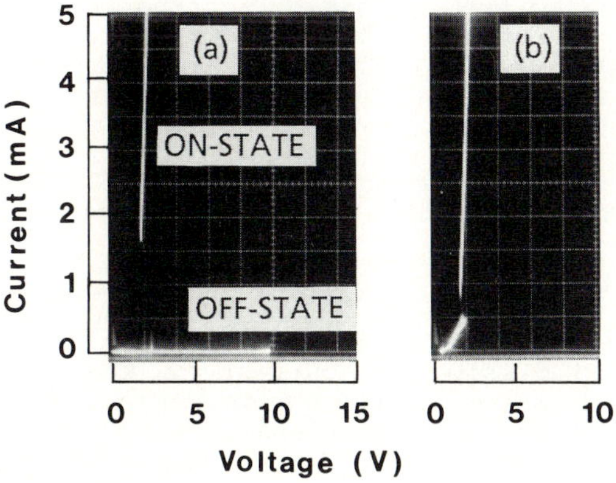

Fig. 3 Voltage-current characteristics of a single switch with an InP collector and a thick (∿ 0.25 μm) base: (a) without input light and (b) with an input light of 175 μW.

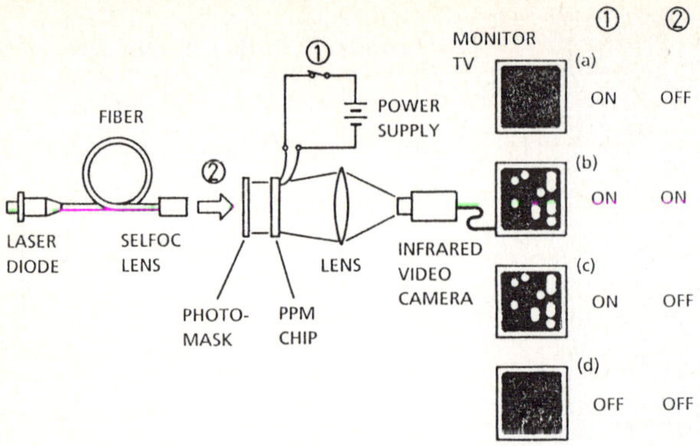

Fig. 4 Experimental setup for operating a PPM chip together with a schematic diagram of memory operation: (a) initial state, (b) write-in operation, (c) memory operation, and (d) reset.

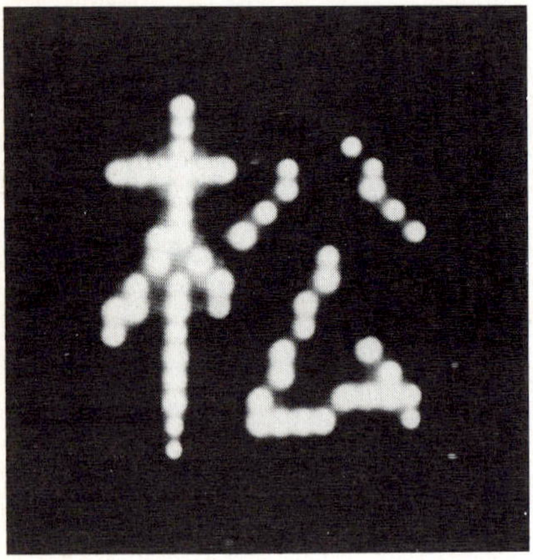

Fig. 5 Infrared image of a PPM chip in memory operation. A Chinese character meaning "a pine tree" is memorized and read out.

applied to all the switches without input light. They are in the off-state and no light emission is observed. Then the light is incident on selected switches, which are turned on and emit output light. Selection of the switch is done by setting photomask between the light source and the chip. The on-state is maintained after removing the input light and the selected switches keep on emitting light. All the switches are turned off by making the supply voltage zero. An example of a memorized pattern on a PPM chip observed with an infrared video camera is shown in Fig. 5. A Chinese character is displayed with 56 selected switches which are turned on and emit light.

4. Reduction of Dissipation Current

Considering operation of the PPM, it is desirable to decrease both turn-on power and dissipation current in the on-state. The minimum current to maintain the on-state (holding current) can be found from voltage-current characteristics and is 1.6 mA for the switch shown in Fig. 3. The characteristics

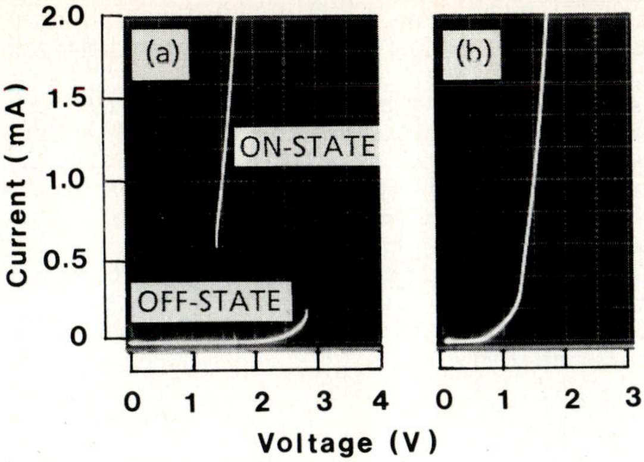

Fig. 6 Voltage-current characteristics of a single switch with a quarternary collector and a thin (∿0.15 μm) base: (a) without input light and (b) with input light of 20 μW.

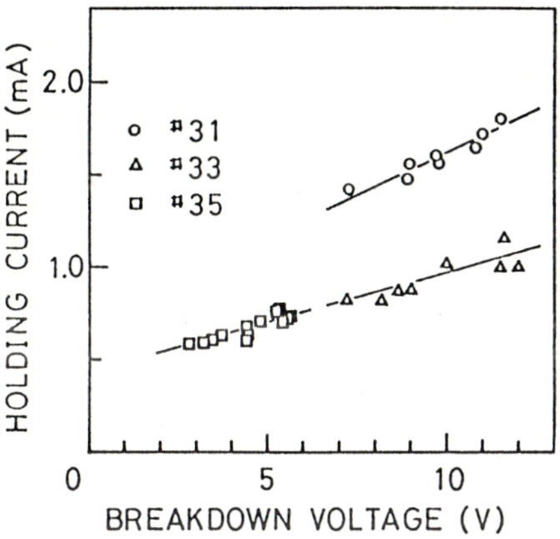

Fig. 7 Relations between breakdown voltage without input light and holding current. The switches are selected from three different wafers: with an InP collector and a thick base (#31), with a quarternary collector and a thick base (#33), and with a quarternary collector and a thin base (#35).

for another switch with a holding current of 600 μA are shown in Fig. 6. The input optical power necessary to obtain a diode-like curve has also been reduced to be 20 μW. For this switch, the HPT collector was changed to InGaAsP and the base was made thinner (∿0.15 μm). As can be seen in Fig. 6(a), the breakdown voltage without input light is also reduced since breakdown caused by punch-through occurs at a lower bias voltage for a HPT with a thinner base.

To isolate the cause of reduction of holding current, the relations between breakdown voltage without input light and holding current have been measured for three types of switches which were selected from three different wafers. They are the wafer with an InP collector and a thick (∿0.25 μm) base (#31), with a quarternary collector and a thick base (#33), and with a quarternary collector and a thin (∿0.15 μm) base (#35). The results are shown in Fig. 7. The breakdown voltage depends on the thickness of the base layer and the various values of breakdown voltage observed in switches on the same wafer is attributed to non-uniformity of layer thickness. The switches with a quarternary collector have smaller holding current than those with an InP collector, and the switches with lower breakdown voltage, which corresponds to a thinner base, have smaller holding current.

The holding current is related to the optical gain of HPT since it is thought to be the current where the product of the efficiency of the LED and the gain of the HPT becomes unity. The gain of the HPT is the product of optical absorption efficiency and current gain. The switches with a quarternary collector have smaller holding current than those with an InP collector because the quarternary collector increases the absorption efficiency of the HPT. The switches with a thinner base have a smaller holding current since a thinner base increases the current gain of the HPT.

5. Dynamic Characteristics

To evaluate the processing rate of the PPM, turn-on and turn-off speed were measured as follows. For turn-on speed, light pulses with various pulse widths were incident upon one of the integrated switches and turn-on power was measured. The results are shown in Fig. 8. The measurement was done for three types of PPM chips on wafers #31, #33, and #35, mentioned in the previous section. The light pulse with a width of 5 ns can turn on the switches on wafers #33 and #35, though the turn-on power is increased. The product of turn-on power and pulse width for the short pulse region is calculated to be 1.5 pJ for the switch on #35.

In this figure the turn-on power for the long pulse region coincides with that in the static characteristics. The turn-on power for the switch with a thin base (#35) approaches that for the switch with a thick base (#33) in the short pulse region because the current gain of the HPT in the high frequency region is limited by the gain-bandwidth product. The gain-bandwidth product has little relation to base thickness since it is not limited by base transit time but by CR time constant.

The results shown in Fig. 8 were obtained using a supply voltage of 1.9 V, which is sufficiently larger than the minimum

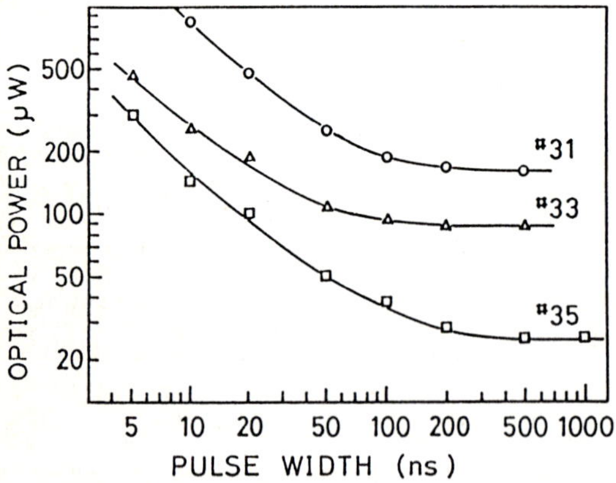

Fig. 8 Relations between pulse width of input light and turn-on power for single switches integrated in PPM chips on wafers #31, #33, and #35.

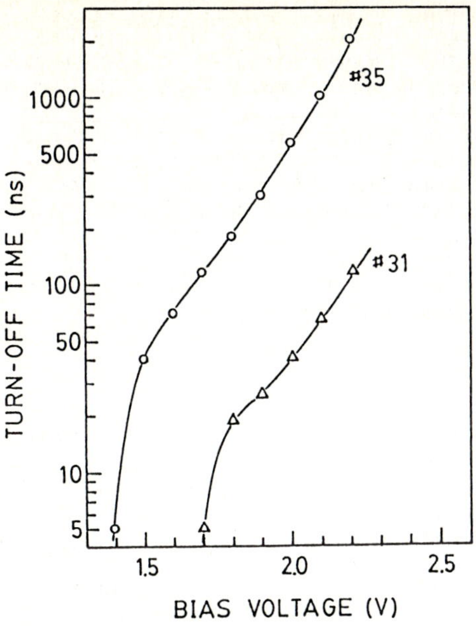

Fig. 9 Relations between bias voltage in the on-state and turn-off time for wafers #31 and #35.

voltage to maintain the on-state (holding voltage). The turn-on power depends only slightly on the supply voltage when the difference between the supply voltage and the holding voltage is larger than 0.2 V. When the difference is less than 0.2 V, the turn-on power increases as the supply voltage approaches the holding voltage.

To measure the turn-off speed, the switches were first turned on and then the supply voltage was set to zero for a short period. The length of period necessary to turn off the switches (turn-off time) was measured for wafers #31 and #35. The results for various bias voltages in the on-state are shown in Fig. 9. The turn-off time strongly depends on the bias voltage in the on-state and a value less than 5 ns was obtained for both wafers in the case of setting the bias voltage close to the holding voltage.

6. Conclusions

A 1 kbit PPM which is an array of 32 x 32 optoelectronic bistable switches has been fabricated successfully. The memory operation of PPM with functions of optical write-in and read-out has been demonstrated as expected. The minimum current to maintain the on-state and the input optical power necessary to turn on the switch were 1.6 mA and 175 μW, respectively, for the switch with an InP collector and a thick (∿0.25 μm) base. They were improved to 600 μA and 20 μW by using a quarternary collector and a thin (∿0.15 μm) base. A light pulse with a width of 5 ns was able to turn on the switch and the product of turn-on power and pulse width for the short pulse region was 1.5 pJ. The turn-off time depended strongly on the bias voltage in the on-state and a value less than 5 ns was obtained in the case of setting the bias voltage close to the holding voltage.

References

[1] H.M. Gibbs, S.L. McCall, T.N.C. Venkatesan, A.C. Gossard, A. Passner, and W. Wiegmann, Appl. Phys. Lett. **35**, 451 (1979).

[2] T. Nakai, N. Ogasawara, and R. Ito, Japan. J. Appl. Phys. **22**, L310 (1983).

[3] H. Kawaguchi and G. Iwane, Electron. Lett. **17**, 167 (1981).

[4] D.A.B. Miller, D.S. Chemla, T.C. Damen, T.H. Wood, C.A. Burrus, Jr., A.C. Gossard, and W. Wiegmann, IEEE J. Quantum Electron. **QE-21**, 1462 (1985).

[5] Zh.I. Alferov, V.I. Korol'kov, V.G. Nikitin, and D.N. Tret'yakov, Soviet Physics-Semicond. **5**, 1314 (1972).

[6] G.W. Taylor, J.G. Simmons, A.Y. Cho, and R.S. Mand, J. Appl. Phys. **59**, 596 (1986).

[7] A. Sasaki, K. Matsuda, Y. Kimura, and S. Fujita, IEEE Trans. Electron Devices **ED-29**, 1382 (1982).

[8] A. Sasaki, M. Taneya, H. Yano, and S. Fujita, IEEE Trans. Electron Devices **ED-31**, 805 (1984).

[9] K. Matsuda, K. Takimoto, J. Shibata, and T. Kajiwara, in Tech. Dig., 7th Int. Conf. Integrated Optics and Optical Fiber Communication (Kobe, 1989), paper 20C3-2.

[10] Y. Ogawa, H. Ito, and H. Inaba, Japan. J. Appl. Phys. **20**, L646 (1981).

[11] K. Takimoto, K. Ohnaka, and J. Shibata, Appl. Phys. Lett. **54**, 1947 (1989).

Realization of a Graded-Threshold Function Using an AlGaAs/GaAs pnpn Optical Switch

K. Hara, K. Kojima, K. Mitsunaga, and K. Kyuma

Central Research Laboratory, Mitsubishi Electric Corporation,
1-1 Tsukaguchi-honmachi 8-chome, Amagasaki, Hyogo 661, Japan

Abstract A graded-threshold function using an AlGaAs/GaAs pnpn optical switch is proposed and demonstrated, which is very useful to optical neural networks.

1. Introduction

The massive parallelism of optical processing is very attractive in the implementation of neural networks [1][2]. In most of the neural network models [3], the activity of a neuron can be modeled by graded-thresholding as shown in Fig.1(a), which is often represented by a sigmoidal function because of the ease of mathematical treatment. Previously this operation was achieved by the combination of several electronic devices. The realization with an optical device is more suitable in terms of 2-D monolithic integration which is essential to parallel processing. However, the response of conventional optical devices is step-thresholding as shown in Fig.1(b). In this paper, we propose and demonstrate a graded-threshold function using a pnpn optical switch [4][5] for the first time. The slope of the graded-threshold function is variable, which is useful for a learning process such as simulated annealing with different annealing temperatures.

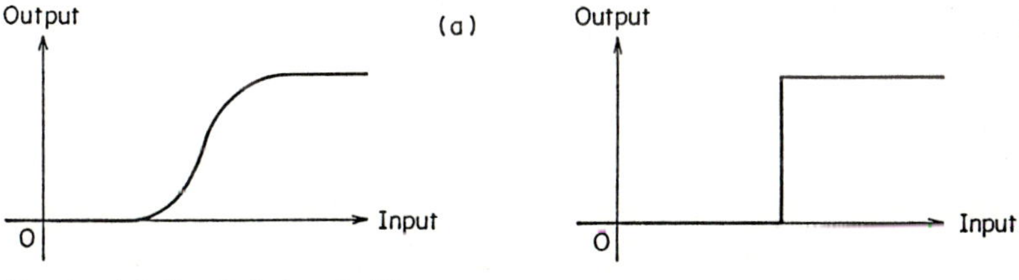

Fig. 1 (a) Graded-thresholding (b) Step-thresholding

2. Operating Principle

When a pnpn optical switch is biased slightly below the switching voltage and is illuminated by an optical input exceeding a threshold level, it is electrically turned on from the high impedance state to the low impedance state. Then, the optical output starts to increase. The total response time T is expressed as the sum of the delay time t_d necessary for the electrical switching and the optical rise time t_r. The rise time t_r is determined by a turn-on response in spontaneous emission, which is independent of the optical input power P_{in}. On the other hand, the delay time t_d is a function of P_{in}. The equivalent circuit of the pnpn device is based on a two-transistor model [6]. We

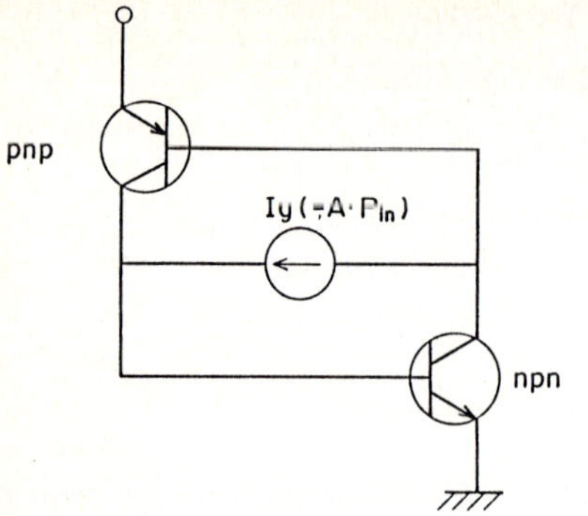

Fig. 2 Two-transistor approximation of pnpn optical switch

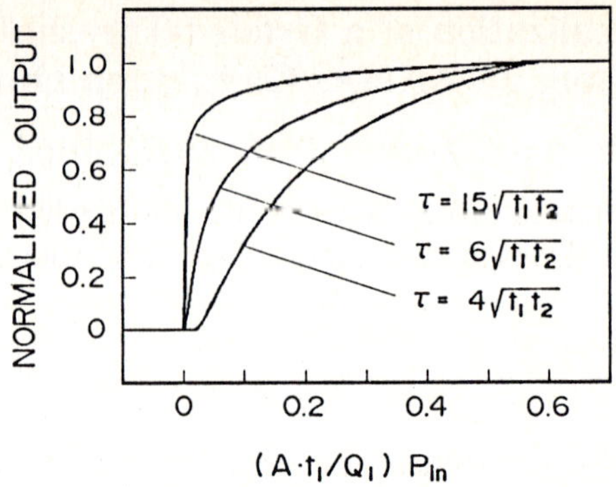

Fig. 3 Graded-thresholding function ($t_{RC} = \sqrt{t_1 t_2}$ $t_r = 2\sqrt{t_1 t_2}$)

have modified this circuit, as shown in Fig.2, by including a constant current source proportional to P_{in}. The collector currents at the high-injection region are given by

$$I_{c1} \simeq Q_1/t_1 \tag{1}$$

$$I_{c2} \simeq Q_2/t_2, \tag{2}$$

where Q_1, Q_2 are the numbers of carriers stored in the pnp and npn transistor and t_1, t_2 are the base transit time, respectively. Assuming the ideal condition that $dQ_1/dt = I_{c2} + I_g$ and $dQ_2/dt = I_{c1} + I_g$, we obtain the following equation:

$$\frac{d^2Q_1}{dt^2} - \frac{Q_1}{t_1 t_2} = \frac{I_g}{t_2}. \tag{3}$$

The solution is given by

$$Q_1 = t_1 I_g(exp(\frac{t}{\sqrt{t_1 t_2}}) - 1) + Q_{const}. \tag{4}$$

This shows that the stored carriers increase exponentially by positive feedback, and the increase is limited by a load resistor. Setting Q_1 in the *on*-state to $Q_{on} + Q_{const}$, the delay time is given by

$$t_d = \sqrt{t_1 t_2}\ln(1 + \frac{Q_{on}}{At_1 Pin}). \tag{5}$$

Therefore, the response time T decreases with increasing P_{in} and finally is limited by the RC constant. We utilize this phenomenon to realize the graded-threshold function. As an output signal we use the integration of optical output power in an integration time τ. The integration time can be set by the time window in the cascaded light-detector or the pulse width of the voltage applied to the switch. With increasing P_{in} above the threshold level the switch turns on faster ; the output signal increases gradually and saturates finally. The relationship between P_{in} and the output signal is shown in Fig.3 with three different values of τ. The longitudinal axis is normalized by the saturation value. This figure shows that the graded-threshold function can be realized, and the slope can be varied by changing the integration time.

3. Experimental Results

The fabricated pnpn optical switch is shown in Fig.4. The pnpn structure was grown on an n-GaAs substrate by molecular beam epitaxy in the following order: an n-GaAs buffer layer ($0.1\mu m, 2 \times 10^{18} cm^{-3}$), an n-AlGaAs layer ($80nm, 5 \times 10^{17} cm^{-3}$), a p-GaAs channel layer ($10nm, 1 \times 10^{19} cm^{-3}$), an n-GaAs active layer ($1\mu m, 1 \times 10^{17} cm^{-3}$), a p-AlGaAs layer ($0.5\mu m, 5 \times 10^{18} cm^{-3}$), and a p-GaAs cap layer ($0.15\mu m, 1 \times 10^{19} cm^{-3}$). A circular mesa (diameter $90\mu m$) was etched down to the substrate. Contact to the mesa was made using a lift-off process of Au/Cr in the form of a ring. The open area for optical access had a diameter of 20μm. The switching voltage was 2.8V and the switching current was 10μA. The optical output power was 350μW at a 350mA injection current. By varying the optical input power from a laser diode (λ=0.84μm), the optical switch-on response was measured. Figure 5 shows that the response time decreases with increasing input power.

To realize a graded-threshold function, the electrical pulse was applied as the biasing voltage. The optical input pulse with a $1\mu s$ width was injected under biasing. The time between the rising edge of the optical input and the falling edge of the electrical

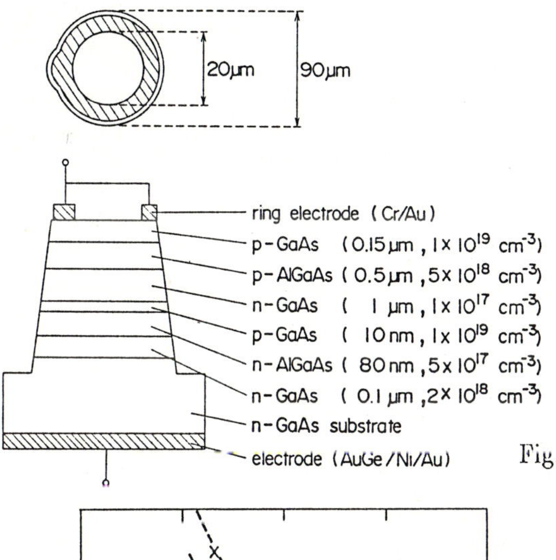

Fig. 4 Structure of fabricated device

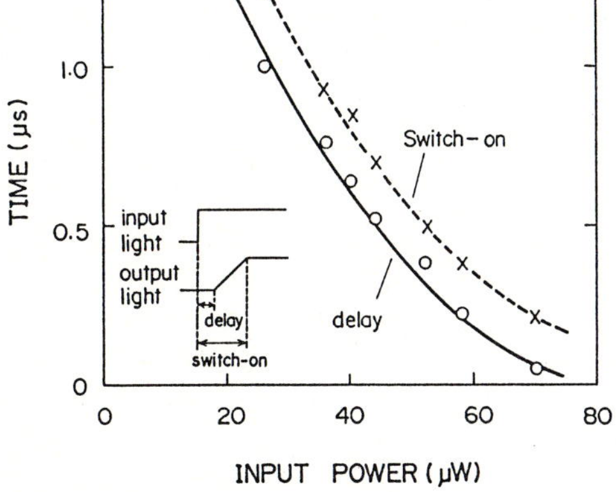

Fig. 5 Measured switch-on response as a function of input power Pin

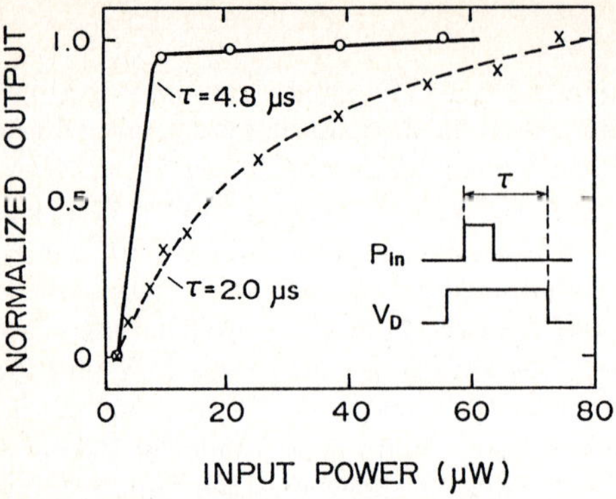

Fig. 6 Input-output response

biasing pulse was the integration time τ as shown in the inset of Fig.6. The output signal was obtained directly by monitoring the average of optical output power. The dependence of the output signal on P_{in} was measured under the integration times of $4.8\mu s$ and $2.0\mu s$. The experimental results as shown in Fig.6 indicate that the graded-threshold function is obtained and its slope is varied by changing the integration time. With increasing integration time, the slope becomes steeper, which corresponds to decreasing the temperature in simulated annealing.

4. Conclusion

The graded-threshold function using an AlGaAs/GaAs pnpn optical switch has been demonstrated for the first time. By changing the integration time, its slope can be varied. The expansibility to 2-D monolithic arrays is very attractive for optical neural networks.

References

[1] N.H. Farhat et al.,Appl. Opt. 24, 1469 (1985).
[2] K. Kyuma et al.,Proc. Soc. Photo-Opt. Instrum. Eng. 963, 475 (1988).
[3] D.E. Rumelhart et al.,PDP, MIT Press (1986).
[4] G.W. Taylor et al.,J. Appl. Phys. 59, 596 (1986).
[5] K. Hara et al.,Electron. Lett. 25, 433 (1989).
[6] S.M. Sze, Physics of semiconductor devices, 2nd ed., Wiley, 214 (1981).

An Injection Controlled Optically Bistable Switch

D. Ronaldson, H.A. MacKenzie, J.I.L. Hughes, and A. Iltaif

Department of Physics, Heriot-Watt University, Edinburgh EH144AS, UK

Abstract The characterisation of a composite bistable device (OBOE) which utilises either electrical carrier injection or photogenerated carriers to achieve switching is reported. The device can be switched in the presence of a holding beam by either a 3V electrical pulse or by an optical power increment of 50 µW.

1. Introduction

Optical bistable devices have an inherent parallelism which may be useful for image and information processing. However, such systems require an electro-optic interface for communication with conventional computing systems. In a basic architecture, the input interface may be a spatial light modulator or a laser diode array with a corresponding detector array or image scanner at the output of the bistable system. In this paper we describe the further development of the OBOE device [1] which has both all optical and electro-optic modes of operation. This integral capability of electrical control and monitoring of the optical logic states (0,1) results in a device with the potential to combine the input, processing and output requirement of an optical logic plane.

The operation of the OBOE device is based on the control of the refractive nonlinearity of a semiconductor etalon, by incrementing the conduction band population by **either** photogenerated or electrically injected carriers in the presence of a holding beam. Thus the device may be used as a 'normal' optically bistable switch or, a holding beam may be used to bring the device close to switching after which an electrical pulse may be used to switch ON. The device also has a transient 'hold' capability, continuous electrical monitoring of the logic state and electrical switch OFF.

Indium antimonide has the largest electronic refractive nonlinearity of any semiconductor material. The carrier recombination time is also sufficiently long ($\tau_R \sim$ 500ns) to allow the creation of dynamic carrier populations with c.w. laser excitation or by carrier injection. The switching energy of fully optimised InSb bistable devices is sufficiently low (75pJ) as is the thermal dissipation per channel (< 50 µW) so that array devices for image processing are compatible with available laser sources and heat sink specifications.

2. Experimental

Fully optimised InSb etalons have been constructed with a thickness of 50 µm and reflectivity coatings of 85%. Using a cw CO laser, bistable switching can be achieved at 500 µW input power with a 50 µm spot diameter [2]. The switching energy of this system is ~ 75 pJ and for bistable switching from a hold point which is within 1% of the switch point, an increment of approximately 10^{10} electrons cm^{-3} will achieve bistable switch-ON. Theory indicates that electrical switching can be achieved with a current of 0.6 mA and applied bias of 0.14 V.

The basic OBOE device consists of an etalon of pure n-type InSb with a pair of surface electrodes formed by the alloy regrowth method. An ohmic contact is formed by melting pure indium into the semiconductor surface and the injecting contact material

Editors: K. Tada · H.S. Hinton © Springer-Verlag Berlin, Heidelberg 1990

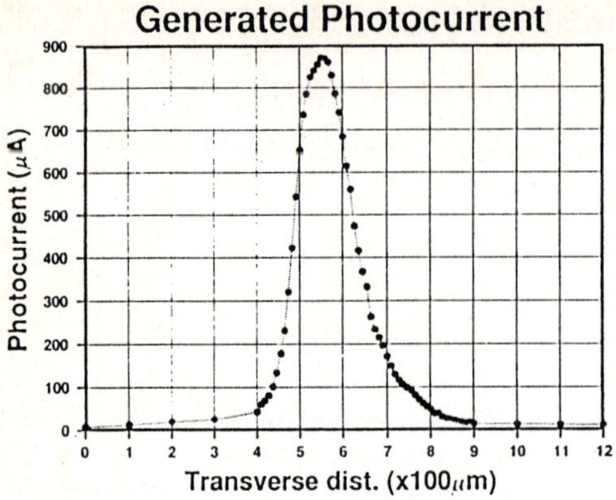

Fig. 1 Variation in photocurrent with probe beam position from injecting electrode.

consists of an indium zinc eutectic alloy composition which when melted on to the semiconductor surface creates a very highly doped p^+n junction. Under forward bias, holes are injected through the junction into the n-type material and establish an excess electron-hole pair population. This population increment creates a sufficient refractive change to induce bistable switching from the holding point [3]. The transverse junction characteristic of the device is shown in Fig. 1.

Electrical switch ON: The bistable characteristic in this mode of operation is shown in Fig. 2(a). The solid line indicates the normal transmission response of the etalon. On generating additional carriers, by using a short voltage pulse, the initial detuning is altered leading to a new bistable characteristic (the dotted line).

Thus if we use a holding power anywhere in the hold range shown and apply a suitable voltage pulse, the etalon will switch from a low transmitting state (OFF) to a high transmitting state (ON). For example, holding at point (a) and introducing a short voltage pulse will switch the system to point (b) and subsequently the transmission will increase to point (c) (the ON point), by bistable switching. The etalon will then stay ON as the operating point is still within the bistable characteristic of the original loop.

Present performance using alloyed injecting contacts has demonstrated the ability to switch the device ON using a 3*V*, 3μs pulse as shown in Fig. 2(b). Also, for different

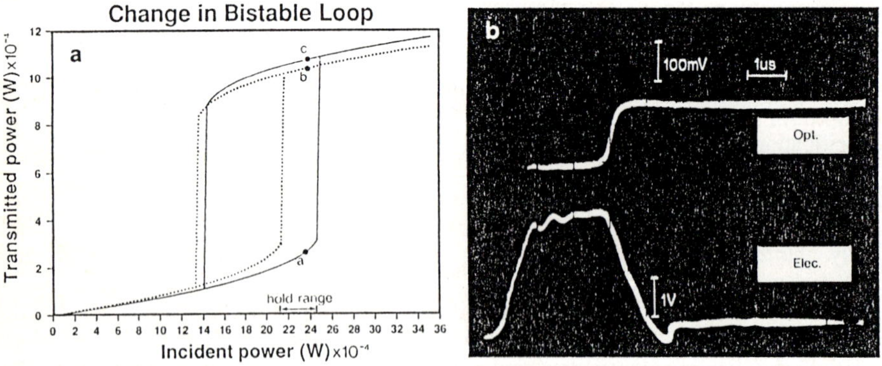

Fig. 2 (a) The bistable characteristic for electrical switch ON. The full line is the characteristic before electrical injection; the dotted line is the characteristic after electrical injection.

(b) Electrical switch ON. The upper trace is the optical signal, the lower trace is the electrical input.

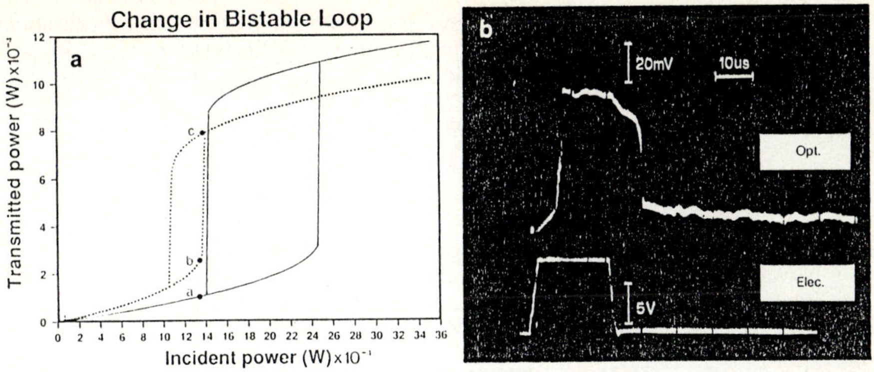

Fig. 3 (a) The bistable characteristic for electrical switch OFF. The full line is the characteristic of the electrically unbiased etalon; the dotted characteristic represents the unbiased system.

(b) Electrical switch OFF. The upper trace is the optical signal, the lower trace is the electrical input.

voltage pulses and holding powers other effects, which mirror the all optical switching situation (i.e. bimodality and critical slowing down) have been demonstrated, illustrating the equivalence of optical and electrical carrier generation.

Electro-thermal Switch OFF: The principal optical nonlinearity in InSb is electronic in origin but in common with many semiconductor materials its refractive index increases with temperature. This competing nonlinearity may be used to switch the bistable element OFF. We have shown [1] that the etalon can be reproducibly switched OFF by utilising an electrical pulse of sufficient energy to induce a thermal change in refractive index.

Electrical Switch OFF: The holding carrier population in this case is achieved by a combination of both optical and electrical carrier generation. To switch the device ON additional optical or electrical bias can be added as described previously and, conversely, a reduction in the optical or electrical bias will lead to a switch OFF.

The bistable characteristic for this regime modelled with practical device parameters is shown in Fig. 3(a), where again the solid curve represents an unbiased etalon (i.e. one with no electrical carrier injection). The switch ON is the same as that shown previously. Thus by holding at point (b) and adding a small voltage pulse in addition to the already existing bias voltage the device can be switched to point (c). In order to switch OFF all electrical bias is momentarily removed, the loop then returns to its original state and the transmission decreases to point (a). If the electrical holding bias is reinstated the transmission will then return to starting point (b).

The principle of this mode of operation has been demonstrated and the switching sequence is shown in Fig. 3(b). In this case we are holding at point (a) (Fig. 3(a)) and then apply the pulse. For the duration of the pulse the transmission is held at point (c). When the pulse stops the high transmission can no longer be maintained and, as the loop returns to its original form, the transmission returns to its original state (a).

3. Discussion

The dual mode operation of the OBOE device creates interesting possibilities for new forms of optoelectronic interfacing for infrared image processing. The reflectivity of the etalon may be modulated by an electrical input and, in common with earlier transphasor experiments [4] the response time is limited to ~ 3 MHz by carrier recombination processes.

In the form of a two dimensional array, reflectivity modulation would be the basis for an infra-red spatial light modulator. The feasibility of such devices has been enhanced by the development of holographic array generators [5] which allow the accurate placement of holding beams. The use of ion beam implantation and diffusion processing [6] will reduce the electrical switching energy to the nJ range. The combination of these developments will result in a versatile opto-electronic interface which can be electrically or optically encoded for pattern recognition or image processing.

References

[1] H.A. MacKenzie, A. Iltaif, J.I.L. Hughes, J.J. Hunter and D. Ronaldson Novel optically bistable optoelectronic device for infrared signal processing, OSA Proceedings on Photonic Switching, **3**, March (1989) pp.26-31.

[2] B.S. Wherrett, D. Hutchings and D. Russell JOSA b, **3**, 351 (1986).

[3] H.A. MacKenzie, J. Young, A.K. Iltaif and H.A. Al-Attar The switching dynamics and noise limitations of indium antimonide bistable elements, IEEE Journal on Selected Areas in Communications, **6**, 7 (1988) pp.1152-1159.

[4] H.A. Al-Attar, H.A. MacKenzie, F.A.P. Tooley and A.C. Walker IEEE Jour. Quant. Elec. **22**, No. 5, 663 (1986).

[5] M.R. Taghizadeh, J.I.B. Wilson, J. Turunen, A. Vasara and J. Westerholm Appl. Phys. Lett. **54**, 1492, (1989).

[6] V.A. Bogatyriov, G.A. Kachurin and L.S. Smirnov Rad. Eff. **49**, 45 (1980).

Demonstration of Optical Gating Using a Semiconductor-Doped Glass Etalon

H. Kobayashi, H. Kanbara, and K. Kubodera

NTT Opto-electronics Laboratories, 3-1 Morinosato Wakamiya, Atsugi-shi, Kanagawa 243-01, Japan

Abstract. Optical gating is achieved using a Fabry-Perot nonlinear optical etalon consisting of a CdSeTe-doped glass and two mirrors. A cw probe light and a pulsed gate light are collinearly illuminated on the etalon. The output of the probe light is gated with a fast response time, due to the refractive-index change in the glass. The nonlinear refractive index of the material, based on the carrier generation by the gate light, is precisely evaluated.

1. Introduction

Semiconductor-doped glasses have attracted much attention because of their unique features[1]. Their response times, often up to ten picoseconds[2], are of great advantage in various applications to optical functional devices. Their optical nonlinearities can be enhanced by the quantum-box effect, which has thus far been mainly investigated theoretically[3]. Some applications to optical devices have been attempted such as an optical bistable memory device using a Fabry-Perot etalon[4], and a wave-guide nonlinear directional coupler[5]. In these experiments, the input light is absorbed and generates carriers in the semiconductor particles, and the input light itself is affected by the refractive index change in the medium (degenerate light). The input-light wavelength was determined through a trade-off of high nonlinearity versus low loss.

This paper reports efficient optical gating of a semiconductor-doped glass etalon, using different wavelengths for the probe and gate lights (non-degenerate lights). In the non-degenerate case, a suitable wavelength is selected for the probe light, while a shorter wavelength is used for the gate light to achieve much higher absorption efficiency. This variety is favorable for optically-gated optical switching techniques.

2. Optical Gating Experiment

The experimental arrangement is shown in Fig. 1 (a). The nonlinear optical etalon consisted of a commercial $CdSe_xTe_{1-x}$-doped glass filter (HOYA IR-80) and two external mirrors. One of the mirrors was mounted on a piezo-translator to tune the cavity length. A continuous-wave probe light generated by a laser diode with a wavelength of $\lambda_p = 812$ nm and a pulsed gate light with the wavelength $\lambda_{gate} = 675$ nm were collinearly impinged on the etalon. The gate light was generated by a Q-switched Nd:YAG-laser pumped dye laser. The duration and repetition rate were 6 ns and 10 Hz, respectively. The doped glass had a thickness of $l_s = 300$ μm and a refractive index of $n_s = 1.545$ at λ_p. The cavity length between the mirrors was $l_{cav} = 765$ μm. Therefore, the optical cavity length was $l_{opt} = n_s\, l_s + (l_{cav} - l_s) = 928.5$ μm. The glass had absorption

Springer Series in Electronics and Photonics, Vol. 29 Photonic Switching II
Editors: K. Tada · H.S. Hinton © Springer-Verlag Berlin, Heidelberg 1990

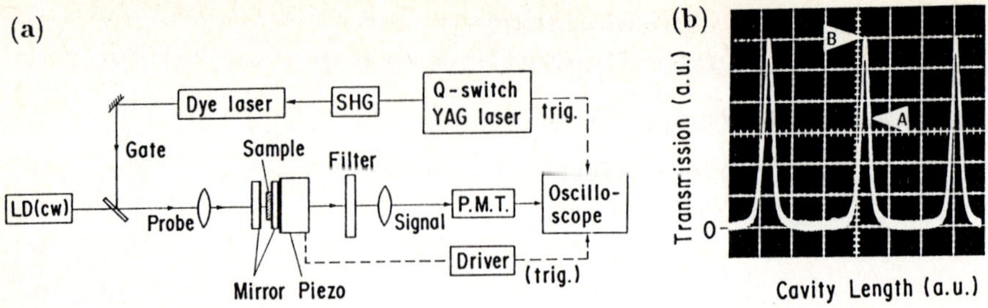

Fig. 1. Experimental arrangement (**a**), and etalon resonance curves for the probe light transmission (**b**).

coefficients of $\alpha_{gate} = 36.2\ \mathrm{cm}^{-1}$ at λ_{gate} and $\alpha_p = 1.1\ \mathrm{cm}^{-1}$ at λ_p. The reflectivities at λ_p and λ_{gate} of the cavity mirrors were $R_p = 0.90$ and $R_{gate} = 0.06$, respectively. The output signal from the cavity was detected by a photomultiplier, followed by an oscilloscope. Transmitted gate light, which accounts for less than 30% of the input gate light, was eliminated by a cut-off filter.

The resonance curve of the etalon transmission for the probe-light input was obtained by changing the piezo-translator voltage, as shown in Fig. 1 (b). An F-value of 10 and maximum transmission $T_{max} = 12$ % were observed. When the cavity length was detuned a little longer than the resonance length, shown as "A" in Fig. 1 (b), the output of the probe light remained low, that is, "OFF" state. The gate light changed the output to the T_{max} transmission, that is, "ON" state. The etalon effect for the gate light could be neglected due to the low reflectivity of the mirrors and a high single-pass loss of 65% in the glass. Oscilloscope waveform traces for the gate light and the output signal are shown in Figs. 2 (a) and (b), respectively. When the cavity length was initially tuned to the resonance length, shown as "B" in Fig. 1 (b), an inverter operation was clearly observed as shown in Fig. 2 (c).

The time responses of the gate light and the output signal are almost the same in the nanosecond time range, as shown in Fig. 2. The resolution time of the detection system was more than 10 ns, though the photomultiplier response time was about 3 ns. A more precise measurement was made using a Si-PIN detector and observing the response to the fine structure of the gate pulse. Experiments showed the switching time to be less than the detecting limit of 1 ns. Considering the reported relaxation time of 7 ps for the CdS_xSe_{1-x}-doped glass[2] and the present cavity life time of 15 ps, the switching time is theoretically expected to be much faster than 1 ns. The exact response time can be measured by using shorter optical gate pulses.

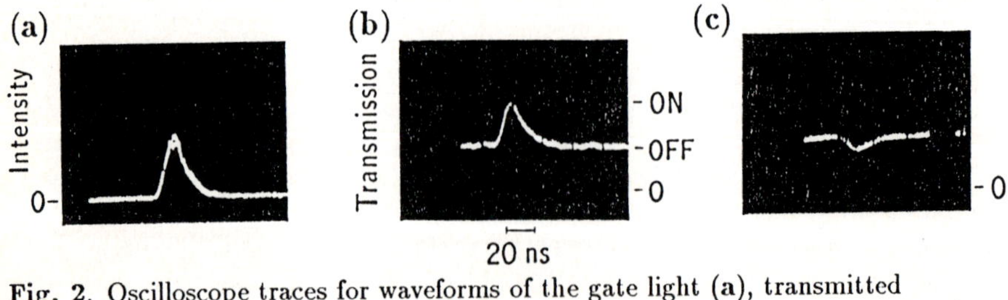

Fig. 2. Oscilloscope traces for waveforms of the gate light (**a**), transmitted probe light signal under gate-light illumination (**b**), and the output signal in inverter operation (**c**), detected by a photomultiplier.

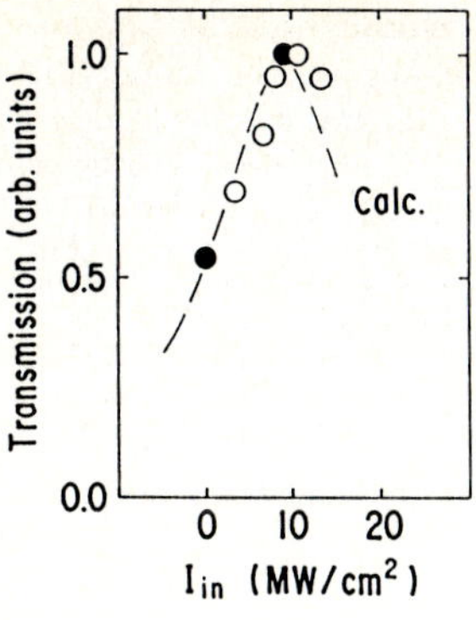

Fig. 3. Output signal dependence on the input gate power density. Data are shown by open and filled circles. Filled circles were used for calculation.

The probe-light transmission in the ON state depends on the input gate-light intensity I_{in} as shown in Fig. 3. The experimental data agree with the calculation mentioned in the following section, corresponding to the etalon resonance curve.

Contrast for the gating was about 2 to 1 as shown in Fig. 2 (b). In a different experiment which used a HOYA IR-83 glass filter, the contrast improved to 3 to 1 because of the larger nonlinear refractive index, as mentioned below. Wavelengths for probe and gate lights were 830 nm and 675 nm, respectively, and the finesse of the etalon was 4. A cavity with a higher F-value will improve this contrast.

3. Analysis and Discussion

The nonlinear refractive index n_2 of the IR-80 glass was evaluated from the experiment as follows. The probe-light transmission T was calculated using the equation:

$$T = \frac{I_{out}}{I_{in}} = \frac{(1-R_p)^2\, e^{-\overline{\alpha}\, l_{cav}}}{(1-R_p\, e^{-\overline{\alpha}\, l_{cav}})^2 + 4\, R_p\, e^{-\overline{\alpha}\, l_{cav}} \sin^2(\delta/2)} \quad .$$

Here, $\delta = 4\pi\, l_{opt} / \lambda_p$ is a phase difference factor, and $\exp(-\overline{\alpha}\, l_{cav}) = (1-\beta)\exp(-\alpha_p\, l_s)$. β represents additional optical loss of a single pass in the cavity. This was due to scattering and so on, and estimated as 15 % from the experimental F-value and T_{max}. The calculated resonance curve is shown in Fig. 4.

The OFF and ON states on the resonance curve in Fig. 4 were obtained by plotting their signal levels. Their levels were evaluated from the results of Fig. 2 (b) which correspond with the filled circles in Fig. 3. The change in optical cavity length Δ was

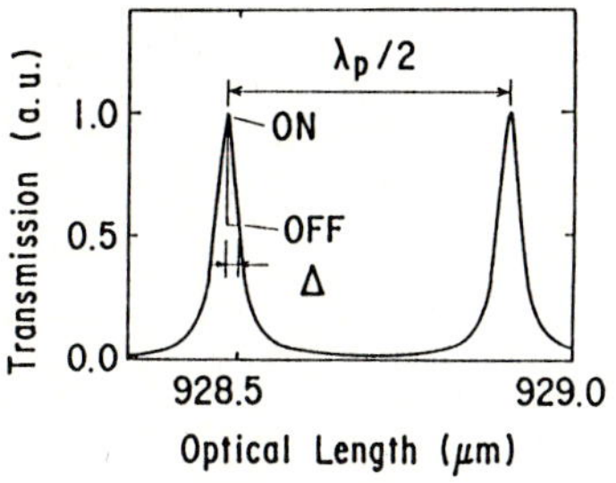

Fig. 4. Calculated etalon resonance curve used for the analysis.

evaluated as $-0.022\ \lambda_p$. This value was negative because the initially-tuned length was longer than the resonance length. Δ is the refractive index change in the sample due to the electrons generated by the gate light. The gate-light intensity was not uniform along the optical path due to the absorption, therefore, Δ can be expressed as follows:

$$\Delta = \int_0^{l_s} n_2\, I_{in}\, e^{-\alpha_{gate}\, Z}\, dZ = n_2\, I_{in}\, \frac{(1 - e^{-\alpha_{gate}\, l_s})}{\alpha_{gate}} \equiv n_2\, I_{in}\, l_{eff} \quad ,$$

where $l_{eff} = 183\ \mu m$ is an effective absorption length for the gate light. The input gate-light intensity of $I_{in} = 9\ MW/cm^2$ gave the value of $n_2 = -1.1 \times 10^{-11} cm^2/W$.

As the imaginary part of third-order optical susceptibility $\chi^{(3)}$ can be neglected in the probe-light wavelength, $\chi^{(3)}$ value was also evaluated as

$$\chi^{(3)} = 3 \times 10^3\ n_s{}^2\ n_2\ /\ (4\pi)^2 = -5.0 \times 10^{-10}\ \text{esu} \quad .$$

Similar analysis was carried out for the IR-83 doped glass. The estimated values are $n_2 = -7.7 \times 10^{-11} cm^2/W$, and $\chi^{(3)} = -3.6 \times 10^{-9}$ esu. These values are consistent with the reported n_2 values in the degenerate case[4, 5]. As the n_2 value has a fairly strong wavelength dispersion around the absorption edge, tuning of the probe-light wavelength is important for an efficient gating operation.

As a means of evaluating n_2 values, the present experiment has higher detection accuracy than the interferometer method[6]. This is due to the use of multiple reflections for the probe light in a Fabry-Perot etalon. The present experiment also has higher accuracy than the bistable operation experiment[4], because multiple reflections of the gate light can be neglected. This latest experiment is considered to be a simple method for evaluating nonlinear optical constants of materials.

4. Conclusion

Optical gating was observed in the nonlinear etalon including a fast response material of CdSeTe-doped glass. The gating was operated efficiently by using the non-degenerate lights. The nonlinear refractive index of the material was precisely evaluated. The high-speed and efficient optical gating should be useful in applications to time-division multiplexing optical communications and optical computing systems.

Acknowledgements. The authors would like to thank Dr. Takayuki Sugeta and Dr. Tetsuhiko Ikegami for their continuous encouragement.

References

1. R. K. Jain and R. C. Lind: J. Opt. Soc. Am. **73**, 647 (1983).
2. M. Mitsunaga, et al.: J. Opt. Soc. Am. B **5**, 1448 (1988).
3. For example, see: T. Takagahara: Phys. Rev. B **36**, 9293 (1987).
4. J. Yumoto, et al.: Opt. Lett. **12**, 832 (1987).
5. N. Finlayson, et al.: Appl. Phys. Lett. **53**, 1144 (1988).
6. G. R. Olbright and N. Peyghambarian: Appl. Phys. Lett. **48**, 1184 (1986).

InGaAs/InP MQW Modulators with Enhanced Performance: Symmetrical Coupled Wells and Double-Sided Epitaxy

M.A.Z. Rejman-Greene and E.G. Scott

British Telecom Research Laboratories, Martlesham Heath, Ipswich IP5 7RE, UK

Abstract. Two novel InGaAs/InP MQW modulator arrays are demonstrated, one using symmetrical coupled wells with improved contrast at 5V, and the other as a packaged 2x2 array of double-sided devices.

1. Introduction

Fast, efficient spatial light modulators are a key component of photonic switching systems and optical computing designs. Of particular interest to applications engineers are modulators with the potential for monolithic integration with either silicon or III-V (opto)electronic devices. Recently, there has been widespread interest in the use of Multiple Quantum Well (MQW) - based materials, utilising the steep absorption edge and near band-gap excitonic features. For example, by using *pin* diodes made from MQW epilayers, a 3dB increase in absorption has been reported in planar GaInAs/InP modulators by applying a reverse bias of 30V [1]. Furthermore, 4x4 arrays of such 60μm diameter devices (Fig1) have been used to demonstrate transmission through a 100MBit/s x 16 channel, parallel interconnect system [2]. The use of this materials system allows for operation in a transmission mode, facilitating a simpler design of optical bus architectures; in contrast with GaAs-based designs which often need to operate in reflection, so as to avoid substrate absorption. In addition, InP-based MQW modulators are compatible with second generation fibre transmission systems, operating at either 1.3 or 1.5μm, by appropriate selection of the quantum well width.

Although MQW modulators have been shown to operate at high speed [3], implementation of two-dimensional arrays in systems has been limited by the combination of

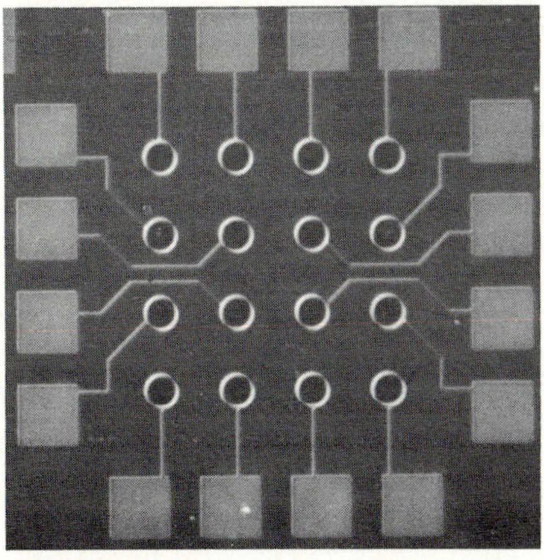

Fig 1 4x4 modulator array

Springer Series in Electronics and Photonics, Vol. 29 **Photonic Switching II**
Editors: K. Tada · H.S. Hinton © Springer-Verlag Berlin, Heidelberg 1990

low contrast ratio and a high operating voltage. (Fabrication of MQW waveguides - which overcome some of these limitations - is a route to *one*-dimensional arrays with improved characteristics [4]). More recently, an asymmetric Fabry-Perot design has been proposed with projected contrast ratios greater than 20dB [5] - improving the performance of GaAs/GaAlAs reflection modulators. In this report, we demonstrate two alternative and independent approaches to enhancing the performance of transmission type modulators.

2. Symmetrical Coupled Wells

Enhanced modulation at low voltages is realised by replacing the thin InGaAs wells (typically ~60Å in width for operation at 1.5μm) with coupled quantum wells (CQW). The separation of two "sub-wells" of thickness 30Å, by a thin (20Å) InP barrier, renders the quantum well wavefunctions more sensitive to an applied field . However, the growth of material with the larger numbers of abrupt interfaces is a severe test of epitaxial methods. We report results on devices made from 200 period MCQW material (i.e. 800 interfaces) grown by Gas-source Molecular Beam Epitaxy (GS-MBE) [6], and fabricated by the planar technology described in [1]. Electroabsorption data through such 60μm diameter modulators, as measured by a micro-Fourier Transform spectrometer [1], is presented in Fig 2: an increase in transmission of 1.2dB for a bias of -6V. The high quality of the epilayers is confirmed by dark current measurements of 8100 diodes at a pitch of 125μm (Fig 3); almost 99% of devices in the central area show less than 10nA leakage at -10V. It is clear that 10x10 spatial light modulators may be produced using this approach.

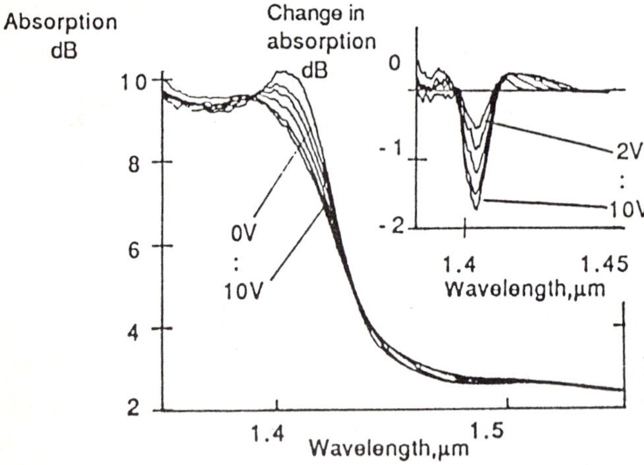

Fig 2 CMQW modulator characteristics

3. Double-sided Epitaxy

A further improvement in the performance of MQW modulators has been realised using the novel approach of double-sided epitaxy (DSE). Fabrication of modulators on *both* faces of substrates (as indicated in Fig 4) would enable a doubling of the maximum number of quantum wells for optimum performance - a limit set by the effects of the residual doping within the wells. However, substrate polishing using conventional

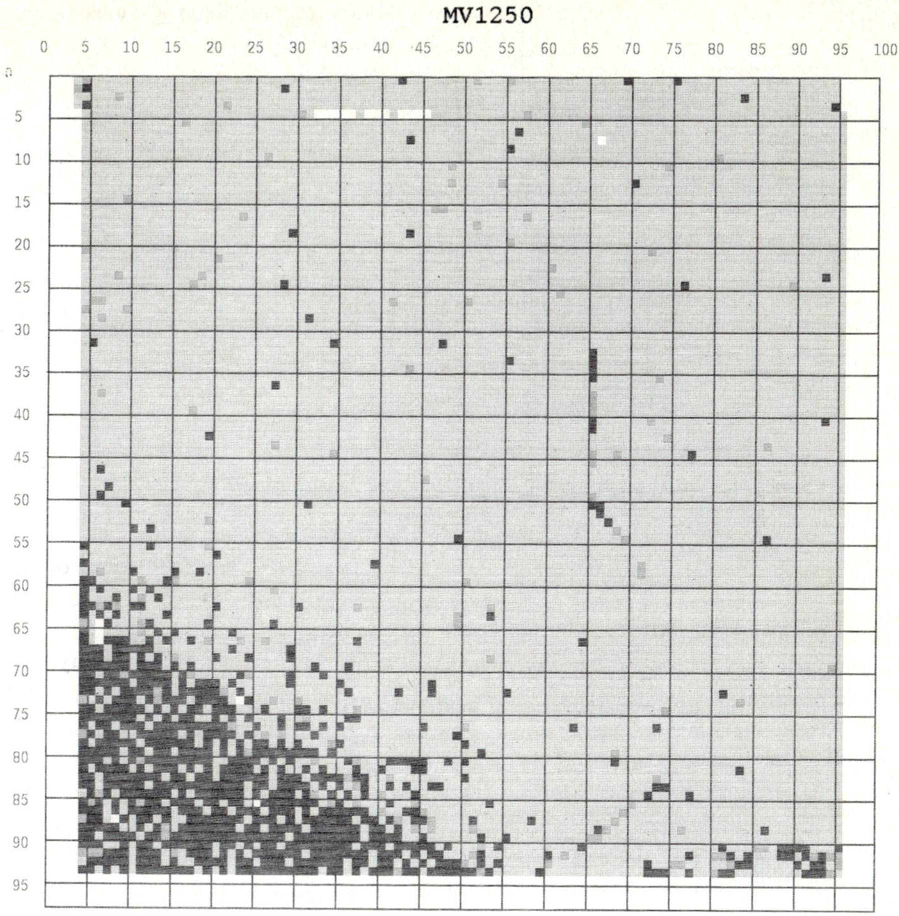

KEY			
	MFANING	TALLY	%
	Cont fail	27	.31
	<10nA	7699	88.94
	10nA-100nA	161	1.85
	>100nA	769	8.88

Fig 3 Leakage currents of CQW modulators at -10V

chemo-mechanical methods invariably results in one face being visibly inferior - with a higher defect density and a significant number of scratches. We show that, in spite of this constraint, matched pairs of modulator arrays can be fabricated.

The polished InP wafer is mounted in a GS-MBE system using an Indium-free method, with a backing wafer in close proximity; during growth, the assembly is rotated to retain compositional uniformity. After completing the growth on the face with the better finish (side 1), the wafer is reversed for epitaxy on side 2. Prior to device fabrication, a piece of the wafer is processed to demonstrate the similarity of the trans-

Fig 4 Double-sided modulator device

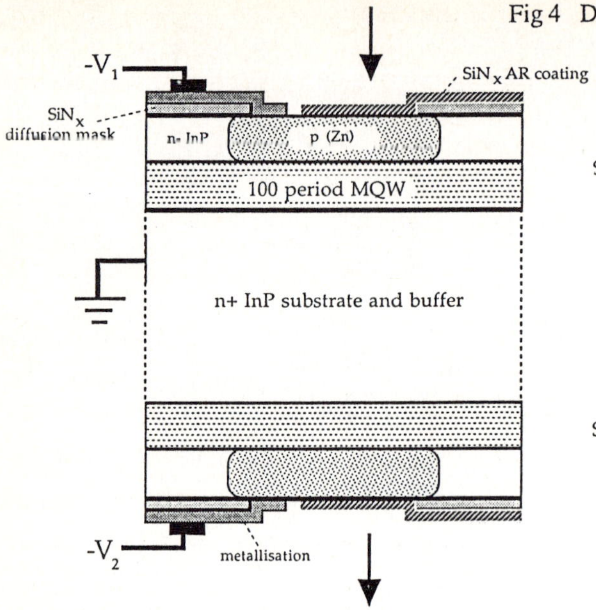

mission spectrum through each of the MQW stacks separately, by removal of each stack in turn [6].

Silicon nitride layers are deposited on both faces using a plasma method, the layer on side 2 being patterned using conventional photolithographic techniques. A *p-n* junction is then formed in the side 2 epilayer by a 1μm deep zinc diffusion at a low temperature and at atmospheric pressure. A similar method was used to form the junction in side 1, defining the pattern in the dielectric by double-sided alignment. Anti-reflection coatings and metallisation layers are deposited and patterned to complete the fabrication process.

Electroabsorption spectra from 250μm diameter devices indicate an optimum standing bias of -10V;Fig 5 shows that by applying a further -10V to corresponding modulators on both sides of the substrate, a modulation of 2.2dB is measured for devices on side 1, and 1.6dB on side 2 - a difference which may reflect a "push-on" of the first zinc diffusion (i.e. the diffusion on side 2).

A 2x2 array of 60μm diameter modulators was packaged as shown in Fig 6. Two Kyocera 18-lead chip carriers were soldered back-to-back, with holes pre-drilled to allow operation in transmission. Each device was then bonded, using a thermocompression technique, to the pads set in the inner lip of each side of the carriers, care being taken to avoid projection of the wires above the level of the outer perimeter of each carrier. Commercially-available sockets are used in the assessment of arrays and their subsequent application in systems.

This successful demonstration of DSE suggests numerous novel applications. For example, in neural network implementations, devices in the two arrays could be separately addressed, yet be close-coupled optically; furthermore, by operating devices on one side of the substrate as a detector array, an interface for InP-based electronic processing of signals is provided on the one chip. Other possibilities include three-level logic, and the fabrication of some complex optical circuits which have, up till now, been precluded by incompatible epitaxy or processing stages.

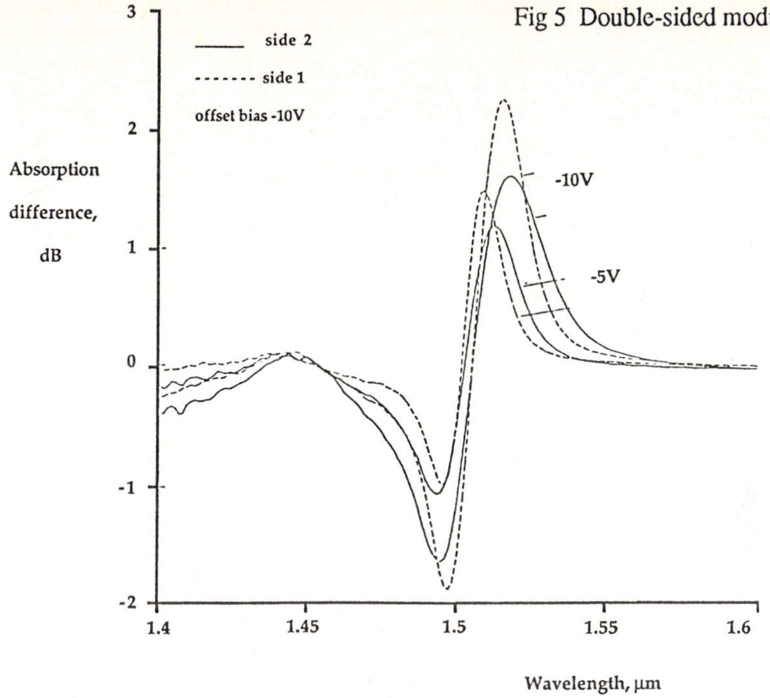

Fig 5 Double-sided modulator results

Fig 6 Packaged double-sided modulators

4. Acknowledgements

The authors thank P. Healey for useful discussions, E. Marsh for the zinc diffusion, J. Turner and M. Learmouth for assistance in the device processing, R. Gooding for electrical probe data and L. Johnston for the device bonding.

References

1. M A Z Rejman-Greene, E G Scott and E McGoldrick:*Electron. Lett.* **24**, 1583 (1988)
2. R P Webb and M A Z Rejman-Greene: *1990 International Topical Meeting on Optical Computing*, Kobe, April 8-12, 1990
3. G Boyd, J E Bowers, C E Soccolich, D A B Miller, D S Chemla, L M F Chirovsky, A C Gossard and J H English: *Electron. Lett.* **25**, 558 (1989)
4. I Kotaka, K Wakita, O Mitomi, H Asai and Y Kawamura: *IEEE Photonics Tech Lett* **1** (5), 100 (1989)
5. M Whitehead and G Parry: *Electron. Lett.* **25**, 566 (1989)
6. E G Scott and M A Z Rejman-Greene: to be published.

Photonic Switching and SEED Effect in AlGaAs/GaAs DFB Structures Grown by MOVPE

S. Zumkley[1], *G. Wingen*[1], *F. Scheffer*[2], *W. Prost*[2], *and D. Jäger*

[1]Institut für Angewandte Physik der Universität Münster, Correnssstr. 2/4, D-4400 Münster, Fed. Rep. of Germany
[2]Fachbereich Elektrotechnik, Universität Duisburg, D-4100 Duisburg, Fed. Rep. of Germany

Abstract. We present theoretical and experimental results of photonic switching and electrooptical modulation in novel AlGaAs/GaAs distributed feedback (DFB) structures grown by MOVPE.

1. Introduction

Nonlinear optoelectronic components and optically bistable devices may be the key elements in future photonic switching and information processing systems. Main emphasis is laid upon hybrid structures where photodetector and modulator characteristics are combined in a suitable manner. With respect to the degree of integration, two basic approaches may be distinguished. Using common OEIC-technology, smart pixels can be fabricated whereas a superintegration of photodetector and modulator functions leads to SEED devices. In any case, however, the realization of the electrooptical modulator is the key problem. In this paper, we study electrooptical modulation and photonic switching in hybrid distributed feedback (DFB) structures made of III-V semiconductor materials. These novel devices are similar to the epitaxially grown microresonators which have been studied recently[1,2,3]. In the present case, however, the thickness of the internal resonator is reduced to a quarter of a wavelength so that the whole structure merely consists of two adjacent, vertically oriented mirrors. It should be emphasized at this point that in contrast to previous work[4,5], the grating mirrors are themselves used as an artificial photonic material with electrooptic and optoelectronic properties. Such periodic devices have theoretically been shown to exhibit optical bistability if nonlinear optical properties are included[6,7], and dissipative structures such as solitons and kinks have been observed and theoretically been described on the basis of the nonlinear Schrödinger equation[8].

 © Springer-Verlag Berlin, Heidelberg 1990

2. Device structure

The experimental DFB structures are grown by metal-organic vapour phase epitaxy (MOVPE). Fig. 1 shows a sketch of the device. On top of the GaAs substrate a buffer layer and the $Al_{0.5}Ga_{0.5}As/GaAs$ superlattice consisting of 20 pairs are grown. The thicknesses are $d(Al_{0.5}Ga_{0.5}As)=65$ *nm* and $d(GaAs)=59$ *nm*. Electrical contacts are integrated to provide an electric field perpendicular to the multilayer. The optical input power P_{in} of a Ti:Sapphire Laser pumped by an Argon Ion Laser is incident on the front Au contact and the reflected power P_{ref} is measured using a beam-splitter.

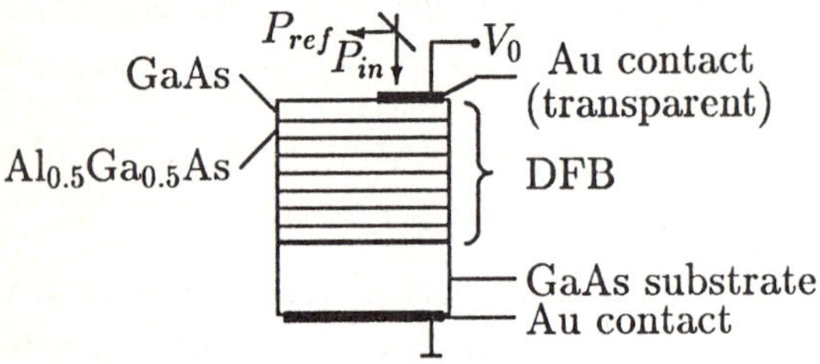

Figure 1. Device structure

3. Experimental results

In a first experiment, the optical reflectivity of the DFB structure is studied as a function of the wavelength of the laser beam. The results are compared with a theoretical analysis on the basis of a common transfer matrix method. A good agreement with the measurements is obtained. In a second step, the electrooptical properties of the device are investigated in detail. For that purpose, the reflectivity $R = P_{ref}/P_{in}$ of the device has been measured using different bias voltages (Fig. 2(a)). In Fig. 2(b) the experimental results of the reflection change are plotted. As a result, a maximum value of nearly 18% is obtained at a wavelength of 873 *nm*, and the on/off ratios are as high as 3:1. Theoretically, it

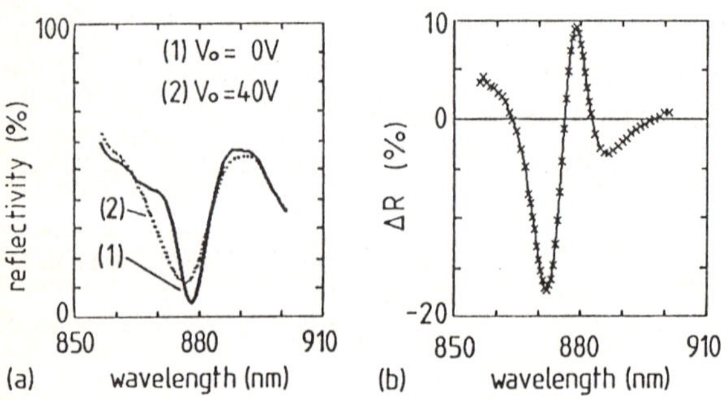

Figure 2. Measured reflectivity for different bias voltages (a) and difference in reflection (b) with $\Delta R = R(40V) - R(0V)$ versus wavelength.

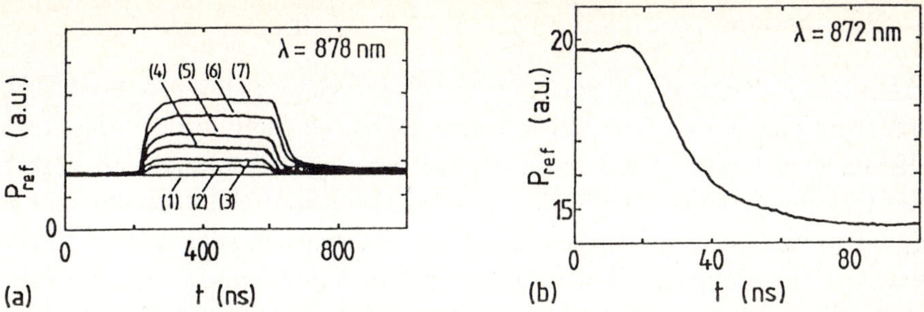

Figure 3. Measured electrooptical switching properties of the DFB structure. (a) Reflected signal for electrical square pulses with amplitudes 0,3,5,10,15,25,35 V for curves (1) to (7), respectively. (b) Reflected signal for an applied voltage step of 26 Volts at $t = 20ns$.

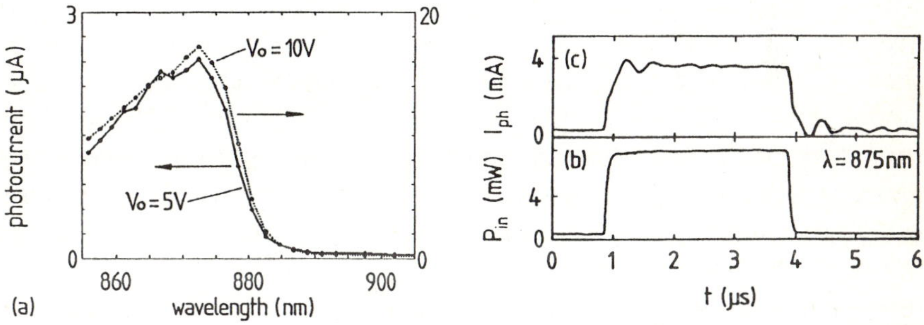

Figure 4. Measured optoelectronic properties of the $Al_{0.5}Ga_{0.5}As/GaAs$ DFB structure: (a) photocurrent as a function of the wavelength for two bias voltages, (b) optical input power P_{in} versus time and (c) photocurrent I_{ph} versus time at a bias voltage of 53 V.

is found that the modulation can be traced back[9] to electrorefraction and electroabsorption in the GaAs layers due to the Franz-Keldysh effect (FKE). In Fig. 3 the results of an electrooptical switching experiment are plotted. As can be seen, a clear modulation of about 60 % and switching times as low as 20 ns are obtained where, however, the wavelength of the laser beam plays an important role, see Fig. 2(a).

In a final, but preliminary measurement, the optoelectronic behaviour of the DFB structure is analysed in the desired wavelength range. The measured photocurrent as a function of wavelength (Fig. 4(a)) reveals that the DFB structure exhibits a clear absorption near the bandgap energy of GaAs. Furtheron, from the results in Fig. 4(b) and (c) it is obvious that the device behaves also as a fast photodetector with a high quantum efficiency.

4. Conclusion

From the above investigations it is concluded that (i) the proposed DFB structure can be used as a distributed reflection modulator with a high efficiency in terms of reflection change per applied voltage and with a short response time. It should, however, be noted that the device has by no means been optimized until now. Theoretical results have shown that a $\lambda/4$-shifted DFB structure exhibits a pronounced transmission peak which can be quenched by an electric field of 200 kV/cm due to the FKE. This leads to a reflection change of more than 50 %. Further improvements are imaginable when the DFB structure consists of periodic pn-heterostructures with built-in electric fields within the active layers. (ii) It is foreseen that the reflection modulator can easily be combined with an external photodetector to optically control the internal electric field and therefore the optical output signal. In particular, the photodetector can be integrated vertically so that a novel self-electrooptical effect device (SEED) arises. The photodetector, which can be a photoconductor, photodiode, or phototransistor serves as the optically controllable external resistance of the modulator, see Fig. 5 for example. (iii) Moreover, if the intrinsic optoelectronic properties of the AlGaAs/GaAs mirror are used, the photodetector and modulator characteristics are spatially integrated and the superlattice behaves as a material with an artificial optical nonlinearity[10]. As a result, for $\Delta R > 0$ optical bistability due to self-induced reflection is expected similar to the properties of the periodic bilayers of linear and nonlinear films[6,7,8]. Again it is anticipated that the optical nonlinearity can become extremely large when using a periodic arrangement of grating resonator filters, i.e. DFB structures with periodic $\lambda/4$-shifted regions. And a further enhancement can be foreseen with MQW material instead of pure GaAs. In any case, a novel kind of an - artificial - optically nonlinear material is generated which can be switched between a transmitting and a reflecting state[8]. It should finally be noted that the proposed DFB structures are especially suited for 2D-arrays of switching devices provided that the epitaxy process ensures layers with sufficiently homogeneous thicknesses.

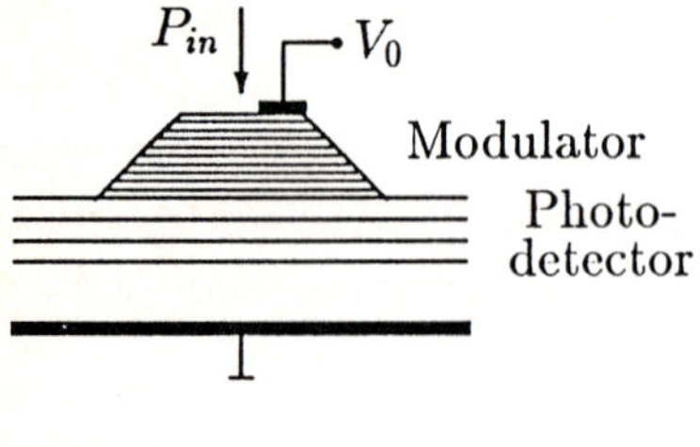

Figure 5. Proposed SEED device consisting of a vertical integration of a modulator and a photodetector. Necessary condition is $\Delta R < 0$.

Acknowledgments. We are indebted to Dr. Y. Iyechika (Sumitomo Chemical, Japan) for his contribution. This study is part of a project that has been supported by the German Federal Minister for Research and Technology (BMFT) under TK 0435. This work is also part of the ESPRIT Optical Computing Project launched by the Commission of the European Communities.

References

1. O. Sahlén, U. Olin, E. Masseboeuf, G. Landgren, and M. Rask, Appl. Phys. Lett. **50**, 1559-1561 (1987). O. Sahlén, E. Masseboeuf, M. Rask, N. Nordell, and G. Landgren, Appl. Phys. Lett. **53**, 1785-1787 (1988).
2. J.L. Jewell, A. Scherer, S.L. McCall, A.C. Gossard, and J.H. English, Appl. Phys. Lett. **51**, 94-96 (1987).
3. R. Kuszelewicz, J.L. Oudar, J.C. Michel, and R. Azoulay, Appl. Phys. Lett. **53**, 2138-2140 (1988).
4. R.H. Yan, R.J. Simes, and L.A. Coldren, IEEE J. Quantum Electron. QE-**25**, 2272-2280 (1989).
5. R.H. Yan, R.J. Simes, and L.A. Coldren, IEEE Phot. Techn. Lett. **2**, 118-119 (1990).
6. W. Chen and D.L. Mills, Phys. Rev. Lett. **58**, 160-163 (1987).
7. W. Chen and D.L. Mills, Phys. Rev. B **36**, 6269-6278 (1987).
8. D. Jäger, A. Gasch, and K. Moser, in: "Optical Switching in Low-Dimensional Systems", H. Haug and L. Banyai (Eds.), Plenum Publishing Corporation, 9-23 (1989).
9. S. Zumkley, G. Wingen, G. Borghs, F. Scheffer, W. Prost, and D. Jäger, Proc. SPIE, vol. 1280 (in press).
10. D. Jäger, J. Opt. Soc. Am. B **6**, 558-594 (1989).

Improving the Performance of Large Arrays of Symmetric Self Electro-optic Effect Devices

L.M.F. Chirovsky[1], *L.A. D'Asaro*[1], *J.M. Freund*[1], *R.F. Kopf*[1], *J.M. Kuo*[1], *A.L. Lentine*[2], *and T. Sizer, II*[3]

[1]AT&T Bell Laboratories, 600 Mountain Avenue, Murray Hill, NJ 07974, USA
[2]AT&T Bell Laboratories, 200 Park Plaza, Naperville, IL 60566, USA
[3]AT&T Bell Laboratories, Holmdel, NJ 07733, USA

Arrays of Symmetric Self Electro-optic Effect Devices can function in multi-stage digital optics systems at higher data rates with less external optical power than previously predicted.

The Symmetric Self Electro-optic Effect Device (S-SEED) can serve both as a logic gate or a memory cell [1]. Large arrays of S-SEED's interconnected via free-space optics can enable experiments to develop powerful new systems with parallel architectures. Arrays of 32×64 S-SEED's are now available for the testing of system concepts [2]. Is the performance of these device arrays sufficient, however, for practical systems? An unequivocal answer is impossible at this point; far too few system experiments have been performed. There are, however, two readily available ways to enhance the performance of S-SEED arrays and thereby to increase the probability of a positive answer to the above question. These are: operation in a sense-amplification mode [3], and further device down-sizing [4].

A quick review of the properties of presently available devices will be helpful for a description of these two enhancements. S-SEED's, in general, exhibit the following device properties necessary or very desirable for a multi-stage system:

1) They are three terminal devices with good input/output isolation and no critical biasing;
2) They are cascadable due to time-sequential gain;
3) They perform logic level restoration;
4) They retime signals in the read-out phase of their operation;
5) They are naturally inverting devices;
6) They are tolerant of nonuniformities because they are bistable without critical biasing and operate with differential inputs and outputs;
7) They can operate as a set-reset latch, for information storage, or can perform any of the four basic logic functions.

Another noteworthy point is that S-SEEDs are essentially reverse-biased detectors/modulators of optical power. This is a positive trait, since historically, reverse-biased devices have been more fault tolerant and more reliable than forward-biased devices which generate optical power. This may also be considered a negative trait since S-SEEDs require external optical power sources, complicating the optical system. The external optical power sources, moreover, must be lasers emitting at a wavelength within a nanometer of the S-SEED's exciton peak wavelength, λ_o. However, the external lasers can automatically

Springer Series in Electronics and Photonics, Vol. 29 **Photonic Switching II**
Editors: K. Tada · H.S. Hinton © Springer-Verlag Berlin, Heidelberg 1990

provide retiming, which otherwise would have to be brought in by some other means. The specific S-SEED (hereafter called A100) incorporated into the 32×64 arrays (mentioned above): contains two p-i-n mesa diodes each with an area of 100 μm^2 (hence the name A100); occupies an area of $20\ \mu m \times 40\ \mu m$; has an optical switching energy, $U_o \approx 1pJ$, when reverse-biased to 6 volts; switches states in as little as ≈ 1 ns; holds its state indefinitely with just 200 nW of optical power per mesa; and has two $5\ \mu m \times 5\ \mu m$ optical windows per mesa.

Functionally, the SEED's are very versatile, so if the free-space interconnects can be made, the issue of system practicality will mostly revolve around speed of operation and external optical power requirements. Due to saturation effects, there is a limit to the amount of power, P_o, that can be brought into a S-SEED on a given beam. The minimum device switching time, τ_{min}, is then the optical switching energy, U_o, divided by P_o, i.e.

$$\tau_{min} = U_o/P_o \,. \tag{1}$$

In a cascadable multi-stage system, signals at one stage will be the outputs of devices in a previous stage. The effective power in the signal beams, P_{sig}, will be $\lesssim 0.1\ P_o$, so the write-in time or the device switching time, τ_{sw}, will be:

$$\tau_{sw} \approx U_o/P_{sig} \gtrsim 10\,\tau_{min} \,. \tag{2}$$

To achieve time sequential gain, read-out time must be equal to write-in time, so a full cycle operating time, τ_{op}, will be $\gtrsim 20\,\tau_{min}$. This would be for conventional operation as described in reference [5] where all the switching energy is supplied by the signal beams.

We constructed a photonic ring-counter [6], to test the system performance of S-SEED's. The devices used had a $\tau_{min} \approx 3$ ns, and the apparatus allowed only a $P_{sig} \approx 0.05\ P_o$. For conventional operation therefore, we expected a $\tau_{op} \approx 40\,\tau_{min}$ ≈ 120 ns. The output of the ring-counter, shown in Fig. 1a, confirms that this was experimentally achieved. Even optimization of the test apparatus and the use of improved A100 devices probably will not reduce this τ_{op} below 20 ns.

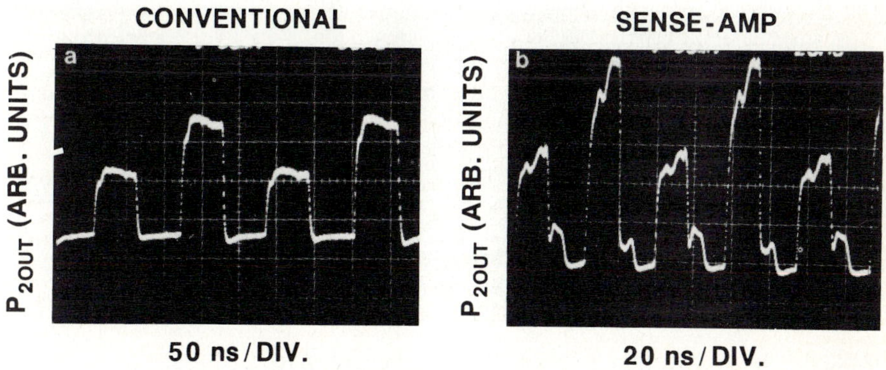

Fig. 1. a) Output of the photonic ring-counter operated in the conventional mode; b) Output of the same ring-counter operated in the sense-amp mode.

The sense-amplification mode of operating S-SEED's [3] takes note of the fact that there is an intermediate metastable state between the two stable states of a S-SEED. Direct external beams at a wavelength, λ_1, several nanometers longer than λ_o (with powers up to the saturation limit, P_o) can set a S-SEED in the metastable state in less than a τ_{min}. Thereafter, signal beams aided by direct external beams at λ_o, (obtained by increasing the duty cycle of the external clock beams beyond 50%), can switch the device to the desired stable state in just over a τ_{min}. Thus the sense-amplification mode of operation can reduce the switching time to a $\tau_{sw} \approx 2\tau_{min}$ and the full cycle operating time to a $\tau_{op} \approx 4\,\tau_{min}$. In the sense-amplification mode of operation the above mentioned photonic ring-counter operated with a $\tau_{op} \approx 37$ ns, as is shown in Fig. 1b. Considering that in the test apparatus the clock beam duty cycle could not be increased beyond 50% and that the temporal width of the λ_1 beams could not be decreased below 7 ns, this was a significant improvement in τ_{op}. With optimization of the apparatus and the use of improved A100 devices, the sense-amp mode operating time should decrease to approximately 4 ns. The disadvantages of this mode of operation, when compared to the conventional mode, are more stringent uniformity requirements and the need for a second external laser source.

Device down-sizing cannot reduce τ_{min} (Eq. 1), because P_o scales with U_o, but can yield devices which require less external optical power for the same τ_{min}. The A100 device was successfully scaled down from an A10,000 device, for a two orders of magnitude reduction in U_o [4]. We have recently fabricated 8×16 arrays of A10 devices, shown in Fig. 2a, for another potential order of magnitude reduction in U_o. Fig. 2b shows all the devices in the array luminescing when in

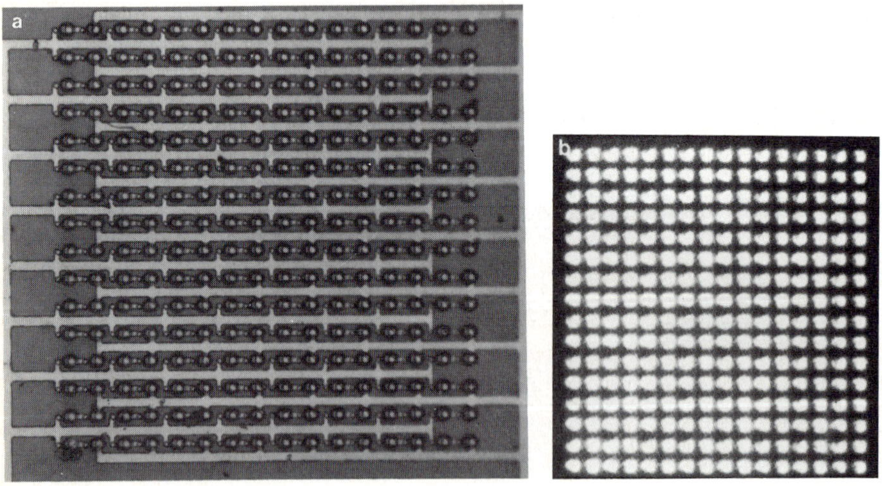

Fig. 2. a) A photograph of an 8×16 array of A10 S-SEED's (cell size: $10\,\mu m \times 20\,\mu m$, mesa size: $2.5\,\mu m \times 4.0\,\mu m$, 2 windows per mesa each $1.25\,\mu m$ square); b) The 256 diodes of an 8×16 array of A10 S-SEED's luminescing when in forward bias.

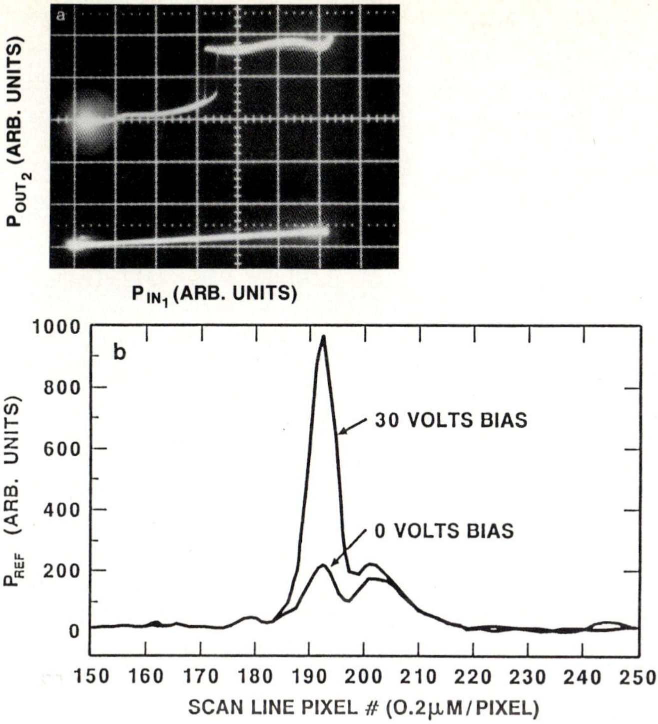

Fig. 3. a) The bistability loop of an A10 S-SEED produced with ~ 3 μm dia. beams and with a reverse bias of 20 Volts; b) Scans of the intensity of light reflected along the 4 μm length of an A10 S-SEED diode when illuminated by an ~ 3 μm dia. beam at 0 Volts and at 30 Volts reverse bias. An optical window output contrast ratio of 5:1 is demonstrated.

forward bias, indicating that all should function uniformly. The data in Fig. 3 demonstrates that the A10 device is bistable and can give an output contrast ratio of up to 5:1. The device's optical switching energy, U_o, was 270 fJ, measured with 10 Volts of reverse bias. The data was taken with oversized diameter beams and so was somewhat degraded. These results indicate, nevertheless, that the fabrication of arrays of much smaller S-SEED's than the A100 devices is possible.

We have presented two straightforward methods of enhancing the performance of large arrays of S-SEED's. If the potential improvements are fully realized, multi-stage digital optics systems could be constructed, for instance, with a 500 Gigabit/second throughput per stage, where each stage is a 32×64 S-SEED array requiring approximately 1 Watt of external optical power. This performance may already be adequate for a practical system. What is more, the SEED technology is in its infancy. It is likely that methods for obtaining additional performance improvements will become apparent with more research, progressively increasing the utility of digital optics systems based on SEED's.

References

[1] A. L. Lentine, H. S. Hinton, D. A. B. Miller, J. E. Henry, J. E. Cunningham, and L. M. F. Chirovsky, IEEE J. Quantum Electron., QE-25, pp. 1928-1936 (1989).

[2] L. M. F. Chirovsky, L. A. D'Asaro, R. F. Kopf, J. M. Kuo, A. L. Lentine, F. B. McCormick, R. A. Novotny and G. D. Boyd, Paper PD 28, Postdeadline Paper Technical Digest, 1989 Annual Meeting of the OSA.

[3] L. M. F. Chirovsky, A. L. Lentine, and D. A. B. Miller, Paper MJ2, Technical Digest, 1989 Conference on Lasers and Electro-Optics.

[4] A. L. Lentine, L. M. F. Chirovsky, L. A. D'Asaro, C. W. Tu, and D. A. B. Miller, IEEE Photon. Tech. Lett, Vol. 6, pp. 129-131 (1989).

[5] L. M. F. Chirovsky, L. A. D'Asaro, C. W. Tu, A. L. Lentine, G. D. Boyd, and D. A. B. Miller, OSA Proceedings on Photonic Switching, John E. Midwinter and H. Scott Hinton eds. (OSA, Washington, D.C. 1989) pp. 2-6.

[6] A. L. Lentine, L. M. F. Chirovsky and L. A. D'Asaro (paper submitted to Optics Lett.)

Part V

Free-Space Switching

A 3D Crossover Switching Network Based on S-SEED Arrays

T.J. Cloonan, F.B. McCormick, M.J. Herron, F.A.P. Tooley, G.W. Richards, E. Kerbis, J.L. Brubaker, and A.L. Lentine

AT&T Bell Laboratories, 200 Park Plaza, Naperville, IL 60566, USA

Abstract. This paper describes a fully-operational prototype of an optical crossover network that was based on S-SEED arrays. The network provided for 32 input ports and 32 output ports, and it was operated at a maximum channel rate of 55.7 kbps. New node-types (2-modules) and 2D to 3D transformations of crossover networks are also discussed.

1. Introduction

One of the more promising interconnection schemes proposed for use in photonic switching is the crossover network, which is topologically equivalent to the shuffle network.[1] [2] To efficiently utilize the parallelism offered by free-space optics, the crossover interconnections must provide 3D connectivity between planar (2D) arrangements of optical logic devices. As a result, we will describe 2D to 3D transformations of crossover interconnections along with a new type of switching node (a 2–*module*) that is easily implemented with optical logic devices. We will then discuss a fully operational 32-input optical crossover network.

2. 3D Optical Crossover Networks Based on 2–modules

An optical crossover network is comprised of multiple node-stages interconnected by optical links. Assuming one input per node in the first stage and one output per node in the last stage, a crossover network with N inputs requires $\log_2(N)+1$ node-stages for full connectivity. Within this paper, a 2*D crossover network* will be defined as a crossover network in which the nodes of a node-stage are arranged in a linear (1D) fashion and the links occupy 2D space. A 3*D crossover network* will be defined as a crossover network in which the nodes of a node-stage are arranged in a planar (2D) fashion and the links occupy 3D space. The link connectivity required by a 2D crossover network with N=16 inputs and N=16 outputs is shown in Fig. 1a. The basic optical hardware required to implement an optical 2D crossover interconnection is shown in Fig. 2.[1]

Transformation of the 2D crossover network into a topologically equivalent 3D crossover network can be accomplished by *fan–folding* the linear node-stages across the fold lines within Fig. 1a. This operation produces node-stages which have a planar arrangement of nodes numbered in serpentine fashion (Fig. 1b). If the connectivity provided by the links is maintained during this fan-folding operation, then the resulting link-stages have either *vertical crossover connections* or *horizontal crossover connections* (but never both). Thus, the 2D crossover hardware in Fig. 2 can be used to implement the 3D crossover connections if the orientation of the prismatic mirror array in each stage is rotated to provide the appropriate connections. Mathematically, the mapping of nodes from the 2D domain into the 3D domain can be described by the transformation

$$\beta_{2D \to 3D}(P) \to (R, C) = (INT(P/J), P_{MOD\ J}) \quad (1)$$

if INT(P/J) is an even number, and it can be described by the transformation

Springer Series in Electronics and Photonics, Vol. 29 **Photonic Switching II**
Editors: K. Tada · H.S. Hinton © Springer-Verlag Berlin, Heidelberg 1990

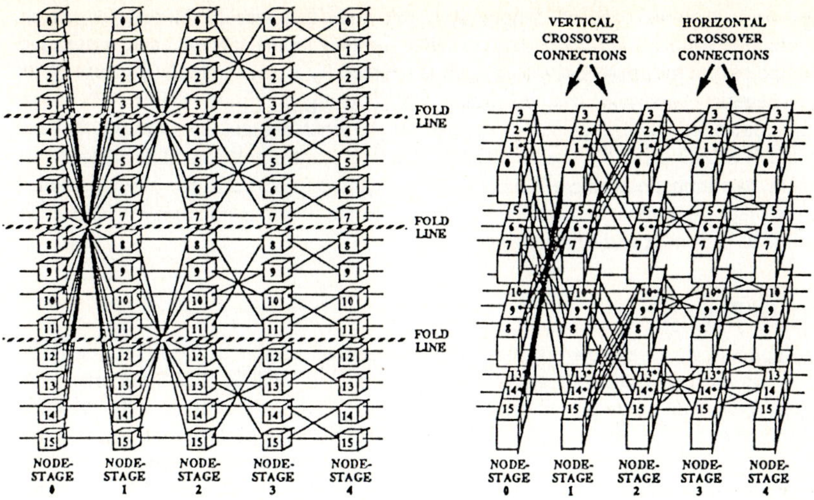

(a) (b)

Fig. 1. Crossover network: (a) 2D implementation and (b) 3D implementation.

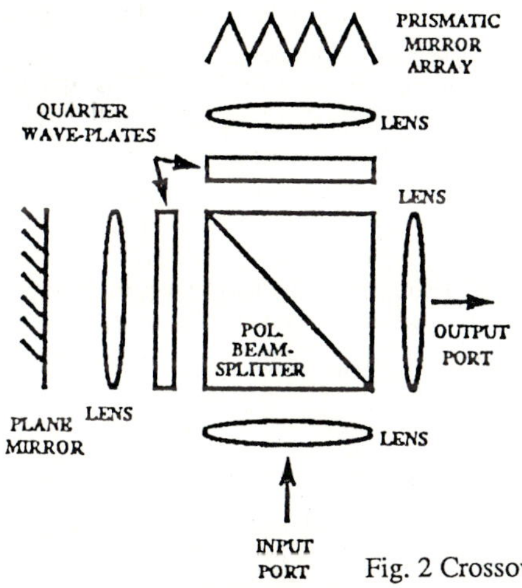

Fig. 2 Crossover hardware.

$$\beta_{2D \to 3D}(P) \to (R, C) = (INT(P/J), J-1-P_{MOD\ J}) \quad (2)$$

if INT(P/J) is an odd number. In the expressions above, P represents the physical address of the node in the 2D crossover network, (R,C) represents the row number and column number, respectively, of the same node in the 3D crossover network, J represents the number of columns of nodes in the resulting planar node-stage, INT(P/J) represents the integer part of the quotient P/J, and $P_{MOD\ J}$ represents the integer remainder of the quotient P/J. System designers can customize the resulting dimensionality of the node-stage array by using more or less fold lines- i.e., by varying the number of columns (J) in the resulting planar node-stage. In general, any 2D crossover network with $I=2^i$ nodes per node-stage (i = positive integer) can be converted into a 3D crossover network with rectangular node-stages containing $J=2^j$ rows and $K=2^k$ columns (j,k = positive integers). An N-input, 3D crossover network based on 2-input, 2-output nodes is required to have I=N and I=JK.

A 2*–module* is a novel type of 2-input, 2-output node that was developed for use in photonic switching applications. A 2-module OR's the two inputs together and passes the results on to the two output ports only if the gate is enabled by a clock signal. Because of its simplicity, a single 2-module can be implemented using a single S-SEED[3] if an electronically-controlled SLM is placed in the path of the clock signals (spot arrays) that are directed at the S-SEED array.[4] The electronic processor controlling the SLM can permit only one of the two inputs entering a 2-module to be enabled (clocked in the previous node-stage), and the other input must be disabled (unclocked in the previous node-stage). This adds new constraints to path hunt and routing algorithms.

3. Experimental Prototype

A four-stage optical crossover network (with interconnections similar to the first three link-stages in Fig. 1b) was constructed using 8x4 S-SEED arrays to produce a 32-input crossover network. Since only four node-stages were constructed, full connectivity is only possible between partitioned subsets of the inputs and outputs (such as the eight inputs and outputs with asterisks in Fig. 1b). Fig. 3 shows the hardware required to implement the network.

Bit-synchronous data is carried into and out of the network through fiber bundles, and each fiber bundle carries 32 dual-rail signals. Thus, each bundle contains sixty-four 100 μm core, multi-mode fibers that are bundled into a square array on a 400 μm pitch. Each input fiber is connected to an 850 nm LED which is driven by a word generator. Distinct data patterns are transmitted through each of the 64 input fibers. The data signals from these fibers are combined with the SLM-controlled clock signal via space-multiplexing.[5] This technique uses transparent optical flats with gold mirrors patterned on it that reflect the data signals and transmit the clock signals. The data signals are imaged onto the first S-SEED array, setting the state of the devices. A 10 mW, 850 nm laser diode which is passed through a pair of crossed binary phase gratings generates an 8-by-8 array of clock beams.[6] The clock signals pass through a mask simulating a SLM before being directed onto the S-SEEDs. The S-SEED array contains 5x10 μm windows on 20 μm centers, so the 8x4 array has a 200 μm diagonal field. Upon reflection from the first S-SEED array, the clock signals are modulated by the S-SEEDs, producing re-generated data signals which are routed through crossover hardware that provides a large vertical cross connection (using a right-angle prism) and a

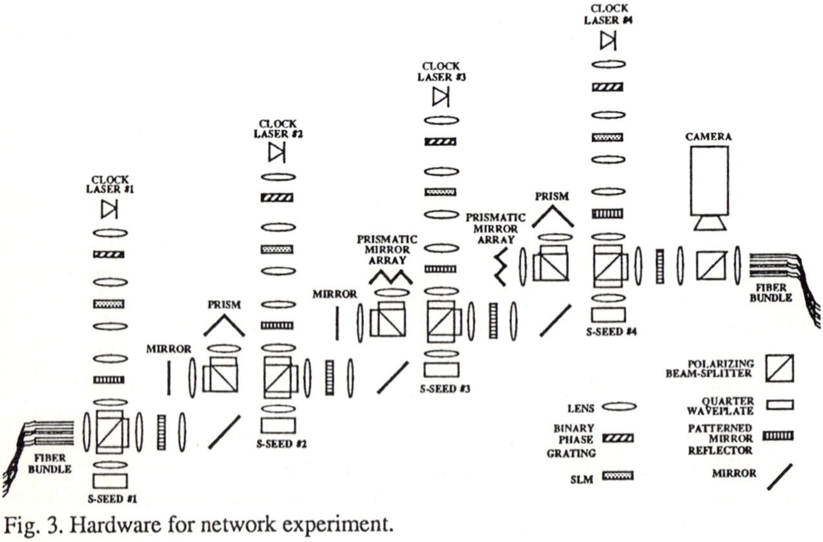

Fig. 3. Hardware for network experiment.

straight connection (using a mirror). The output signals from the crossover interconnection are then combined with the second SLM-controlled clock and imaged onto the second S-SEED array. The signals from the second S-SEED array are similarly re-generated and routed through crossover hardware that provides a small vertical cross connection (using a ruled aluminum prismatic mirror array) and a straight connection (using a mirror) before being imaged onto the third S-SEED array. The signals from the third S-SEED array are then routed through crossover hardware that provides a large horizontal cross connection (using a right-angle prism) and a straight connection (using a small prismatic mirror array to invert the dual-rail data in the straight path) before being imaged onto the fourth S-SEED array. This inversion of data in the straight path is required, because the S-SEED windows are oriented horizontally and the dual-rail data in the cross path is also inverted (the rails are crossed). The output of the fourth S-SEED array is imaged into a camera and into the output fiber bundle, which has photodiodes on selected fibers. The available crossover connections permit the data from each input fiber to be switched to any one of eight output fibers by manually controlling the four SLMs to block (disable) desired clock signals. Error-free switching was verified by oscilloscope traces of the signals arriving at the photodiodes. The maximum system speed of 55.7 kbps is currently limited by the available laser power and by the optical power losses. These losses (which are primarily due to vignetting, grating and S-SEED inefficiencies, Fresnel reflections, and aberrations) permit only 0.6 percent of the initial clock power to be routed to the S-SEEDs in the following node-stage. In future system designs, we will attempt to minimize these effects to increase the network speed, and we will also attempt to increase the S-SEED array size to construct larger networks.

4. Conclusions

2D to 3D transformations for crossover networks have been described. We have shown that S-SEED arrays can be operated as simple 2-input, 2-output nodes (called 2-modules) within a controllable network. A 32-input 3D crossover network using free-space optics to route signals between S-SEED arrays has also been experimentally demonstrated.

The authors would like to thank S. J. Hinterlong, R. L. Morrison, S. L. Walker, L. M. F. Chirovsky, L. A. D'Asaro, and H. S. Hinton for suggestions and components.

References

1. J. Jahns and M. J. Murdocca, "Crossover Networks and their Optical Implementation," Appl. Opt. 27, 3155 (1988).

2. T. J. Cloonan, "Topological Equivalence of Optical Crossover Networks and Modified Data Manipulator Networks," Appl. Opt. 28, 2494 (1989).

3. A. L. Lentine, H. S. Hinton, D. A. B. Miller, J. E. Henry, J. E. Cunningham, and L. M. F. Chirovsky, "The Symmetric Self Electro-optic Effect Device," Applied Physics Letters 52, 1419 (1988).

4. A. L. Lentine, F. B. McCormick, R. A. Novotny, L. M. F. Chirovsky, L. A. D'Asaro, R. F. Kopf, J. M. Kuo, and G. D. Boyd, "A 2 Kbit Array of Symmetric Self Electro-optic Effect Devices," IEEE Photonic Technology Letters, Vol. 2, pp. 51-53 (1990).

5. M. E. Prise, M. M. Downs, F. B. McCormick, S. J. Walker, and N. Streibl, "Design of an Optical Digital Computer," Journal de Physique, Colloque C2, Supplement au No. 6, Tome 49, pp. 15-18, June 1988.

6. R. L. Morrison and S. L. Walker, "Binary Phase Gratings Generating Even Numbered Spot Arrays," OSA Annual Meeting, 1989 Technical Digest Series, Vol. 18, (Optical Society of America, Washington, D.C., 1989), p. 111.

A Polarization-Switched Optically Reconfigurable Interconnection Network

H. Ito, Y. Yamamoto, and T. Nakagami

Fujitsu Laboratories, Ltd., 1015, Kamikodanaka, Nakahara-ku, Kawasaki 211, Japan

Abstract. An application of our new polarization-based optical switch in the construction of an optical Banyan interconnection network is described. The test results of its basic operation are also given.

1. Introduction

Optical free-space interconnection is expected to overcome the bottleneck of electrical interconnection because of the high-bandwidth, high-speed transmission, parallel multi-beam inputs, and the absence of interference among beams. A highly parallel and reconfigurable architecture is possible [1],[2],[3].

In this paper, we describe the principles behind our new polarization-based optical switch and its extension to the optical Banyan interconnection network. We also present the results of the basic channel shift test.

2. A new multi-input multistage polarization-based switch

2.1 Principles of the polarization-based optical switch

The proposed optical switch consists of a polarizing beam splitter (PBS), two quarter-wave plates ($\lambda/4$), a mirror, a prism with reflective coating, and a two-dimensional polarization control device (PCD). This switch has two modes of operation. If the beam having passed through the PCD is p-polarized, the switch is in the bypass mode as shown in Figure 1(a). If it is s-polarized, the switch is in the exchange mode as shown in Figure 1(b). In the exchange mode, the s-polarized beam is reflected by the polarizing beam splitter and the prism with the reflective coating. In this process, it goes through the quarter wave plate twice, and becomes p-polarized. It then passes through the polarizing beam splitter and is reflected back by the mirror. In this process, it goes through the quarter wave plate twice and becomes s-polarized. It is then reflected by the polarizing beam splitter and exits.

The shift distance between the input beam and the output beam is determined by the refraction index and the shape of the prism. The multiple-input beams thus shift an equal amount. For example, consider four beams input in parallel. For the single prism as shown in Figure 2(a),

Springer Series in Electronics and Photonics, Vol. 29 **Photonic Switching II**
Editors: K. Tada · H.S. Hinton © Springer-Verlag Berlin, Heidelberg 1990

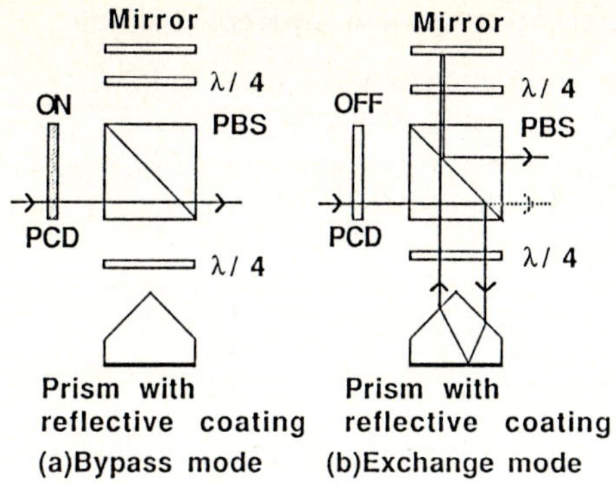

Figure 1. Basic operation of the polarization-based optical switch

PCD: Polarization control device

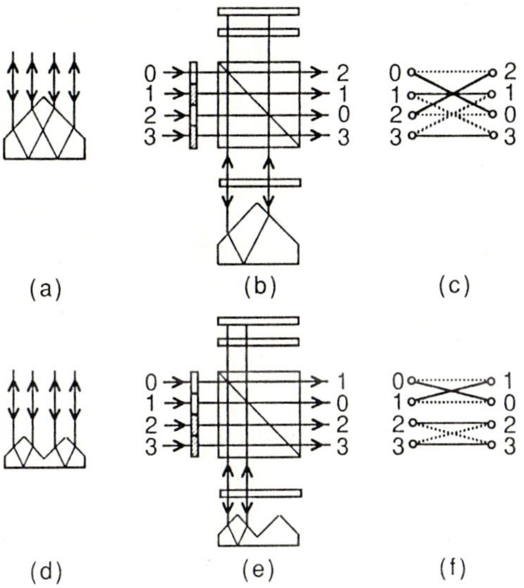

Figure 2. Characteristics of the polarization-based optical switch

The shift distance between the input beam and the output beam is determined by the refraction index and the shape of the prism. The multiple-input beams thus shift an equal amount. For example, consider four beams input in parallel. For the single prism as shown in Figure 2(a), the beam shifts by 2 channels. For a double prism as shown in Figure 2(d), the beam shifts by 1 channel. If a single prism is used and the PCD array is set as OFF,ON,OFF,ON, as shown in Figure 2(b), data points 1 and 3 are in the bypass mode and data points 0 and 2 are in the exchange mode. The equivalent connections are indicated by the solid lines in Figure 2(c). If a double prism is used and the PCD array is set as OFF,OFF,ON,ON, as shown in Figure 2(e), data points 2 and 3 are in the bypass mode, and data points 0 and 1 are in the exchange mode. They are indicated by the solid lines in Figure 2(f).

2.2 Characteristics of the polarization-based optical switch

The characteristics of the proposed optical switch can be summarized as follows:
(1) Beams traveling in opposite directions can take the same paths without interfering with each other. If they can be separated at the output, signal transmission is possible.
(2) Wherever each input beam impinges on to the optical switch , the corresponding output beams shift an equal amount. This implies that multiple beams can be input with good tolerance in the prism's lateral alignment.
(3) This optical switch enables various free-space wiring by changing the shape and size of the prism. In addition, the free-space wiring can be reconfigurated by controlling the ON/OFF states of the PCD array.

3. Application of the optical switches to the Banyan interconnection network

The Banyan network can be optically implemented by cascading our optical switches as shown in Figures 2(b) and 2(c). In this 4x4 optical Banyan network, the cross-connection of input beams can be accomplished by controlling the PCD arrays at each stage. We have confirmed the beam exchange operation through a 1x4 two-stage optical test as shown in Figure 3. In the test setup, a Liquid Crystal Light Valve (LCLV) has been used to control the polarization states of the input beams by adjusting the power of the white light sources S1 and S2. Figure 4 shows the channel shifts and settings of the control light beams S1 and S2. To increase the number of channels to 8, the basic structure of the optical switches does not need to be changed except for the shape of the prism at each stage. This is shown in Figure 5.

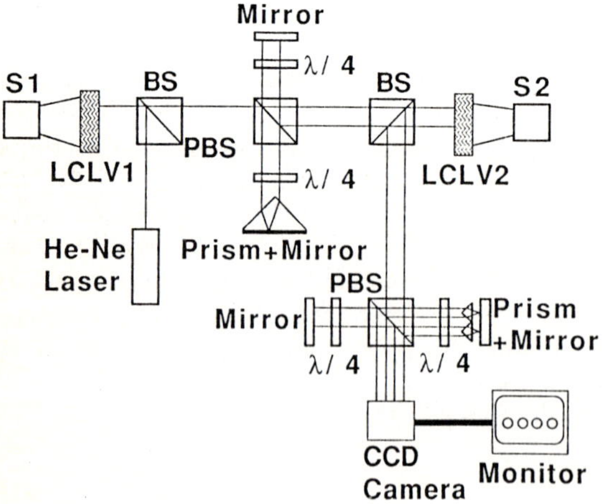

Figure 3. Two-stage optical network experiment

BS: Beam splitter
PBS: Polarizing beam splitter
LCLV 1,2: Liquid Crystal Light Valve

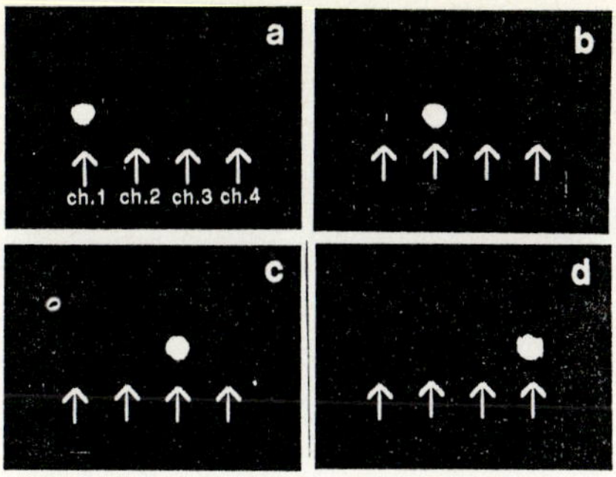

Figure 4. Channel shift results (a)S1=ON,S2=OFF (b)S1=ON,S2=ON (c)S1=OFF,S2=ON (d)S1=OFF,S2=OFF

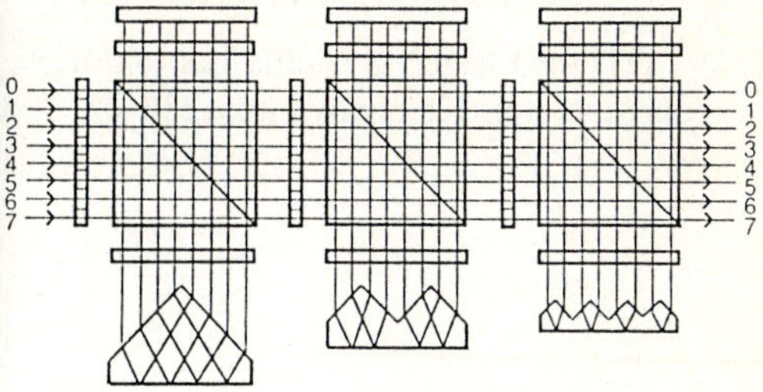

Figure 5. 8x8 optical Banyan network

4. Summary

We have proposed a new polarization-based optical switch and its application to the optically reconfigurable Banyan interconnection network. We also confirmed the channel shift operation. Further research will be done on the non-blocking self-routing algorithm and on the small-scale integration of our optical switch.

Acknowledgment

The authors thank Dr. Takanashi and Mr. Touge for their encouragement and guidance.

References

[1] Healey,P.,"Optical Interconnection Networks Employing Multiplexed Crosspoints",Conference Record of Photonic Switching'89, Salt Lake City, U.S.A.,March 1989.
[2] Johnson,K.M.,M.R. Surette, and J. Shamir ,"Optical Interconnection Network using Polarization-Based Ferroelectric Liquid Crystal Gates",Applied Optics,Vol.27,No.9,1988,pp1727-1733.
[3] Jahns,J.,"Optical Crossover Interconnections",SPIE Proc. on Optical Computing,Vol.963,1988.

Two-Dimensional Multichannel Optical Switch

K. Hogari, K. Noguchi, T. Sakano, and T. Matsumoto

NTT Transmission Systems Laboratories, 1-2356 Take, Yokosuka-shi, Kanagawa-ken 238-03, Japan

Abstract. This paper proposes a new type of two-dimensional multichannel optical switch that has small-size and low-driving-power characteristics. An 8×8 optical switch is demonstrated.

1. INTRODUCTION

The fields of parallel processing and telecommunication demand switches that have an enormous number of channels and high throughput performance. Optical spatial switches using polarization switching devices have the potential to realize such switches[1]. These switches operate using polarization control that changes optical paths[1]-[3]. However, these switches cannot be made sufficiently compact, because they use many aligned optical components, and because these components have to be physically separated to form these switches. Thus, they cause signal delay and skew problems due to the long optical path and the long optical path differences.

This paper proposes a new type of two-dimensional multichannel optical switch, which has small-size and low-driving-power characteristics, by demonstrating an 8×8 optical switch.

2. SWITCH STRUCTURE AND OPERATION PRINCIPLE

An 8×8 bitonic sorting switching network[4] is shown in Fig.1. This network consists of six switching stages and five routing stages. Each switching stage consists of four 2×2 switch elements. The cross or through state in the switch elements is determined according to the bitonic sorting algorithm to establish a data path[4]. The switch structure becomes very simple with a two-dimensional channel arrangement[5]. The proposed optical switch structure is shown in Fig.2. It consists of six cascaded 2×2 switch element arrays which are mentioned below. In this arrangement, the beams are shifted a distance equal to the channel interval vertically in the 2nd and 5th arrays in the cross state, and horizontally in the 1st, 3rd and 6th arrays. Also, the beams are shifted vertically twice the channel interval in the 4th array. Generally, a $2^n \times 2^n$ optical switch is composed of $n(n+1)/2$ switch element arrays that are composed of 2^{n-1} 2×2 optical switch elements[4].

Two kinds of the proposed 2×2 optical switch element array structures are shown in Fig.3. Figure 3(a) shows the switch element array structure (Type 1) which consists of a polarization controller array (P.C. array) and a polarizing beam splitter array. Structure (A) is an array structure where the beams can be shifted equal to the channel interval and structure (B) is an array structure where the beams can be shifted by twice the channel interval. When the transmitted beams of a channel pair are designated as p-polarization by P.C. array, the through state can be achieved. Also, when the transmitted beams of a channel pair are designated as s-polarization, the cross state can be achieved.

Springer Series in Electronics and Photonics, Vol. 29 **Photonic Switching II**
Editors: K. Tada · H.S. Hinton © Springer-Verlag Berlin, Heidelberg 1990

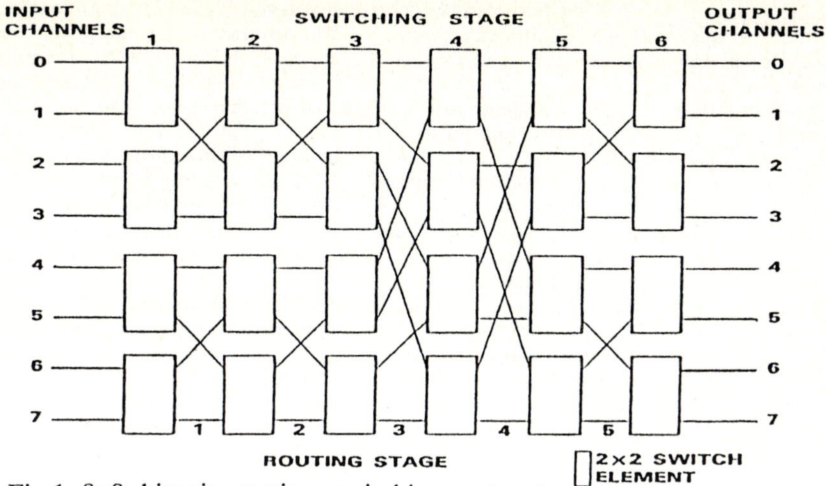

Fig.1 8×8 bitonic sorting switching network

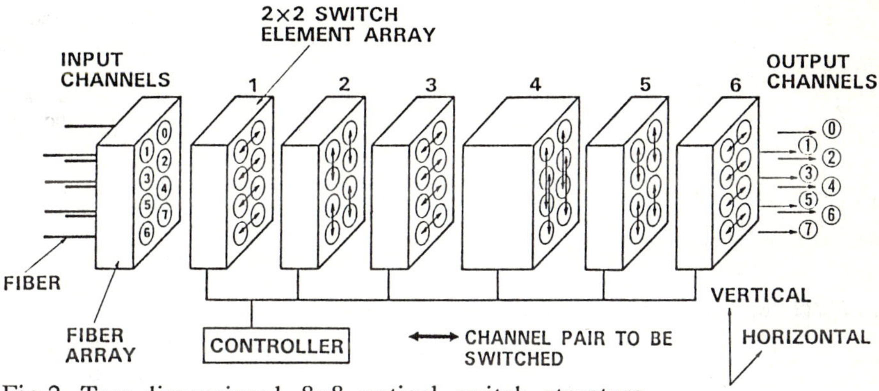

Fig.2 Two-dimensional 8×8 optical switch structure

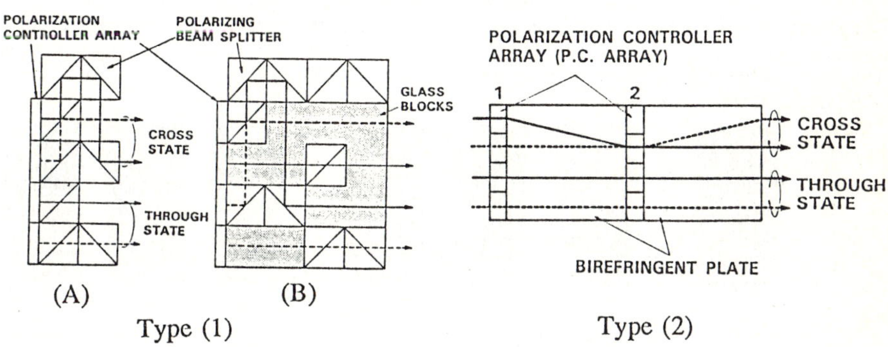

Fig.3 2×2 optical switch element array structures

(a) Type 1 consists of a P.C. array and a polarizing beam splitter array.

(b) Type 2 consists of two P.C. arrays and two birefringent plates.

Figure 3(b) shows another switch element array structure (Type 2) that consists of two P.C. arrays and two birefringent plates. The amount of extraordinary ray (e-ray) shift is equal in the two birefringent plates, and the plates are arranged so that the e-ray can be shifted in exactly the opposite direction. When the transmitted beams of a channel pair are designated as the e-ray and ordinary ray (o-ray) by P.C. array 1 and that designation is reversed by P.C. array 2, a cross state can be achieved. Also, when the transmitted beams of a channel pair are designated as the o-ray by both P.C. arrays 1 and 2, a through state can be achieved. The amount of optical beam shift is controlled by the birefringent plate length. Therefore, a birefringent plate with twice the amount of beam shift for a channel interval is twice as long as a plate in which the amount of beam shift is equal to the channel interval.

These switch element array structures are suitable for a small-size 2x2 switch element array, because they can have a two-dimensional channel arrangement, and because the polarization controller arrays and the polarizing beam splitter arrays or the birefringent plates can be joined in a body. Thus, the small-size optical switch can be constructed by using these 2x2 optical element arrays, and this switch passes high-bit-rate optical signals without delay and skew. As with other optical switches, this switch has the same advantage in terms of its low-driving-power.

3. SWITCHING PERFORMANCE

In order to evaluate the performance of the proposed switch, a two-dimensional 8×8 optical switch is demonstrated. The 2×4 optical fiber array at 2 mm fiber pitch is used to inject the optical beams. Thus, the channel interval is 2 mm. SELFOC lenses are attached to the end of the fiber array to obtain the collimated optical beams. In type 1 as shown in Fig.3(a), each polarizing beam splitter is 10 mm in length, 2 mm in width and 2 mm in height, and the eight polarizing beam splitters are arranged and fixed. The liquid crystal display panels (L.C. panels) are used as the polarization controller arrays. The L.C. panel pixel dimensions are 330×330 μm. Six L.C. panels and six polarizing beam splitter arrays are cascaded as shown in Fig.2. The L.C. panels are controlled by a personal computer. Also, in type 2 as shown in Fig.3(b), the birefringent plate in which the amount of beam shift is equal to the channel interval in the cross state, is 18.5 mm in length, 12 mm in width and 12 mm in height. Twelve L.C. panels and twelve birefringent plates are cascaded.

An optical switch experiment using two kinds of switch element arrays was carried out using a He-Ne laser at a 0.633 μm wavelength. When the optical beams were injected into channels 0 and 1, and the polarization was controlled according to the switching network algorithm, the output patterns were observed using a vidicon camera. The experimental results are shown in Fig.4.

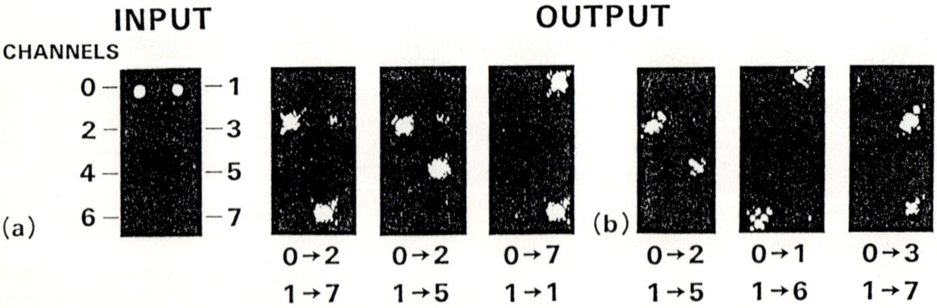

Fig.4 Experimental results
(a) Output patterns for switch element array Type 1.
(b) Output patterns for switch element array Type 2.

It was found that the optical paths could be changed entirely by using the proposed 2×2 optical switch element arrays. The output optical beam is deformed, because the collimated optical beam is larger than a L.C. panel pixel and the collimated optical beam is scattered and diffracted by the border between the L.C. panel pixels. Therefore, if the L.C. panel pixel dimensions are matched to the channel interval, the multichannel optical switch will have more satisfactory characteristics.

4. CONCLUSION

A two-dimensional multichannel optical switch is proposed and an 8×8 optical switch has been demonstrated. A small-size, multichannel and low-driving-power optical switch was obtained. This two-dimensional multichannel optical switch will be used in optical transmission systems and optical interconnections between many processors.

ACKNOWLEDGMENTS

The authors wish to thank S. Shimada and H. Kimura for their support and encouragement.

References

[1] K. M. Johnson et al., Applied Optics, 27, pp.1727-1733, 1988.
[2] M. Yamaguchi et al., ECOC 87, Helsinki, pp.397-400, 1987.
[3] J. Jahns et al., Applied Optics, 27, pp.3155-3160, 1988.
[4] K. E. Batcher, Proc. Spring Joint Comput. Conf., pp.307-314, 1968.
[5] C. D. Thompson et al., Commun. of the ACM, 20, pp.263-271, 1977.

A Rearrangeable Multichannel Free-Space Optical Switch Using a Polarization-Multiplexing Technique

K. Noguchi, K. Hogari, T. Sakano, and T. Matsumoto

NTT Transmission Systems Laboratories, 1-2356 Take,
Yokosuka-shi, Kanagawa-ken 238-03, Japan

Abstract: A rearrangeable multichannel free-space optical switch using a polarization multiplexing technique is described. A 32-channel switch is constructed and its performance is confirmed.

1. Introduction

Multichannel switches are suitable for communication link network reconfiguration in parallel processing computer systems. For electrical switches, however, channel number is limited by signal delay and skew(1). To overcome this problem, free-space optical multichannel switches using holograms(2) or liquid crystal gates(3) have been proposed. To the authors' knowledge, however, rearrangeable free-space optical switches with a large number of channels have not been constructed. This communication describes a rearrangeable free-space optical switch with a large number of channels which uses polarization multiplexing. The performance of a 32-channel prototype switch is examined.

2. Switch Structure

The Benes network configuration is convenient for rearrangeable switches. Figure 1 shows a 32-channel Benes network using 2×2 switch elements which consists of 9 switching stages and 8 routing stages.

Here, we propose a multichannel free-space optical switch using a polarization multiplexing technique according to a Benes configuration. Figure 2 shows the structure of the proposed switch. The switch consists of polarization rotator arrays constructed from liquid crystal (LC) light modulators and routing elements mentioned below. Optical ports for signal input and output are arranged two-dimensionally. Each port carries two optical signal channels which are differentiated by their orthogonal polarization. Each pixel can be discretely controlled so that polarization rotation occurs in steps of 90 degrees.

The routing element exchanges one polarization component of the two signal light beams going into different input ports with each other, while it allows to pass through the other polarization components of the beams. Figure 3 shows the routing elements used in the proposed switch. Half-wave plate stripes, which are arranged so that the gap between the stripes equals the stripe width, are inserted between two

Springer Series in Electronics and Photonics, Vol. 29 **Photonic Switching II**
Editors: K. Tada · H.S. Hinton © Springer-Verlag Berlin, Heidelberg 1990

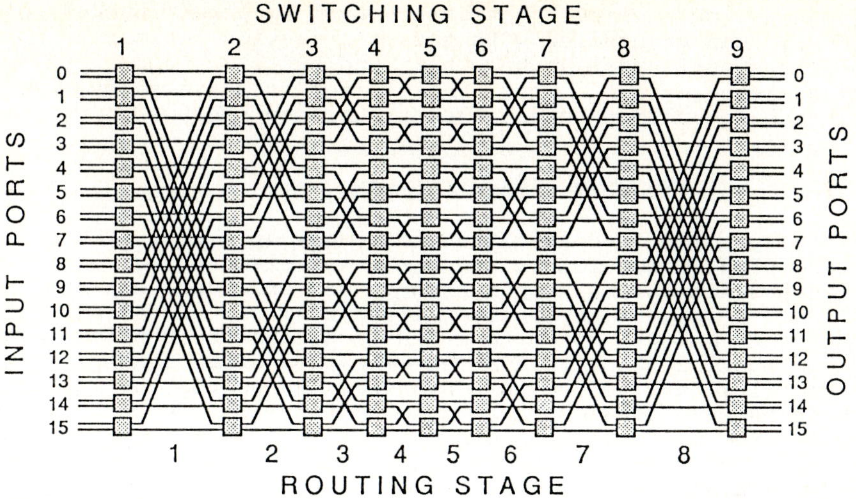

Fig. 1 A 32-channel Benes network configuration

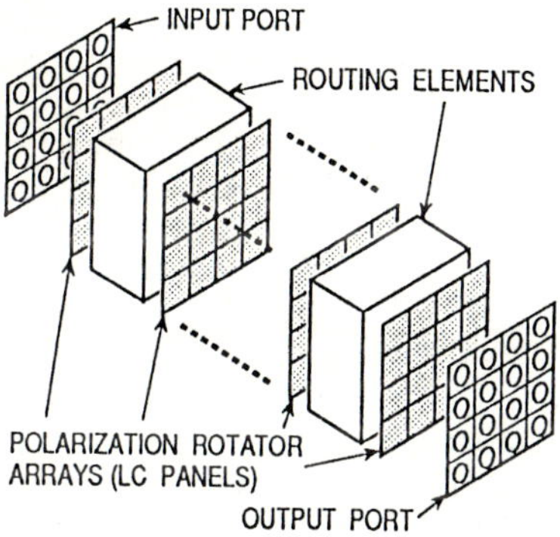

Fig. 2 Structure of the proposed optical switch

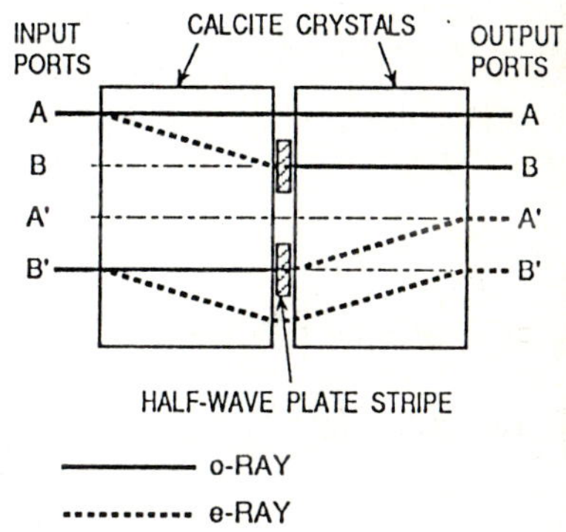

Fig. 3 Structure of the routing element

calcite crystals to exchange the ordinary ray (o-ray) and extraordinary ray (e-ray). The input-side crystal deflects the e-ray downward and the output-side deflects it upward. Both displacements equal the stripe width.

In this arrangement, the e-ray going into the input port A (or A′) and the o-ray going into the input port B (or B′) fall in a half-wave plate stripe. Then, the former is converted to the o-ray and reaches the output port B (or B′), and the latter is converted to the e-ray and reaches the output port A (or A′). As a result, one component of each of the orthogonally polarized light beams going into the ports A and B are exchanged with each other, while the other ones pass without any changes.

This switch passes high-bit-rate optical signals without distortion or skew because optical paths in the switch do not vary with channel permutations. Also, a large number of channels can be handled with the polarization multiplexing, two-dimensional port arrangement, and switching and routing element integration. Thus, this switch is suitable for network reconfiguration in parallel processing systems.

3. Experiment for a 32-Channel Switch

A prototype of a 32-channel rearrangeable optical switch consisting of 9 LC modulators and 8 routing elements was constructed and its switching performance was examined experimentally. A 633-nm He-Ne laser light beam was used as the light source. In order to avoid light beam broadening due to diffraction, lens waveguides were used for transmitting the signal lights to the next stage. Images for the signal beam spots were focused on LC modulator pixels and the plane at half-wave plate stripes and gaps in the routing elements. The pitch of the beam spots passing through adjacent ports was 0.47 mm at the LC modulators and was 0.66 mm at the half-wave plate stripes. The separation of adjacent LC modulators was about 35 cm, into which a routing element and four guiding lenses were inserted.

Figure 4(a) shows the port arrangement for this switch. At the n*th* port, the n*th* channel and $(n+16)$*th* channel are superimposed as an o-ray and an e-ray, respectively. In this arrangement, four types of beam exchanges occur at routing stages. The exchange patterns are shown in Fig. 4(b)-(e). Figure 4(b) is the pattern for the 1st and 8th routing stages, (c) is for 2nd and 7th, (d) is for 3rd and 6th, and (e) is for 4th and 5th.

The switching performance for channels 0, 1, 4, and 5 was examined. Figure 5 shows the beam spot patterns observed at the output ports. Figure 5(a) represents the spot pattern which takes no change ($n \rightarrow n$). Figures 5(b)-(e) represent those corresponding to "reverse" ($n \rightarrow 31-n$), "perfect shuffle" ($n \rightarrow 2n$ $(n<16)$, $n \rightarrow 2(n-16)+1$ $(n>16)$), "multiplied by three" ($n \rightarrow 3n$ $(n<11)$, $n \rightarrow 3n-32$ $(11 \leq n<22)$, $n \rightarrow 3n-64$ $(22 \leq n)$), and "subtract one" ($n \rightarrow n-1$ $(n>0)$, $n \rightarrow n+31$ $(n=0)$) permutations, respectively. In the figures, two channels assigned to n*th* and $(n+16)$*th* are duplicated in the n*th* port as o- and e-rays. Regarding this, the photographs show that the light beams are entirely connected to the desired output channels.

The insertion loss of all channels in the switch in the reverse permutation was examined. The average of the measured loss was 17.0 dB. All of the measured loss values fell in the 15.7-18.9 dB range. The results show that the insertion losses are relatively small and stable.

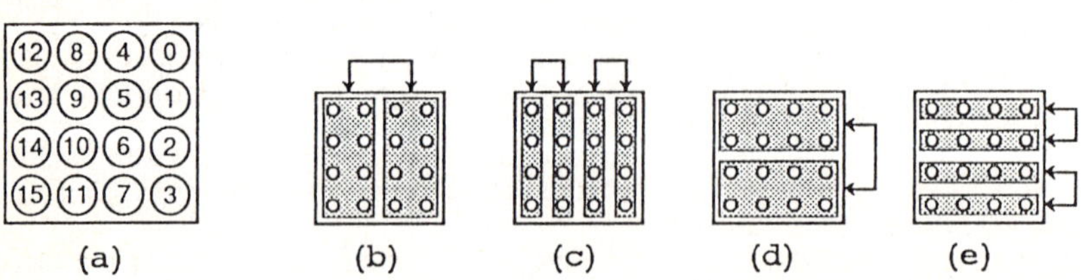

Fig. 4 Port arrangement and channel exchange pattern

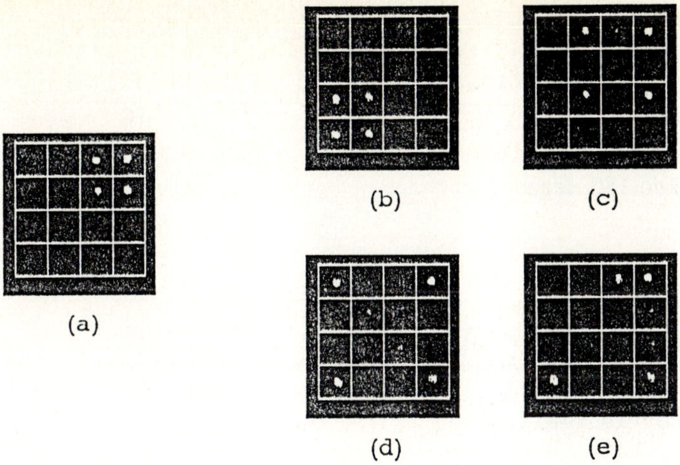

Fig. 5 Photographs of the switched beam spot patterns

4. Conclusion

A rearrangeable free-space optical multichannel switch using a polarization multiplexing technique is proposed. This switch consists of polarization rotator arrays and routing elements. It is shown that the proposed switch is suitable for network reconfiguration in parallel processing systems. A 32-channel prototype switch was constructed, and excellent switching performance and stable insertion loss were confirmed.

Acknowledgement: The authors would like to thank Dr. Hidetoshi Kimura for his continuous encouragement.

References:

(1). J. W. Goodman, F. I. Leonberger, S-Y. Kung, R. A. Athale, Proc. IEEE 72, 850(1984).

(2). J. B. McManus, R. S. Putnam, and H. J. Caulfield, Appl. Opt. 27, 4244(1988).

(3). K. M. Johnson, M. R. Surrete, and J. Shamir, Appl. Opt. 27, 1727(1988).

Scale Estimation of a Two-Dimensional Optical Concentrator Using Position Shifters

M. Yamaguchi

NTT Communication Switching Laboratories, 3-9-11, Midori-cho, Musashino-shi, Tokyo 180, Japan

Abstract. Maximum scale, limited by birefringent plate characteristics, is estimated for a two-dimensional concentrator. The results indicate that a compact 1024-input-port concentrator is realizable.

1. Introduction

Optical parallel processing technology is expected to increase the performance of optical switching systems [1],[2]. An impressive application of this processing would be to achieve a compact optical subscriber-line concentrator that can pick out live lines from among all the input lines and connect those to the output lines [3]. For such a concentrator, the author proposed a two-dimensional optical concentrator using birefringent-plate-based position shifters and confirmed its basic operation with an experimental one having twenty-five inputs and nine outputs [4].

This paper estimates the maximum scale and dimensions, limited by birefringent plate characteristics, of such a two-dimensional concentrator to evaluate the possibility of realizing a compact and large-scale concentrator.

2. Two-Dimensional Optical Concentrator

2.1 Structure and Principle

The structure of the proposed two-dimensional concentrator is shown in Fig. 1. This concentrator consists of cascaded two-dimensionally-arrayed parts: i.e. optical position shifters, coupling lens arrays, and input and output optical fiber arrays. The key device is the position shifter which has N^2 (equal to the number of input fibers) shift cells and shifts optical signals from one cell to another in the x- or y-direction. Shift direction is defined at each stage of the position shifters.

Optical signals from the input fiber array are collimated by the lens array and are input to the first stage of the position shifters. Each signal beam is shifted to the adjacent shift cell, if the output port of the adjacent cell is not being used. The beam is then transferred to the next position shifter. At that point, the signal beams are again shifted in the same manner as in the

Springer Series in Electronics and Photonics, Vol. 29 **Photonic Switching II**
Editors: K. Tada · H.S. Hinton © Springer-Verlag Berlin, Heidelberg 1990

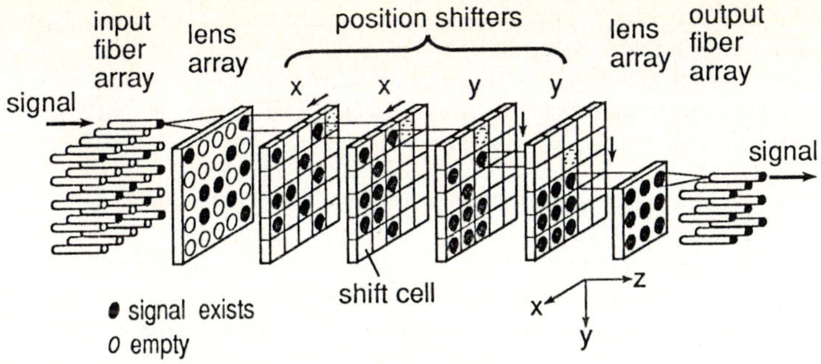

Fig. 1 Structure and principle of the two-dimensional concentrator

first stage. As a result, the signal beams are gathered in the corner of the last stage and they are then coupled into the output fiber array.

2.2 Optical Position Shifters

A shift cell is made up of a birefringent plate, a polarization controller, a photo detector and a processing circuit, as shown in Fig. 2. The incoming optical beam is launched into a cell through a circular window and goes through the polarization controller and the birefringent plate. Polarization of the incoming beam is rotated (0 or 90 degrees) by the polarization controller; e.g. if the incoming beam is an ordinary ray, it is changed into an extraordinary ray by 90-degree polarization rotation. In the birefringent plate, the beam goes straight or slantwise according to the beam polarization; i.e. the ordinary ray goes straight and the extraordinary ray goes slantwise.

For example, if the incoming ordinary ray requires shifting, it is changed into the extraordinary ray by the polarization controller and then shifted by the birefringent plate to exit from the adjacent cell. If the incoming ordinary ray requires traveling

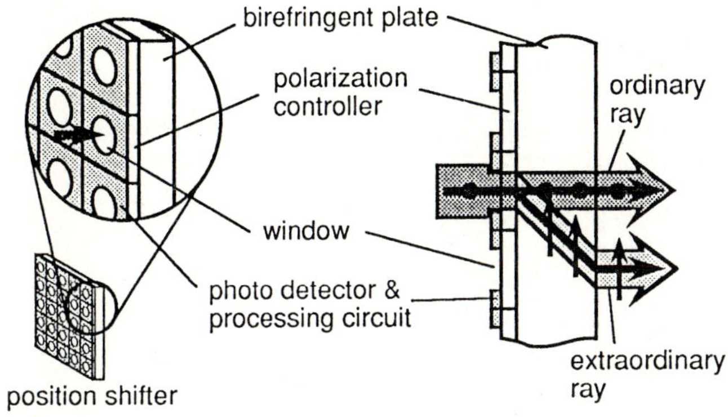

Fig. 2 Position shifter structure

straight, its polarization is unchanged. In that case, the ray goes straight through the birefringent plate and exits from the same cell. The control signals for the polarization controller are generated from the input beam level and the adjacent cell states by the photo detector and processing circuit.

3. Estimation of Scale & Dimensions

To evaluate the possibility of realizing a compact and large-scale concentrator, it is important to estimate the maximum available scale and dimensions of it. The scale and dimensions are limited by the characteristics of the birefringent plates in the position shifters. To estimate the limits by these birefringent plates, a model (Fig. 3) is adopted in which the optical beams go through a single thick birefringent plate whose thickness L_z is equal to the total thickness of the birefringent plates in all the stages. Two basic limitations are imposed by the birefringent plate. The first limitation arises from the relationship between separation angle ϕ of an ordinary ray and an extraordinary ray and expansion angle θ of the signal beam that is based on diffraction. The second limitation arises from the size restriction of the birefringent plate. These limitations are represented by Equations (1) and (2), respectively, where D is window diameter, L_c is cell size, and L_{xy} is the size (x-y direction) of the birefringent plate:

$$N = 2\frac{\phi}{\theta} = \frac{2\phi D}{2.44\,\lambda}, \tag{1}$$

$$N = \frac{L_{xy\_max}}{L_c} = L_{xy\_max}\,\frac{a}{D}. \tag{2}$$

Figure 4 shows the relationship between the number of shift cells N in a single column on the position shifter and window diameter D, where line (1) and curve (2) are calculated by Equations (1) and (2), respectively. The number of shift cells allowed in a column for any window diameter is shown by the hatched

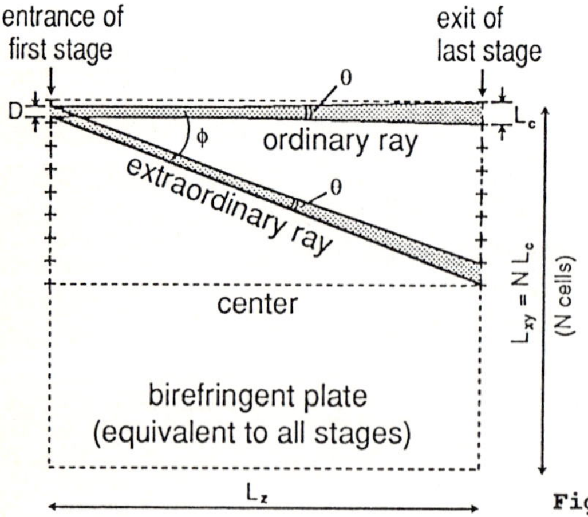

Fig. 3 Model for estimation

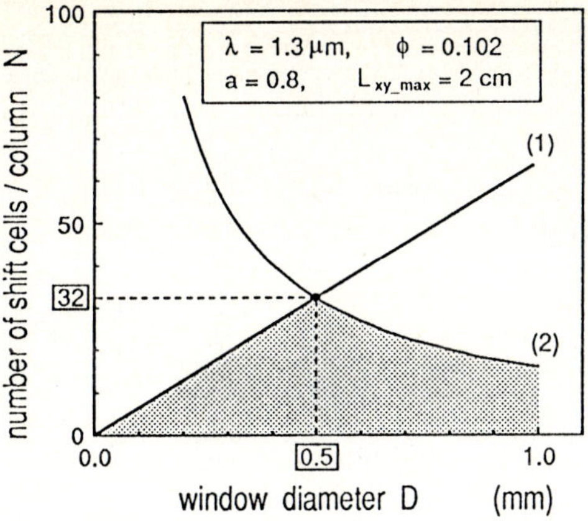

Fig. 4 Calculation results

area. The maximum available number of shift cells, given by the line and curve intersection, is thirty-two with a window diameter of 0.5 mm. This corresponds to a 1024-input-port concentrator with dimensions of 20 mm x 20 mm x 400 mm. Thus, a compact and large-scale concentrator can be realized by adopting a two-dimensional-array structure.

4. Conclusion

The scale and dimensions of the proposed two-dimensional optical concentrator were estimated. The results of the estimation indicated that a compact 1024-input-port concentrator is realizable with dimensions of 20 mm x 20 mm x 400 mm.

References

1. T. Yasui, K. Kikuchi: In *Photonic Switching, Electronics and Photonics 25* (Springer-Verlag, 1988) pp.24-35
2. M. Yamaguchi, K. Kanayama, K. Kikuchi: 13th European Conf. on Opt. Commun., Helsinki (1987) **1**, pp.397-400
3. I. Kobayashi: Topical meeting on Photonic Switching, Salt Lake City (1989) FA1
4. M. Yamaguchi, I. Kobayashi: 15th European Conf. on Opt. Commun., Gothenburg (1989) **1**, pp.272-275

Free-Space Multistage Optical Interconnection Networks Using Micro Lens Arrays

S. Kawai

C&C Information Technology Research Laboratories, NEC Corporation, 4-1-1, Miyazaki, Miyamae-ku, Kawasaki 213, Japan

Abstract. New optical interconnection networks with a self-routing function using lens arrays, whose optical systems have low aberration and high light power efficiency, are presented. Interconnection experiments are successfully demonstrated using two sets of planar micro lens arrays.

1. Introduction

Following the advances achieved with regard to high-density and high-speed for electronic devices, as the number of signal paths increases, problems with clock skew and bandwidth limitations are occurring. Light beams can be used to cross free-space noninductively, having a high frequency handling bandwidth and capability for high-speed communications. If free-space interconnections that satisfy these characteristics are realized, several problems inherent in electronic wiring can be solved. Some free-space optical interconnection techniques for these purposes have already been proposed[1]-[4]. An interconnection technique using lens arrays[2] is one appropriate method. It has higher light power efficiency through using imaging and the possibility of chip-like integration due to its simpler architecture.

New optical interconnection networks with a self-routing function using lens arrays are presented and their principles are successfully demonstrated, using planar micro lens (PML) arrays[5].

2. Multistage Interconnection Networks

Multistage interconnection networks with a self-routing function are widely used for connections between processors or computer memories. These networks include basic interconnections, such as shuffle exchanges. If these interconnections can be extended to 3-D arranged systems, the number of processors or computer memories connected with each other can be increased tremendously. Free-space optical interconnections, which do not have fixed pathways, are suitable for 3-D interconnections. Shuffle exchanges have already been realized using lens arrays [2]. The omega network and the baseline network can be realized by combining shuffle exchanges. Figure 1 shows some techniques used to realize a 4-shuffle,

Springer Series in Electronics and Photonics, Vol. 29 Photonic Switching II
Editors: K. Tada · H.S. Hinton © Springer-Verlag Berlin, Heidelberg 1990

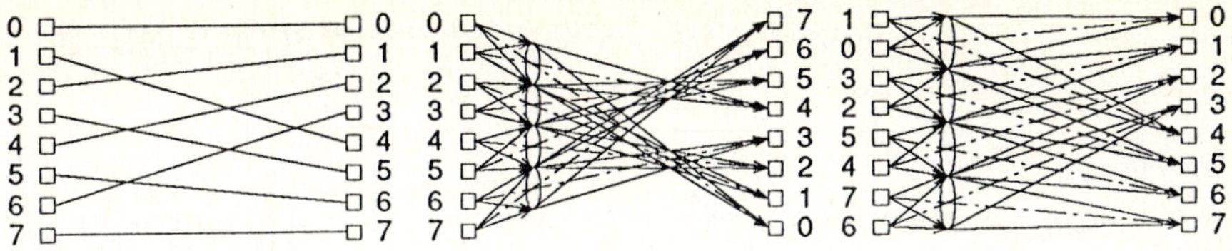

(a) Ordinary Pathway (b) Ordinary Optical Pathway (c) New Optical Pathway

Figure 1. Shuffle Exchange Pathway

which shuffles 8 elements into 2 groups of size 4. Figure 1(a) indicates an ordinary electric wiring technique. The optical axes in an ordinary optical system arrangement, whose outputs are inversely arranged versus the inputs, are shown in Fig. 1(b).

When these interconnections are applied to self-routing networks, they are connected to each other through 2x2 switches. Therefore, new light paths, shown in Fig. 1(c), are realized by exchanging the output port for each switch. In some cases, this kind of optical system can decrease any aberration in the signal path. Using such an optical system, butterfly exchanges, used to construct a banyan network, and interconnections which form a full crossover network can also be realized.

3. New Optical Interconnection Network

The interconnection networks mentioned above are for electric wiring, and are not always suitable for use in optical systems. In optical interconnections using lens arrays, it is important to construct an optical system whose aberrations are low and whose light power efficiency is high. Figure 2 shows a new network with a self-routing function which satisfies these conditions. The lateral magnification for each stage optical system is always exactly double.

Commonly, since an aberration is proportional to the light source point heights for the same radius lens, an optical system whose point heights are low is desired. Figure 3 shows the ratio of the point heights to the lens radius for the networks mentioned. Since the ratio for the presented network is smaller than for the omega and crossover networks, a low aberration optical system can be constructed.

The image size is proportional to a lateral magnification size square for the system. The signal light incident on the desired detector and the crosstalk light incident on the neighboring detectors are changed according to the magnification size. Figure 4 shows the light power efficiency for the networks. Since the lateral magnification for the omega network and the proposed network is always exactly double, no light beams are incident on neighboring photodetectors and light power efficiency does not decrease, while for the other networks S/N ratios decrease to near 1 for more than 1000 (32x32) interconnections.

When the same numerical aperture size lenses are used, the optical system size is almost proportional to the lateral magnification size for the system. Figure

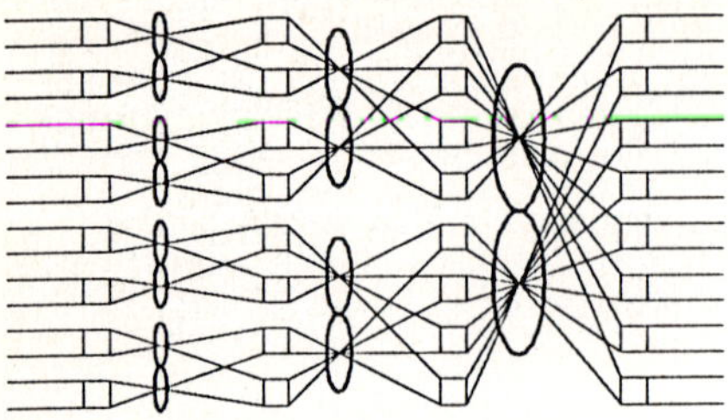

Fig.2 New Self-routing Network

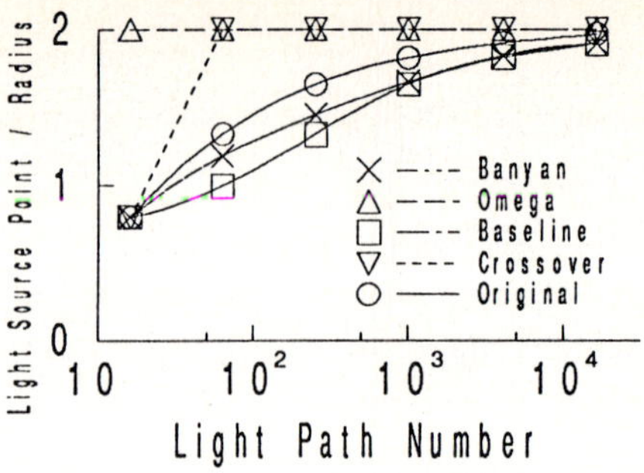

Fig.3 Comparison for Aberration

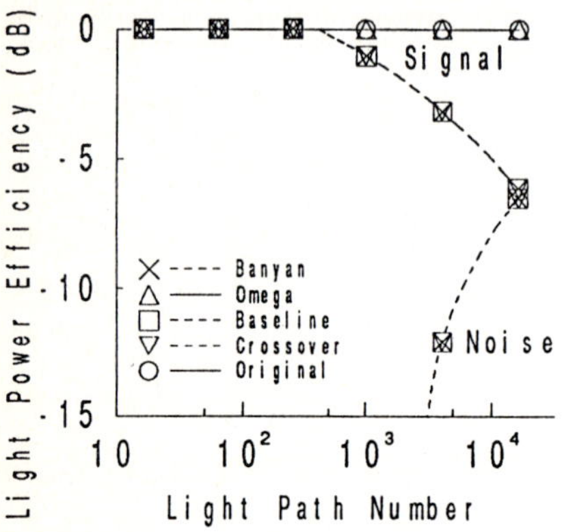

Fig.4 Light Power Efficiency

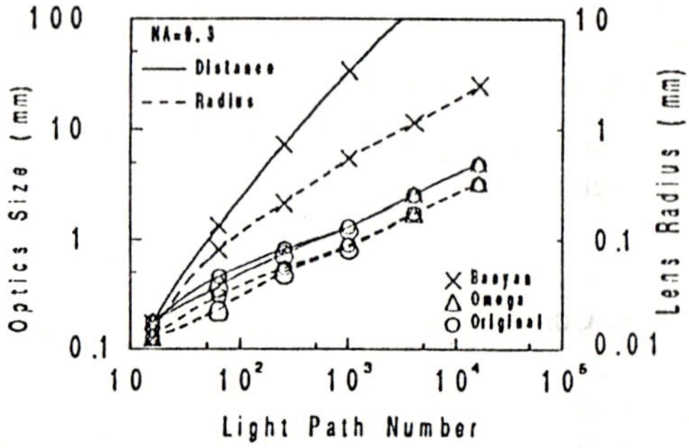

Fig.5 Optical System Size

5 shows the optical system size and lens radius. The distance between an input and an output plane in the omega and the proposed networks is about 1mm, while in the other networks, the distance exceeds more than 100mm. Therefore, the omega and the proposed networks have the possibility of highly dense interconnection.

In order to confirm the light path number limit for this network, ray tracings using biconvex spherical lenses have been carried out. As a result, using this network, more than 16000 (128x128) processors or computer memories, which is ten times the number in previous equivalent networks, can be connected to each other. If aspherical lenses are used, the number of light paths can increase further.

4. Experimental Results

In order to confirm interconnection principles, imaging experiments using PML arrays were carried out. They have an aspherical refractive index distribution and can converge laser beams to near the diffraction limit. They have almost ideal convergence characteristics and are suitable for interconnection purposes. In this experiment , assuming 31.25μm pitch light sources and photodetectors, two kinds

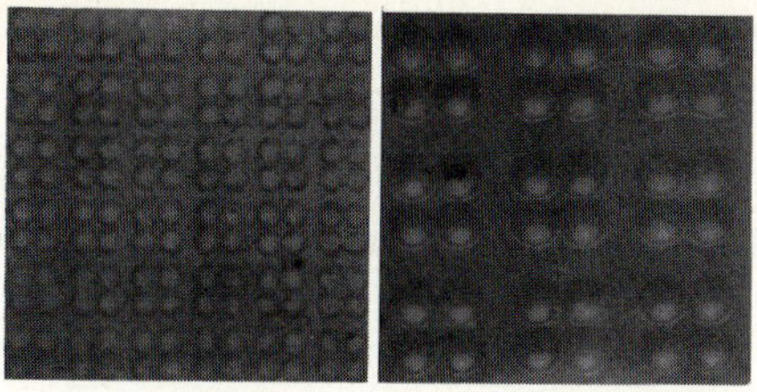

1st Stage 2nd Stage
Fig.6 Planar Micro Lens Arrays

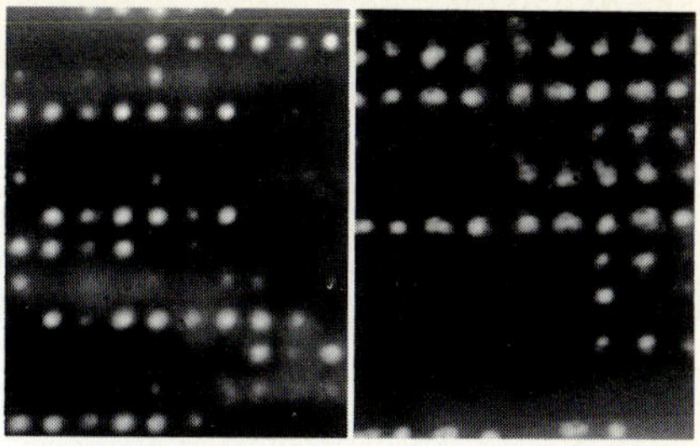

1st Stage 2nd Stage
Fig.7 Imaging Patterns

of PML arrays were made up to construct an 8x8 network. Figure 6 shows 1st stage PML arrays, whose average pitch is 62.5μm, and 2nd stage PML arrays, whose average pitch is 125μm. Using these lenses, circular input mask patterns arranged at 31.25μm pitch are imaged as shown in Fig. 7. It is confirmed that the desired optical light handling pathways have been realized with low aberration and high light power efficiency.

5. Conclusion

New multistage networks, which are suitable for interconnections using lens arrays, are presented. Their principles are successfully demonstrated, using PML arrays. If a PML array is integrated with surface-emitting lasers and photodetectors, a free-space optical interconnection chip can be obtained.

The author gratefully thanks Dr. M. Oikawa of Nippon Sheet Glass Co., Ltd. for supplying the planar micro lens arrays used in these experiments. He also thanks Drs. K. Iinuma and T. Temma as well as Mr. M. Mizoguchi for their suggestions and encouragement.

References

1. J.W. Goodman et al., Proc. IEEE 72, 850 (1984)
2. A.A. Sawchuk and I. Glaser, Proc. SPIE 963, 270 (1989)
3. J. Jahns and M.J. Murdocca, Appl. Opt. 27, 3155 (1988)
4. A. Guha et al., Appl. Opt. 29, 1077 (1990)
5. M. Oikawa et al., Proc. SPIE 898, 3 (1988)

Multistage Interconnection Networks and d-Dimensional Architectures

J. Giglmayr

Advanced Technology System Analysis GmbH,
Rüdesheimer Platz 5, D-1000 Berlin 33, Fed. Rep. of Germany

Abstract. A model for optically interconnecting d-dimensional data cubes is presented. For d=3, 4 a realization is proposed and discussed. Further applications of d-dimensional architectures are reported.

1. Introduction

The present paper assumes that photonic switching networks are the first step towards a digital optical processor. In photonic switching networks processing is applied to find a path from any idle input to any idle output through the 3-D MIN. For a digital optical processor further computation capabilities have to be added to the nodes.

There exists a wide variety of different approaches to the architecture of photonic systems and no systematic methodology seems yet to exist. The aim of the ongoing work is a systematic design of photonic systems. Mapping parallel algorithms efficiently on 3-D architectures is part of such a systematic design. In particular, mapping the multidimensional problem graph onto a d-dimensional architecture which finally is layed-out in the 3-D physical space is central within the ongoing work.

2. Model of Interconnections

For the upcoming digital optical technology the modification of common concepts from electronics is desired. The main work in electronics was devoted to 2-D MINs with 2×2-switches. In contrast, free space digital optics is characterized by 3-D systems with at least 4×4-switches. For the non-blocking multistage interconnection of any idle input with any idle output the least number of stages is $2\log_k(N^2) - 1$ with the number of N^2/k switches of size k×k in each stage [1]. But if rearrangements have to be considered we may need $3\log_k(N^2) - 4$ stages [2] and even more.

For this 3-D MIN a distributed control by a routing tag is assumed. The routing tag for the MIN is divided into the binary representation of the mixed radix numbered (MRN) inputs to the MIN and into the routing tag complement which is centrally computed at the arrays in front of the MIN (Fig. 1).

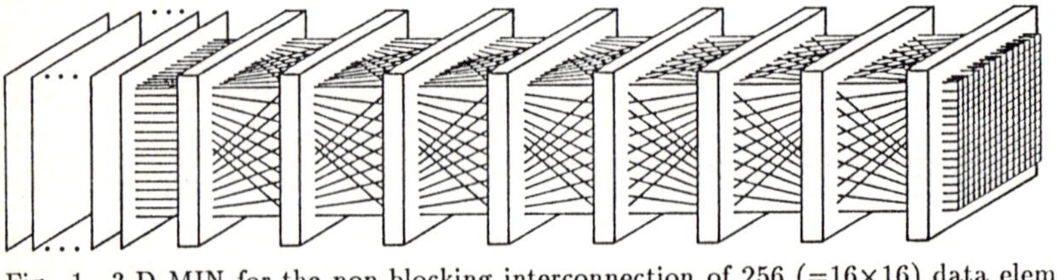

Fig. 1. 3-D MIN for the non-blocking interconnection of 256 (=16×16) data elements. [The computation of the routing tag complement is placed in front of the MIN. The switches are of size 4×4 and the 2-D interstage pattern is $S_{2-D}^{SGPS}(2,8;2,8)$ and its inverse (SGPS for symmetric generalized perfect shuffle).]

Springer Series in Electronics and Photonics, Vol. 29 Photonic Switching II
Editors: K. Tada · H.S. Hinton © Springer-Verlag Berlin, Heidelberg 1990

A main advantage of the digital optical technology is the massive parallelism which has been applied on the logic level. But the massive parallelism may also be applied on the process and algorithm level. In particular, mapping data arrays onto equivalent d-dimensional ($d \geq 3$) data cubes may increase the parallelism of the algorithms. The multistage interconnection of the d-dimensional data cubes by d-dimensional regular patterns [3] and by d-dimensional switching and processing cubes [4] provides a d-dimensional architecture. Some of the elements of these d-dimensional architectures and their applications are described throughout the paper.

3. Data Cubes and Regular Interconnections

In free space digital optics one may extend the dimension of the interconnected data set by at least one coordinate if additionally the frequency domain is considered [3]. In particular, optical frequency division multiplexing (OFDM) is assumed on each link and in the frequency domain the pattern is generated by means of frequency conversion (FC). Starting with a data vector of length N^2 (representing a row of lasers) we may generate a data array of size $N^2 = N^{(1)}N^{(2)}$ and starting with an equally sized data array (representing an array of lasers) we may generate a 3-D data cube of size $N^2 = N^{(1)}N^{(2)}N^{(3)}$, respectively. Assuming the factorization $N^{(i)} = n_1 n_2$ (≥ 4) the regular pattern defined on a 4-D cube is expressed by ($\otimes$ for the Kronecker product)

$$S_{4-D}^{SGPS}(n_1, n_2) = S(n_1, n_2) \otimes S(n_1, n_2) \otimes S(n_1^2, n_2^2). \tag{1}$$

But the existence of d-dimensional ($d = 3,4$ etc.) data cubes of size $N^{(1)}N^{(2)} \ldots N^{(d)}$ is also a question of their organization rather than of the existence of appropriate physical phenomena. Generally, higher-dimensional geometric objects may be generated by reconfiguring lower-dimensional geometric objects according to the laws of projective geometry [5].

In particular, we may reconfigure the arrays and 3-D cubes and obtain the requested 3-D and 4-D data cubes [Fig. 2 (a)] and the related regular patterns [Fig. 2 (b)]. The topology of these patterns is independent of the initial dimension of the data set (vector or array, respectively) but their evaluation provides different results which are shown throughout the paper (Figs. 3).

4. Switching and Processing Cubes

The d-dimensional switching and processing cubes interconnected by d-dimensional patterns are d-dimensional geometric objects. The inputs and outputs are tensorial quantities and the transformation by the switches is described by tensor equations [4]. The ordering of a d-dimensional cube may be described by transformations on the binary representation of MRN d-tuples. In particular, for a d-dimensional cube with the elements numbered by a i-binary d-tuple (X_i^d) the shuffled column major port ordering is described by

$$\overline{X_d^i} = S^{-1}(d,i)X_i^d \tag{2}$$

[cf Fig. 2 (c) for d=4] and other orderings are obtained by similar transformations [4]. This framework which we call Horizontal Coding Scheme [4] prescribes (1) the location of the switches within the cube, (2) the geometry of the port configuration, (3) the size of the switches and (4) provides a distributed control of the MIN based on d-tuple addresses.

The architecture for the computation of the routing tag complement in front of the MIN is similar to the architecture of a switching and processing cube. On this architecture a parallel algorithm has to be implemented for the decomposition of the requested permutations. In particular, during the decomposition procedure the maximum of the particular source and destination addresses has to be computed [6].

The chosen paradigm for the implementation of the algorithm is the divide-and-conquer-principle. This means that the solution of a problem in d-space is obtained by the decomposition into small parts and their parallel solution and the solution of the problem in the d-1 space [7]. Note, the ordering of the ports of the switches of a 4-D cube contains the port ordering of a 3-D cube and the latter contains the port ordering of a switching array [Fig. 2 (c)]. Thus by

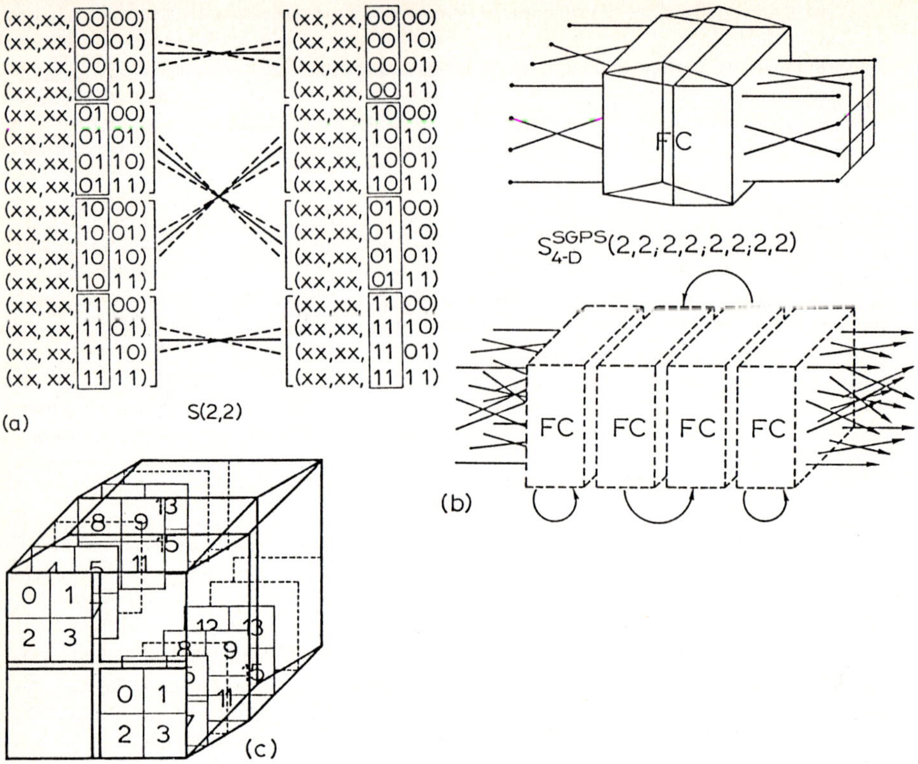

Fig. 2. (a) Scheme for the generation of a 4-D cube of size 4×4×4×4 and the related 4-D pattern by reconfiguring a 3-D cube of size 4×4×16. [The S(2,2) pattern shuffles the order of the 3-D cubes (solid lines) as well as the order of the arrays (dashed lines). The 3-D cubes are represented by the binary digits surrounded by the rectangles, the arrays are represented by the residual two binary digits and ×× means 00, 01, 10, or 11.] (b) Abstract model of a 4-D cube and the 4-D SGPS pattern (above) and scheme for the realization (below) (FC for frequency conversion). (c) 4-D switching cube with 16×16-switches and the port numbering (relative addresses) in shuffled column major order.

means of the presented concept and the chosen algorithm paradigm one may run the complete algorithm on a d-dimensional processing cube.

5. Evaluation of Regular Interconnections

For the evaluation of regular d-dimensional patterns a quantity is introduced which we call the maximal displacement d_{max} of a pattern (more precisely, the maximal displacement of the end of the graphs with respect to their beginning) [5]. The introduction of this quantity is reasonable: (1) It has a physical meaning (the maximal path length of a light beam if interconnecting two arrays by a regular pattern may be deduced which bounds the speed of future optical processing systems), (2) it measures the capability of a regular pattern to implement algorithms (for the 1-D case d_{max} is called the skew of a pattern) and (3) it has convex properties which provide a minimum of d^c_{max} with respect to the size and dimension of the interconnected data set, respectively.

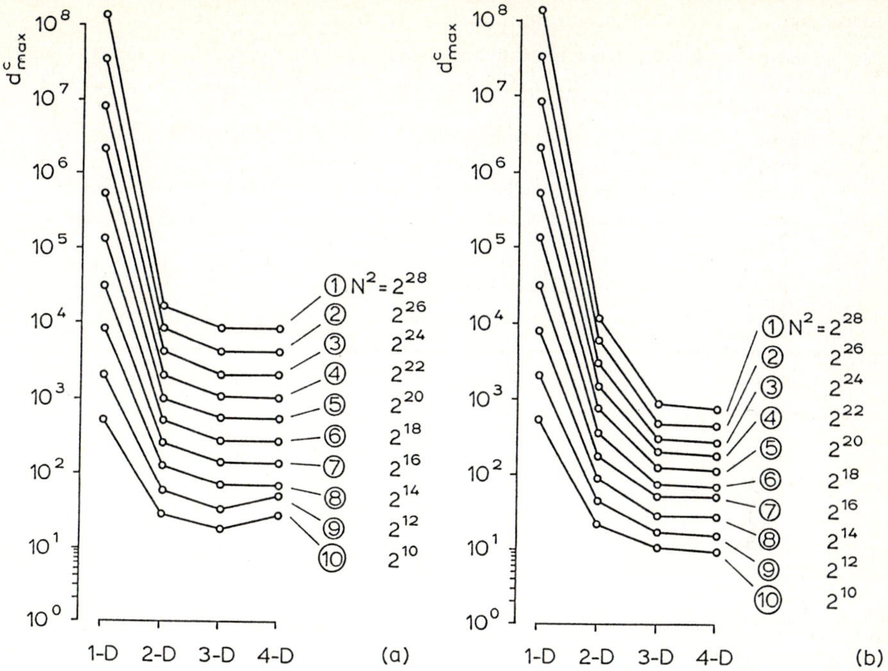

Fig. 3. d^c_{max} vs dimension of the pattern arising from multiplexing on (a) data vectors and (b) data arrays.

In particular, d^c_{max} may be expressed by the algebraic bases of the 1-D and 2-D patterns (s for space)

$$d^s_{max} = (n_1^{(1)} - 1)(n_2^{(1)} - 1) \tag{3}$$

$$= \sqrt{(n_1^{(1)} - 1)^2(n_2^{(1)} - 1)^2 + (n_1^{(2)} - 1)^2(n_2^{(2)} - 1)^2} \tag{4}$$

and the considered d-dimensional extension is (c for complexity or costs and f for frequency)

$$d^c_{max} = d^s_{max} + d^f_{max} \tag{5}$$

whereby d^f_{max} is expressed by a weighted superposition of Eq.(3) [5]. [For an application of Eqs.(3) to (5) cf. Figs. 3.]

Figs. 3 show in both cases (starting with data vectors and arrays, respectively) the dramatic decrease of d^c_{max} from 1-D to 2-D and an additional decrease if 3-D data cubes are interconnected. The least d^c_{max} indicates the best partition of N^2 into arrays and 3-D and 4-D cubes, respectively (Section 3). The d-dimensional patterns are proposed to model the interstage patterns of OFDM based switching systems whereby the local minimum of d^c_{max} with respect to the size of the data set may be assigned to the optimal number of frequency channels [5]. The extension of this multiplexing scheme provides the global minimum of d^c_{max} and is called subdivided multiplexing [5] because the d-dimensional data cubes are obtained by the subdivision of the arrays and 3-D cubes addressed above.

References

[1] J. Giglmayr, "Spatial Extension of Multistage Interconnection Networks," in *Proceedings of the OSA Topical Meeting on Photonic Switching*, March 1-3, 1989, Salt Lake City, Utah, USA, pp.170-179.

[2] H. S. Hinton, "Photonic Time-Division Switching Systems," IEEE Circuits and Devices Magazine, July, 39-43 (1989).

[3] J. Giglmayr, "On d-Dimensional Shuffle Interconnections" (submitted for publication).

[4] J. Giglmayr, "Organization of k×k-Switches Interconnected by d-Dimensional Regular Optical Patterns" (submitted for publication).

[5] J. Giglmayr, "Generation and Evaluation of d-Dimensional Regular Optical Patterns" (submitted for publication).

[6] H. R. Ramanujam, "Decomposition of Permutation Networks," IEEE Trans. Comput. **C-22**, 639-643 (1973).

[7] J. L. Bentley, "Multidimensional Divide-and-Conquer," Communications of the ACM **23**, 214-229 (1980).

Part VI

Wavelength-Based Devices

Frequency Conversion by Four-Wave-Mixing in LD Amplifiers

G. Großkopf, R. Ludwig, R. Schnabel, N. Schunk, and H.G. Weber

Heinrich-Hertz-Institut für Nachrichtentechnik Berlin GmbH,
Einsteinufer 37, D-1000 Berlin 10, Fed. Rep. of Germany

Abstract. Nearly-degenerate four-wave-mixing in traveling wave laser diode amplifiers enables all-optical frequency conversion of an optical data signal. The frequency conversion range is limited by the gain bandwidth of the amplifier, which is about 4000 GHz in the experimental and theoretical work reported here. In general, the conversion efficiency is larger than 1 due to the internal gain of the amplifier. Moreover, an optical data signal can be converted simultaneously to several ($\geq$10) different frequencies with use of one semiconductor laser amplifier only. The frequency converter operates for data signals up to 1 Gbit/s independently of modulation scheme and bit rate. It can preferably be applied to angle modulation schemes.

1. Introduction

Frequency conversion of optical data signals is a key operation of switching nodes in coherent optical multicarrier systems. The function of a node is to switch a given data signal at the carrier frequency f_x to a desired carrier frequency f_y. The most straight forward way for carrier frequency conversion is opto-electronic conversion with use of a heterodyne receiver and a transmitter [1, 2]. All-optical methods for frequency conversion are the induced frequency shift of a DFB-laser by light injection [3] and four-wave mixing in a semiconductor laser optical amplifier [4 - 6, 18].

Nearly-degenerate four-wave-mixing (NDFWM) in semiconductor laser optical amplifiers (SLA) was studied in detail in recent years experimentally and theoretically [7 - 19]. The physical mechanism behind NDFWM is carrier density modulation at the difference frequency between two incident light waves. This carrier density modulation generates two new light waves, which are frequency shifted from the two input waves by the beat frequency. If one of the two input signals is a modulated wave, i.e. an optical data signal, one of the newly generated waves represents a frequency shifted replica of this data signal. This offers the possibility of all-optical frequency conversion of an optical data signal. However, in the conventional scheme of NDFWM efficient frequency conversion is restricted by the carrier lifetime to a conversion range of a few GHz [8, 9].

Recently, a frequency conversion scheme based on NDFWM was demonstrated, which extends the conversion range up to several THz [4]. The physical mechanism behind this frequency converter is NDFWM in the SLA together with homogeneous gain saturation of the SLA. The carrier density modulation associated with NDFWM affects homogeneously the entire gain spectrum of the SLA. If a second pump wave (converter wave) at frequency $\approx f_y$ is passed through the SLA, it experiences this carrier density modulation and generates two side bands. These side bands are also frequency shifted replicas of the input data signal. Turning on and off the converter wave enables switching of the data signal from a given carrier frequency f_x to the desired carrier

Springer Series in Electronics and Photonics, Vol. 29 Photonic Switching II
Editors: K. Tada · H.S. Hinton © Springer-Verlag Berlin, Heidelberg 1990

frequency f_y. Simultaneous switching to several channels is possible by application of several converter waves.

Carrier density modulation is also possible without NDFWM if the optical data signal at the input of the SLA is intensity or amplitude modulated [20]. In order to simplify the discussion we consider here only modulation schemes of the optical data signals for which the optical power is constant as for instance frequency and phase modulation.

2. Frequency Conversion Scheme

The experimental arrangement of the considered frequency conversion scheme is shown in Fig. 1. Three light waves, the signal wave E_s, the pump wave E_p and the converter wave E_c are coupled via optical fibers and fiber couplers into a SLA. Fiber polarization controllers (PC) adjust the polarization of E_s and E_p to obtain maximum NDFWM efficiency [9]. For the SLA used in our experiment, maximum NDFWM efficiency is achieved if both waves are TE polarized input waves. The conversion efficiency is zero if E_s and E_p have perpendicular polarizations. The polarization of E_c can be arbitrary with respect to E_p or E_s. But maximum conversion efficiency is obtained if E_c is a TE polarized input wave because of the higher TE gain.

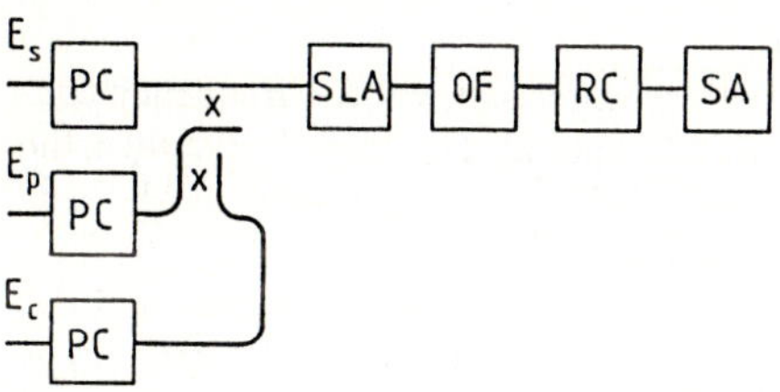

Fig. 1: Frequency converter experiment
PC: polarization controller
SLA: semiconductor laser amplifier
OF: optical filter
RC: heterodyne receiver
SA: spectrum analyzer

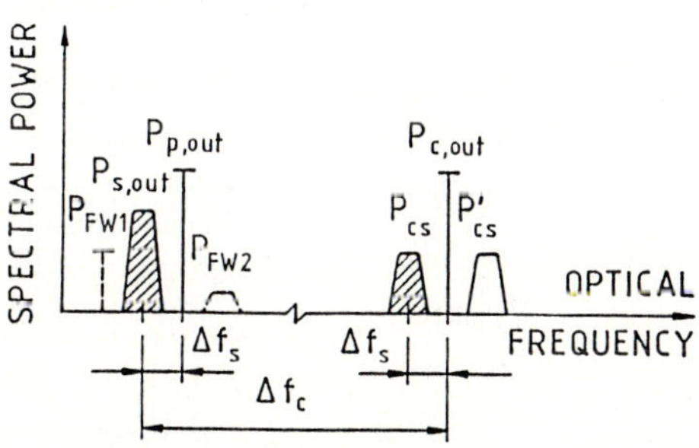

Fig. 2: Schematic of the light spectrum at the output of the optical amplifier as frequency converter

At the output of the SLA we obtain an optical spectrum as depicted in Fig. 2 schematically. To discuss this spectrum we introduce the following definitions for the signal wave E_s, the pump wave E_p and the converter wave E_c, respectively: frequencies (wavelengths) $f_s(\lambda_s)$, $f_p(\lambda_p)$, $f_c(\lambda_c)$, SLA input powers: $P_{s,in}$, $P_{p,in}$, $P_{c,in}$ and SLA output powers: $P_{s,out}$, $P_{p,out}$, $P_{c,out}$. Furthermore, we introduce the frequency spacings $\Delta f_s = | f_p - f_s |$ and $\Delta f_c = | f_s - f_c |$. The spectrum shown in Fig. 2 is valid under the following conditions: $\Delta f_s < 5$ GHz is required for efficient NDFWM interaction between E_s and E_p and $\Delta f_c > 10$ GHz is required to avoid direct NDFWM interaction of E_c with E_p or E_s. Direct NDFWM interaction between E_s and E_p generates the two sidebands E_{FW1} and E_{FW2} with the SLA output powers P_{FW1} and P_{FW2}. Only one of these two sidebands namely E_{FW2} represents a replica of input signal. The spectrum of E_{FW1} is more complicated, because it results from a product of E_s with itself. In case of phase modulation with a phase deviation of π, this spectrum degenerates to a δ-function. The sidebands E_{FW1} and E_{FW2} are also present if the converter wave E_c is off. If the converter wave is on, there are also sidebands associated with this wave at a frequency spacing Δf_s on each side of E_c. These sidebands, designated E_{cs} and E'_{cs} with the powers P_{cs} and P'_{cs} respectively, represent both replicas of the input wave E_s.

The experimental set up for the above described frequency conversion scheme requires stable and tunable light sources for E_s, E_p and E_c. In our experiments we used three tunable external cavity semiconductor lasers, each having a spectral width of less than 100 kHz and a tuning range of up to 10 THz. Modulation of the signal wave E_s, was accomplished by a $LiNbO_3$ phase modulator. The SLA was an AR-coated (SiO_x) Hitachi HLP 5400 laser exhibiting an internal unsaturated single pass gain of approximately 21 dB in the gain maximum, a gain ripple of less than 2 dB, and a 3 dB bandwidth of about 4000 GHz. The optical waves at the output of the SLA are in general measured with use of a heterodyne receiver (RC) and a spectrum analyzer (SA), as Fig. 1 shows. An optical filter (OF) at the SLA-output suppresses all light components near f_s and f_p.

3. Transmission Experiment

In a first experiment we demonstrated optical data transmission via the frequency converted signal E_{cs} with E_s being a 140 Mbit/s DPSK optical data signal and with the frequency spacings $\Delta f_s = 1$ GHz and $\Delta f_c = 1500$ GHz. The local laser at the heterodyne receiver RC (see Fig. 1) was tuned to select only the converted signal wave E_{cs} while suppressing $E_{c,out}$ and E'_{cs} electronically by a seven-stage Chebyshew filter. All optical fields around $E_{p,out}$ and $E_{s,out}$ were suppressed by the optical filter OF. A BER $<10^{-9}$ was obtained for the converted signal E_{cs}. The SLA input powers were $P_{s,in} = 1.2\ \mu$W, $P_{p,in} = 14.5\ \mu$W and $P_{c,in} = 28\ \mu$W. At the output of the SLA we measured the converted signal power $P_{cs} = 3\ \mu$W. Thus, we obtained a conversion efficiency $\eta = P_{cs}/P_{s,in} \approx 4$ dB. However, taking into account coupling losses at the SLA, splitting losses in the fiber coupler at the SLA and losses caused by the optical filter OF we obtained an overall conversion efficiency of -7 dB. We will see in section 4 that it is possible to increase η strongly, so that higher overall conversion efficiencies can be expected.

4. Conversion Efficiency

The conversion efficiency is characterized by the ratio $\eta = P_{cs}/P_{s,in}$, where P_{cs} and $P_{s,in}$ are the powers of the converted signal and of the input signal, respectively. η depends on the characteristics of the SLA and on the following parameters: the frequencies (wavelengths) $f_s(\lambda_s)$, $f_c(\lambda_c)$ of the signal wave and the converter wave, the frequency spacing Δf_s between the signal wave and the pump wave and on the powers $P_{c,in}$ and $P_{p,in}$ of the converter wave and pump wave. We obtained the expression [18]

$$\eta(\lambda_c, \lambda_s) = k \left(\frac{\partial n(\lambda_c)}{\partial N} \right)^2 \cdot G^2(N, \lambda_s) P_{p,in} G(N, \lambda_c) P_{c,in} \quad (1)$$

$$k = \frac{1}{1 + (2\pi \Delta f_s \tau_e)^2} \cdot \left(\frac{\tau_e \Gamma}{h f_s A \lambda_c} \right)^2 \quad (2)$$

with the following parameters of the SLA: the lifetime τ_e and the density N of the free carriers, the derivative $\partial n/\partial N$ of the index of refraction n with respect to the carrier density N at the wavelength λ_c, the confinement factor Γ, the active layer cross section A and the single pass gain G. The quantity k describes the low pass behaviour of the NDFWM efficiency. The carrier density modulation does not follow the beat frequency Δf_s if $2\pi\Delta f_s\tau_e \gg 1$. The gain G depends on the wavelength λ and on the density N. In eq. 1 we assumed $G(\lambda_s) = G(\lambda_p)$ because of the restriction for Δf_s. In general N is a function of the light intensity in the active region. Consequently, via N the gain is a complicated function of $P_{p,in}$, $P_{c,in}$ and $P_{s,in}$. A calculation of η requires therefore numerical methods. This calculation was performed in [19] for the conventional scheme of NDFWM with only two copropagating light waves. We used a similar method for the conversion scheme described here. For low light powers $P_{p,in}$, $P_{c,in}$, $P_{s,in}$ and low gain (i.e. for negligible saturation) G (N, λ) is only weakly dependent on the light intensity. Under these conditions the quantity $P_{cs}/(P_{s,in}G(\lambda_s)P_{p,in}G(\lambda_c)P_{c,in})$ is independent of the injected light intensities to a good approximation. In previous work [18] this quantity was designated as conversion coefficient.

Fig. 3 shows calculated contours of constant conversion efficiency η (solid lines) and measurements of η (dots) in the λ_c, λ_s-plane. The measurements were performed with an unmodulated signal wave E_s. The frequency spacing Δf_s was kept constant with $\Delta f_s = 70$ MHz. This was directly controlled via the SLA injection current because the carrier density modulation causes a corresponding modulation of the injection current at the beat frequency Δf_s. The input powers were approximately $P_{s,in} \approx$-30 dBm, $P_{p,in} \approx -27$ dBm and $P_{c,in} \approx -18$ dBm in all measurements and the unsaturated effective gain of the SLA was about G$\approx$21 dB in the gain maximum. The parameters for the numerical calculation were determined by a detailed investigation of the SLA used in the experiment. The agreement between theory and experiment is within the expected accuracy. It is remarkable that one obtains $\eta \geq 0$ dB in a large area of the λ_c, λ_s-plane and that η shows only small variations in this area. With increasing powers of the pump and converter wave the area with $\eta \geq 0$ dB increases and is shifted towards longer wavelengths λ_c and λ_s due to the higher gain saturation at shorter wavelengths.

If η is considered for fixed values λ_s, λ_p and Δf_s as function of $P_{p,in}$ and $P_{c,in}$ we obtain the following results as shown in Fig. 4. For given $P_{p,in}$ the conversion efficiency η has a maximum versus $P_{c,in}$. This maximum shifts to higher powers $P_{c,in}$ if $P_{p,in}$ is increased. Similar results are obtained if η is calculated versus $P_{p,in}$ for various values $P_{c,in}$. The maxima for η increase and finally decrease again if both $P_{c,in}$ and $P_{p,in}$ are increased. This behaviour is depicted in Fig. 5, which shows calculations of η_{max} versus

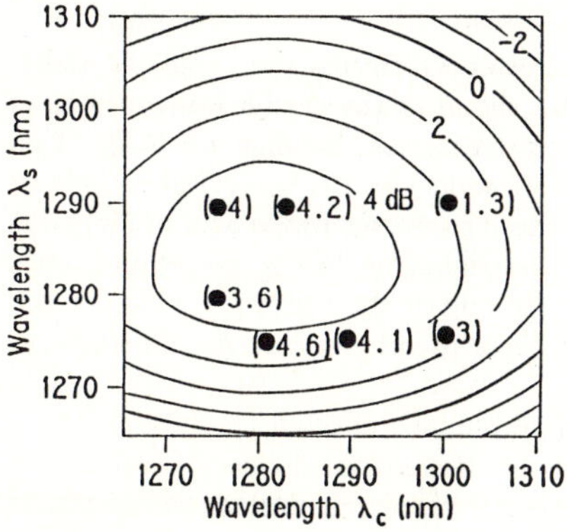

Fig. 3: Calculated contours of constant conversion efficiency η in dB (solid lines) and measurements of η (dots) in the λ_c, λ_s-plane for $P_{p,in}$ = -27 dBm, $P_{s,in}$ = -30 dBm and $P_{c,in}$ = -18 dBm

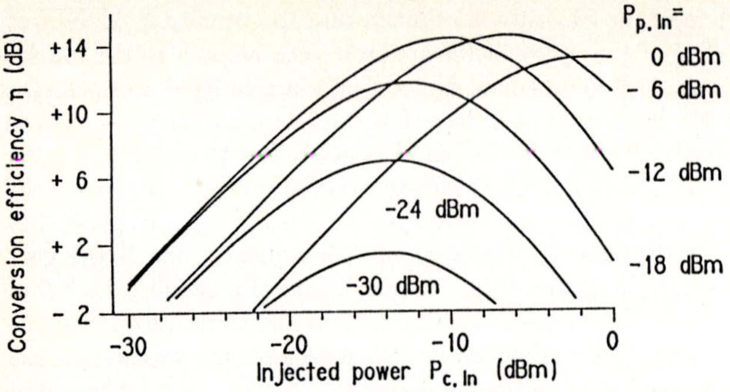

Fig. 4: Conversion efficiency η[dB] versus the injected power $P_{c,in}$ of the converter wave for different pump powers $P_{p,in}$.

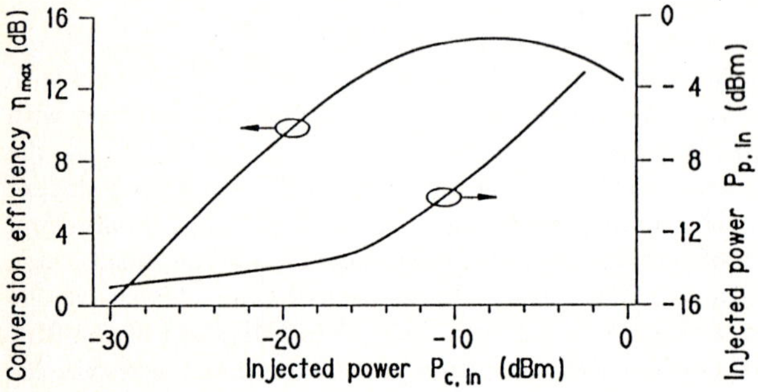

Fig. 5: Maximum conversion efficiency η_{max}[dB] versus the injected power $P_{c,in}$ of the converter wave. The optimized injected pump power $P_{p,in}$ for η_{max} is given on the right scale.

$P_{c,in}$. Here η_{max} is the maximum of η for the given $P_{c,in}$ versus $P_{p,in}$. The corresponding values $P_{p,in}$ for which these maxima are obtained are also depicted in this figure on the right scale. In the calculations for Figs. 4 and 5 the following parameters were used: λ_p = 1280 nm, λ_c = 1300 nm, Δf_s = 360 MHz, $2\pi\Delta f_s\tau_e$ = 0.5, $P_{s,in}$ = -30 dBm and τ_e = 0.22 ns.

5. Point to multipoint switching

To highlight the usefulness of the frequency converter we discuss the "point to multipoint switch" depicted in Fig. 6. An optical data signal at the carrier frequency f_s, is converted into one or several optical data signals at the carrier frequencies f_1 to f_N. This is accomplished if one or several converter waves (frequencies f_{c1}, $f_{c2}, \ldots$) are tuned to the frequencies $f_1 + \Delta f_s, \ldots, f_N + \Delta f_s$. Each of these converter waves acts in the same way as the converter wave E_c described before. The desired switching operation is only effective if the pump wave (associated with f_s) and one of the converter waves at the frequencies f_{c1} to f_{cN} are initiated. Consequently the switching operation represents an AND-gate.

It is of interest to know how the conversion efficiency is affected by the number of converter waves. Fig. 7 depicts a calculation of the conversion efficiency η versus the number N of converter waves all with the same $P_{c,in}$. In these calculations we used λ_s =

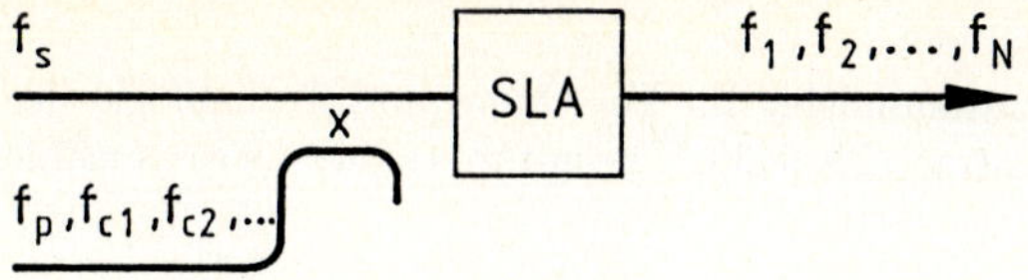

Fig. 6: Point to Multi-point Switch

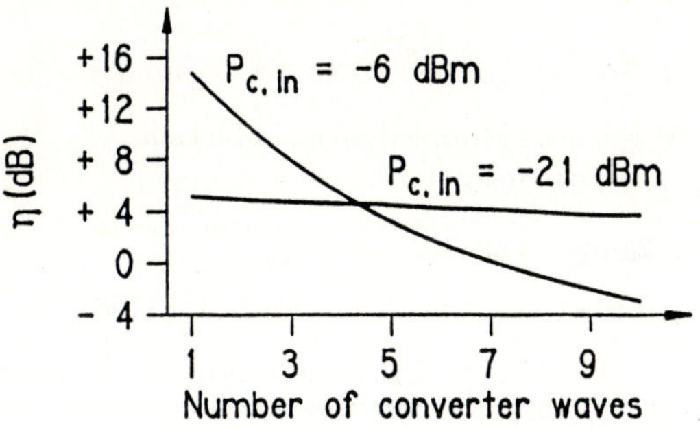

Fig. 7: Conversion efficiency η[dB] versus the number N of converter waves. All converter waves have the same $P_{c,in}$ with $P_{c,in}$ = -6 dBm or $P_{c,in}$ = −21 dBm.

1280 nm, $2\pi\Delta f_s\tau_e = 0.5$, $P_{p,in}$ = -6 dBm, $P_{s,in}$ = -30 dBm and λ_c = 1299 nm + $N\Delta\lambda_c$ with $\Delta\lambda_c$ = 1 nm and N = 1 to 10. In Fig. 7 the curve with $P_{c,in}$ = -6 dBm is optimized to give η_{max} if one converter wave is used only. In this case, increasing the number N of converter waves causes a strong decrease of η due to gain saturation. The second curve is calculated for $P_{c,in}$ = -21 dBm. Under these conditions η is slightly dependent on the number N of converter waves.

It is evident from the results in Fig. 7, that the number N of simultaneously converted signal waves can be larger than 10. An accurate estimation of this number depends on the required output powers P_{cs} of the converted data signals. This output power must exceed the required minimum signal powers at the inputs of the subsequent switching nodes.

6. Discussion

Compared with other known all-optical frequency conversion schemes the conversion efficiency of the frequency converter described here is very high [21]. The investigations showed that conversion efficiencies >4 dB are easily obtainable with today's semiconductor laser amplifiers (SLA). For saturation output powers of the SLA $P_{sat} \gg 10$ dBm an even larger conversion efficiency is expected.

The maximum data rate B of the signal to be converted is limited by the conditions $B < | f_s - f_p |$ and $| f_s - f_p | 2\pi\tau_e \leq 1$, where f_s and f_p are the carrier frequencies of the signal and pump wave, respectively. The term $| f_s - f_p | 2\pi\tau_e$ determines the efficiency of NDFWM and depends on the carrier lifetime τ_e. For smaller values of τ_e the conversion efficiency will decrease, but the flat frequency response will increase. With state of the art optical amplifiers frequency conversion can be achieved for data signals with bit rates up to 1 Gbit/s.

Disadvantages of this frequency conversion scheme are requirements of stable and narrow linewidth lasers, which provide the pump and converter wave. Furthermore,optical filters with high resolution are necessary in order to suppress all undesired frequency components at the output of the SLA and the spontaneous emission noise of the SLA.

Despite these disadvantages the frequency conversion scheme described here is a promising concept for optical switching and routing applications.

Acknowledgement

The work was supported by the Bundesminister für Forschung und Technologie der Bundesrepublik Deutschland under contract TK 0438. The authors alone are responsible for the contents of the paper.

References

[1] Ch.Caspar,E.Großmann, B.Strebel: Techn.Digest ECOC'87, Vol. I, pp. 317-320

[2] M.Fujiwara, S.Suzuki, E.Emura, M.Kondo, I.Mito, K.Kaede, M.Shikada, M.Sakaguchi: Trans. IEICE E.72 pp. 55-62 (1989)

[3] K.Inoue, N.Takato: Electron.Lett. 25 pp. 1360-1362 (1989)

[4] G.Großkopf, R.Ludwig, H.G.Weber: Electron.Lett. 24, pp. 1106-1107 (1988)

[5] E.J.Bachus, R.P.Braun, Ch.Caspar, H.-M.Foisel, K.Heimes, N.Keil, B.Strebel, J.Vathke, M.Weickhmann: Opt. Fiber Comm. Confer., Postdeadline paper PD 13-1-3, 1989

[6] K.Inoue: Electron.Lett. 25, pp. 630-632 (1989)

[7] K.Inoue, T.Mukai, T.Saitoh: Appl. Phys.Lett. 51, pp. 1051-1053 (1987)

[8] R.G.Waarts, G.Großkopf, R.Ludwig, H.G.Weber: Journ.Opt.Soc.Am. A4-13, pp. 129-130 (1987)

[9] G.Großkopf, R.Ludwig, R.G.Waarts, H.G.Weber: Electron.Lett. 24,pp.31-32 (1988)

[10] H.Nakajima, R.Frey: IEEE J.Quantum Electron. QE-22, pp.1340-1354 (1986)

[11] G.P.Agrawal: Optics Lett. 12, pp.260-261 (1987)

[12] G.P.Agrawal: Electron.Lett. 23, pp.1175-1177 (1987)

[13] G.P.Agrawal: J.Opt.Soc.Am B-5, pp.147-158 (1988)

[14] F.Favre, D.Le Guen, J.C.Simon, P.Doussière: Electron.Lett. 25, pp. 272-273 (1989)

[15] R.P.Webb, T.G.Hodgkinson: Electron.Lett. 25, pp. 491-493 (1989)

[16] T.E.Darcie, R.M.Jopson: Electron.Lett. 24, pp.638-640 (1988)

[17] R.Nietzke, P.Panknin, W.Elsäßer, E.O.Göbel: IEEE J.Quantum Electron. 25, 1399-1406 (1989)

[18] N.Schunk, G.Großkopf, R.Ludwig, R.Schnabel, H.G.Weber: submitted to IEE, Proc. J.Optoelectron.

[19]T.Mukai, T.Saitoh: to be published in IEEE J.Quantum Electron.

[20]G.Großkopf, R.Ludwig, H.G.Weber: Electron.Lett. 22, pp.900-902 (1986)

[21] L.Thylén: ECOC'89, We A13-1

Gigabit Operation of a Wavelength-Conversion Laser

K. Kondo[1], *H. Nobuhara*[1], *S. Yamakoshi*[1], *and K. Wakao*[2]

[1]Optoelectronics Technology Research Corporation,
Present address: c/o Fujitsu Laboratories, Atsugi-shi
[2]Fujitsu Laboratories, Atsugi-shi, 10-1 Morinosato Wakamiya,
Atsugi-shi, Kanagawa-ken 243-01, Japan

Abstract. The first 1-Gb/s operation of a wavelength-conversion laser is demonstrated. This performance is achieved with 250-picosecond turnoff response by minimizing input light power which operates the device.

1 Introduction

A tunable wavelength-conversion device has been strongly required for photonic switching and optical processing systems[1]. Recently, we have successfully demonstrated a novel optical-wavelength-conversion laser[2] and also demonstrated its functional operations. This device showed the wide tunability of 3.5 nm[3], large optical gain and large on–off ratio. Furthermore, this device showed such logical functionality as AND, NOT[4] and XOR[5].

This work demonstrates the first 1-Gb/s dynamic operation of the wavelength-conversion laser. To realize this performance, we clarified that the turn-off response strongly depended on input light power and it was important to minimize the input power for fast operation of the wavelength-conversion laser.

2 Wavelength-conversion laser

The schematic structure of the device is shown in Fig. 1. This device is based on a tandem-electrode bistable laser diode with a saturable absorption region in

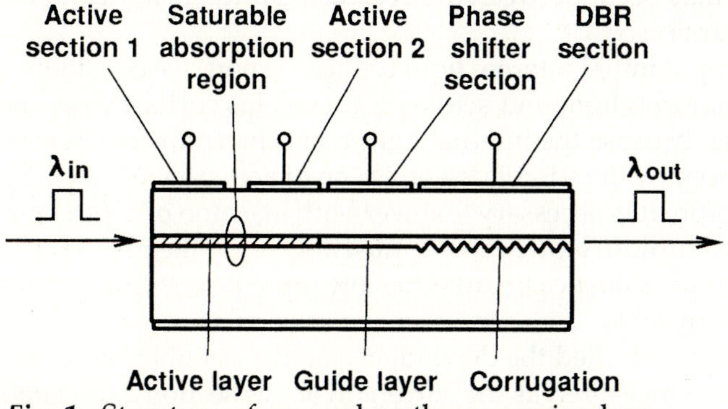

Fig. 1. Structure of a wavelength-conversion laser

Springer Series in Electronics and Photonics, Vol. 29 Photonic Switching II
Editors: K. Tada · H.S. Hinton © Springer-Verlag Berlin, Heidelberg 1990

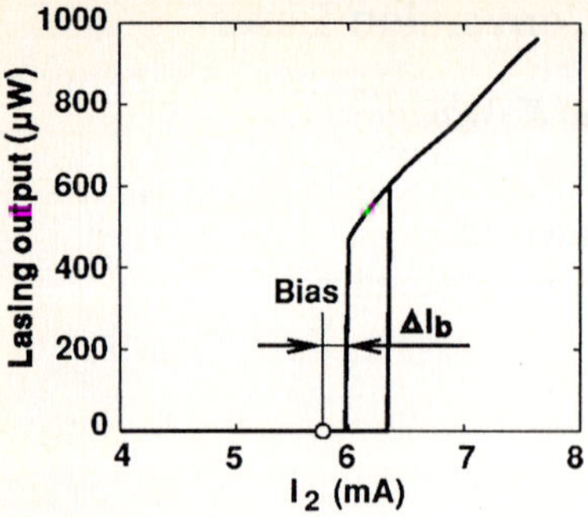

Fig. 2. Lasing characteristics of a wavelength-conversion laser measured against the current in active section 2. The current in active section 1 was fixed at 35.2 mA.

the lasing active layer. Fig. 2 shows hysteresis characteristics of the device. A bias point is set just below the turnoff threshold so that turnon and turnoff of lasing output is driven by signal-on and signal-off of light input, respectively. Light input signal with a wavelength λin induces the lasing state with an independent lasing wavelength λout. We integrated a DBR reflector with a phase shifter so that we obtained tunability of the wavelength of lasing output signal.

3 Turnoff response of wavelength-conversion laser

To achieve gigabit wavelength conversion, we needed to improve the slow turnoff response. Typical turnoff time was two or three nanoseconds, which was slower than a few-hundred-picosecond turnon time.

To improve turnoff response, we studied turnoff mechanism in a wavelength conversion laser. Turnoff starts when injection light stops. At this time, the absorber is saturated and has very low absorption in the large internal power of the lasing light. In this situation, turnoff proceeds very slowly due to the absorber's low absorption rate of internal light power. Therefore, to obtain faster turnoff, we must heighten the absorber's absorption, or lower the power of initial internal light. Furthermore this will make much faster response because feedback acceleration may occur between the decreasing internal light and increased absorber absorption.

To lower the power of initial internal light for fast turnoff, we considered that injection power of input light and setting point of bias current were the most important factors, because the internal light is produced by injection of input light, and enhanced by the stimulated emission of active gain caused by the bias current. Therefore it is necessary to lower both injection of input light and bias current for fast turnoff. However, we must also study the turnon condition as well as turnoff dependence because too low input power and current bias will make turnon unstable.

In our experiment, we studied the dependence on input light power and bias setting. Bias ΔIb was measured as the current in active section 2 deviated from the turnoff threshold shown in Fig. 2.

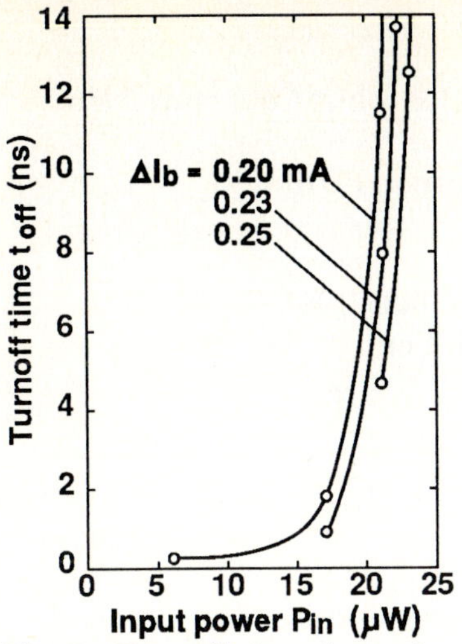

Fig. 3. Turnoff-time dependence on input power

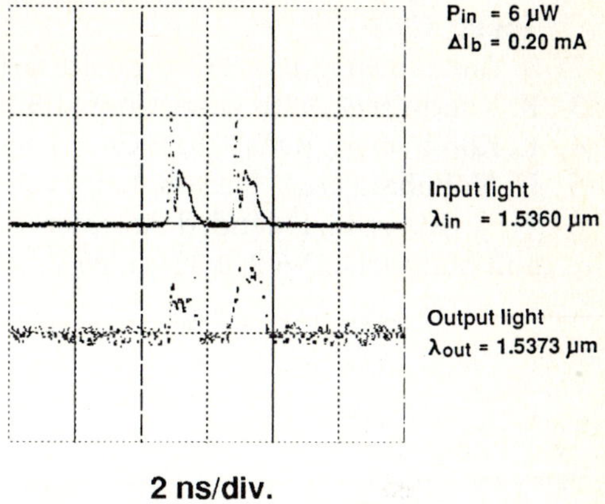

Fig.4. 1-Gb/s operation of wavelength conversion

Fig. 3 shows experimental result that turnoff time of output light response depends on input power of light signal. Under a fixed bias, the smaller input power causes the faster turnoff. However, turnon is unstable with a certain small power of input light in the experiment, and this is shown as the limiting point of the plots. The larger input power drastically increases turnoff time and we consider that this is the effect of large saturation of the absorber absorption. Under a fixed input power, the lager bias deviation causes the faster turnoff, but also the larger limitation on input power.

The plot at 0.20-mA ΔIb has a few-hundred-picosecond turnoff under 10-μW input power. In this experiment, we obtained a minimum 250-ps turnoff at 6 μW of input power and 0.20 mA of ΔIb.

Fig. 4 demonstrates 1-Gb/s dynamic operation of a wavelength-conversion laser. We attained this performance under the condition of the fastest 250-ps turnoff. Light output response with 1 ns width is successfully achieved.

4 Conclusion

In conclusion, 1-Gb/s operation of a wavelength-conversion laser is demonstrated for the first time. This result was realized by minimizing input power which drastically reduced the turnoff response time. We clarified that this dependence was derived from the nonlinear absorption which varied with the internal light power in our optical fuctional device. We also consider that this phenomenon universally applies to other optical nonlinear systems[6].

References

1 M. Sakaguchi *et al.*, Tech. Dig. of 1st Optoelectronics Conf., July 1986, Tokyo, C9-3
2 S. Yamakoshi *et al.*, OFC'88 postdeadline paper, PD10
3 K. Kondo *et al.*, 11th Laser Conf., J-3, 1988
4 K. Kondo *et al.*, IOOC'89, 19C4-1,1989
5 H. Nobuhara *et al.*, Electon. Lett., vol. 25, p. 1485, 1989
6 T. Sanada *et al.*, Extended Abstracts of 21th Conf. on Solid State Devices and Materials, D-6-2, p.321, 1989

Spectral Bistability in Multielectrode DFB Lasers

J. Jacquet, M. Sotom, O. Audouin, D. Leclerc, J.-M. Gabriagues, and J. Benoit

Laboratoires de Marcoussis, C.G.E. Research Center,
F-91460 Marcoussis, France

Abstract : We report the bistable behaviour of tunable multi-electrode InGaAsP DFB lasers under the conditions of electrically- and optically-controlled longitudinal mode hopping.

1.Introduction

Tunable multi-electrode InGaAsP DFB lasers are attractive devices for all-optical fiber networks and switching systems based on Frequency-Division-Multiplexing (FDM) or Wavelength-Division-Multiplexing (WDM). Although their intrinsic operation characteristics are not yet fully understood, useful applications of these devices have already been found and demonstrated, such as tunable coherent FSK transmitters for FDM networks [1] and optically-triggerable tunable wavelength converters for WDM switching networks [2]. Also, optical tristability including spectral bistability using an inhomogeneously excited multielectrode DFB laser has recently been reported, and application of the device as an optical logic gate experimentally demonstrated [3]. In this paper we report new results on electrically- and optically-controlled spectral bistability using a three-electrode DFB laser and propose some preliminary analysis of the phenomena.

2. Description and Characteristics of Multielectrode DFB lasers

The three-electrode DFB lasers [1] we have used for these experiments basically consist of a uniform first-order DFB heterostructure, with 300 μm total length, driven by a non-uniform current distribution, by means of three electrically-isolated electrodes, with identical 100 μm length that determine three sections inside the laser cavity; the two lateral electrodes are connected so that the component is driven by two independent current sources. The facets are not anti-reflection coated. Typical threshold currents, when the three electrodes are strapped, are of the order of 40 mA. This laser diode is continuously tunable over 2 nm around 1.52 μm.

The spectral properties of the light emission of an uniform DFB laser above threshold is determined by both the reflectivity of the chip facets and the relative longitudinal position of the grating with respect to the two facets. Thus, depending on these parameters, the uniform DFB oscillates either in a single-longitudinal-mode regime or in a two-degenerate-longitudinal-mode regime [4]; by using a three-electrode structure and two separate drive currents, the spectral properties of the laser emission can be controlled by inhomogeneous pumping of the component, i.e. by adjusting the longitudinal gain distribution .

Springer Series in Electronics and Photonics, Vol. 29 **Photonic Switching II**
Editors: K. Tada · H.S. Hinton © Springer-Verlag Berlin, Heidelberg 1990

In the case where one section is electrically pumped below threshold, the launching of external radiation into the laser cavity can switch on the component : this effect is interpreted as saturable absorption, and under such conditions the component can be operated either as a wavelength converter [2] or as an optical logic gate [3].

In the case where the set of selected drive currents is imposing that two longitudinal modes have nearly equal threshold gains, but different longitudinal field profiles, mode-hopping can be induced by localised electrical or optical pumping.

We have experimentally studied these effects under the conditions of electrical and optical control.

3. Experimental Results

In order to demonstrate the saturable absorber effect, the set-up represented in figure 1 has been used: the multielectrode DFB sample is optically controlled by an other standard-type DFB laser diode. Figure 2 shows the operation of switching on/off the multielectrode DFB under optical control, with a clear bistable behaviour: a few µW launched power only is required for the switch-on operation under various values of the drive currents (I1 & I2); on the other hand, switching off the component simply by removing the control radiation requires extremely critical adjustment of I1 & I2. This phenomenon does not depend on the polarisation state and wavelength of the control radiation.

The effect of modification of the longitudinal gain distribution is illustrated by the second set of experiments. Figure 3 shows optically-induced mode-hopping between two modes with high (>30 dB) side-mode suppression ratio. The competition between the two longitudinal modes, separated by 2 nm, is modified in a reproducible way by applying an external optical signal. Once again, a clear bistable behaviour is be observed , there is no critical dependence on the wavelength

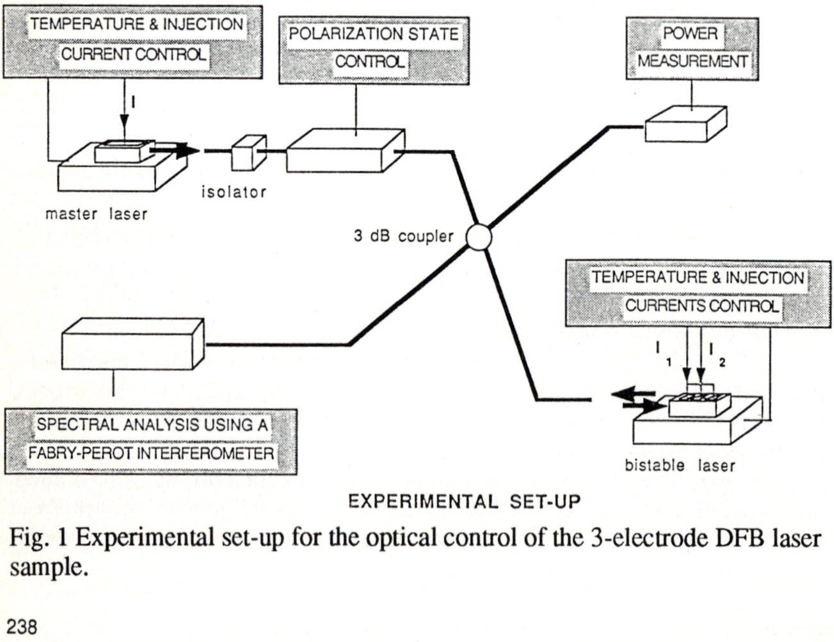

Fig. 1 Experimental set-up for the optical control of the 3-electrode DFB laser sample.

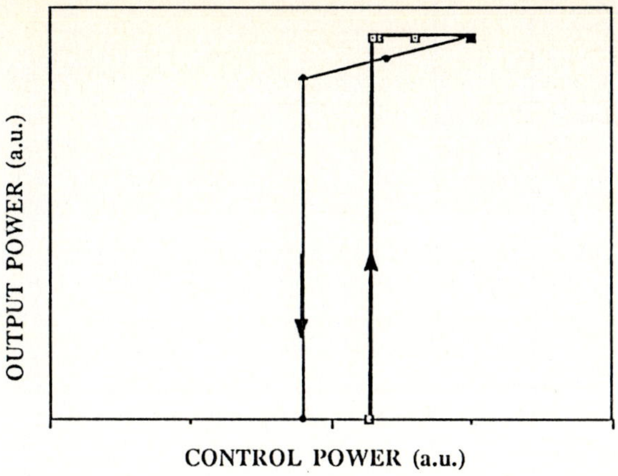

Fig. 2: Optical control of a 3-electrode DFB laser : switch on/switch off sequence

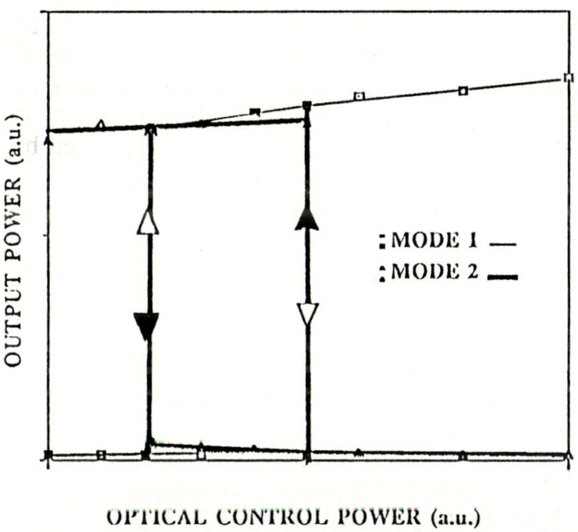

Fig 3: Optically-controlled spectral bistability using a 3-electrode DFB laser.

and polarisation state of the control radiation, and only a few µW launched control power is required for the "wavelength switching" operation. Associated with a simple filter, the device would act as an optical inverter.

This effect has been reproduced under purely electrical control. Figure 4 shows the variation of the lasing wavelength as a function of the center-electrode drive current. The dynamic properties have been investigated by sinusoidally modulating the centre-electrode current (I2) with an average value equal to 42 mA, corresponding to the center of the hysteresis shown in fig. 4. The typical response curve is represented in figure 6, showing a good "wavelength switching" capability up to 500 MHz. The same set of results has been obtained on interchanging the two drive currents (I1 & I2).

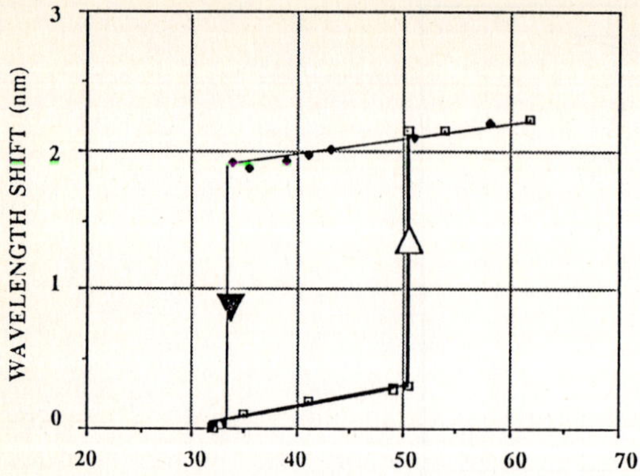

Fig. 4: Electrically-controlled spectral bistability using a 3-electrode DFB laser

4. Conclusion

Spectral bistability has been observed using an electrically- or optically-controlled three-electrode DFB laser. The most attractive features are the low optical control power required (a few μW) and the insensitivity to both the wavelength and polarisation state of the control radiation (1.3 μm and 1.5 μm range control sources have been successfully used). We interpret this phenomenon as mode-hopping between modes with different longitudinal field profiles, induced by a localized modification of the longitudinal gain distribution.

5. References

1. D. Leclerc, J. Jacquet, D. Sigogne, C. Labourie, Y. Louis, C. Artigue, J. Benoit, Electron. Lett. **25**, 45 (1989).
2. H. Kawaguchi, K. Oe, H. Yasaka, K. Magari, M. Fukuda, Y. Itaya, Electron. Lett., **23**, 1088 (1987).
3. M. Jinno, M. Koga, T. Matsumoto, Electron. Lett., **24**, 1030 (1988).
4. W. Streifer, R. Burnham, D. Scifres, IEEE J. Quantum Electron. **QE-11**, 154 (1975).

Tunable Fiber Ring Lasers with an Electrically Accessible Acousto-optic Filter

S. Oshiba, T. Kamijoh, and Y. Kawai

Semiconductor Technology Laboratory, Oki Electric Industry Co., Ltd., 550-5 Higashi-asakawa, Hachioji, Tokyo 193, Japan

Abstract. This is the first report on tunable fiber ring lasers whose wavelengths were widely controlled from 1285 to 1367 nm by changing the frequencies of the acoustic shear waves in acousto-optic (AO) filters. In this tuning range, a very narrow spectral linewidth of 15 kHz was observed.

1. Introduction

For large capacity subscriber networks and multichannel distribution systems based on the wavelength division multiplexing (WDM) technologies, tunable fiber ring lasers would be one of the most suitable light sources and they could be used as optical switching devices because fiber ring lasers can offer extremely narrow spectral linewidth and very wide tuning characteristics. We have been developing fiber ring lasers with wide tuning ranges of about 100 nm [1], [2] by using high-performance travelling wave amplifiers (TWAs) as a gain medium, whose basic structure is the same as the GaInAsP Fabry-Perot lasers, i.e. V-grooved inner stripe lasers on a p-type substrate (VIPS lasers) [3]. In these demonstrations, however, the tuning speed was limited by the mechanical movement of the gratings or the bandpass filter which were used as tuning elements. In terms of the tuning speed and the fine wavelength control, one alternative is using an acousto-optic (AO) filter [4] as a tuning element which can be controlled by electrical signals, promising rapid tuning. Moreover, precise control by the feedback of electrical signals is expected.There has, however, been no report on fiber ring lasers with AO filters. In this paper, we report the electrical tuning characteristics of the fiber ring lasers with AO filters.

2. Structure of the Fiber Ring Laser

Figure 1 shows a schematic diagram of the ring laser whose major components are a TWA, 1-m-long polarization maintaining fiber (PMF), an isolator, a directional coupler and an AO filter. The AO filter made of TeO_2 was inserted in the cavity of the ring laser with a low insertion loss of 5 dB which is comparable to those of the ring lasers with bandpass filters or diffraction gratings. The bandwidth of the AO filter was less than 5 nm and the peak transmission for the center wavelength was approximately 80 %. The generated beam from the TWA was guided into the AO filter, where the beam was filtered by the Bragg diffraction caused by an acoustic shear wave in the filter, through the PMF and a micro lens. The incident beam was

Springer Series in Electronics and Photonics, Vol. 29 **Photonic Switching II**
Editors: K. Tada · H.S. Hinton © Springer-Verlag Berlin, Heidelberg 1990

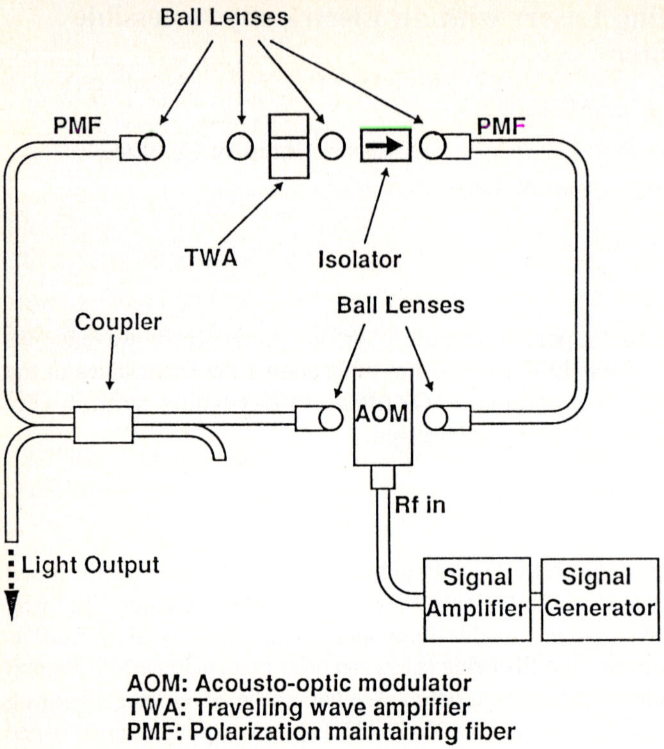

Fig.1 Configuration of a fiber ring laser using an AO filter as a wavelength tuning element.

controlled to polarize extraordinarily because the diffracted direction depends on the polarization of the beam. In order to control the beam from the AO filter, the drive frequencies of the filter were changed from 42 to 46 MHz. The TWA had anti-reflection (AR) coatings (< 0.01 %) on both facets, and was operated at a relatively high current injection level of 120 mA without lasing operation caused by Fabry-Perot resonance. The ripple of Fabry-Perot oscillation was less than 1 dB. No significant gain saturation and a high peak gain of 30 dB were achieved.

For very low reflectivities of the facets of TWA, the lasing condition of the ring laser is determined when the gain $G(\lambda)$ compensates for the total cavity loss $T_{ex}(\lambda)$ of the external ring cavity, as follows,

$$T_{ex}(\lambda)G(\lambda)=1.$$

The total loss is estimated to be approximately 16 dB; the coupling loss between the TWA and the fiber was 3 dB, the directional coupler's loss was 3 dB and the insertion loss of the isolator was 1 dB. The tuning range of the ring laser is estimated as wide as 80 nm from the gain-width of the TWA for 16 dB signal gain.

3. Tuning Characteristics

The lasing spectra of the ring laser tuned by the AO filter are shown in Fig.2, measured at three different acoustic frequencies. The operation wavelengths were controlled from 1285 to 1367 nm by modulating the AO filter: the maximum tuning range was about 80 nm, which is comparable to those of the ring lasers with bandpass filters or gratings. This wide tuning range is ascribed to the low optical cavity loss of 16 dB and the high gain of the 500-μm-long TWA. A very narrow linewidth of 15 kHz was obtained in this wide tuning range, although, as already pointed out, each main mode with a 130 MHz spacing contains external cavity modes.

Figure 3 shows the tuning wavelengths as a function of the acoustic drive frequency of the AO filter. The scanning time from one wavelength to other wavelength was 0.5 msec. When the scanning time exceeds the time of the transit of the acoustic wave crossing the incident beam, broadening is expected because of the large fluctuation in the frequency of the acoustic wave. No spectral broadening,

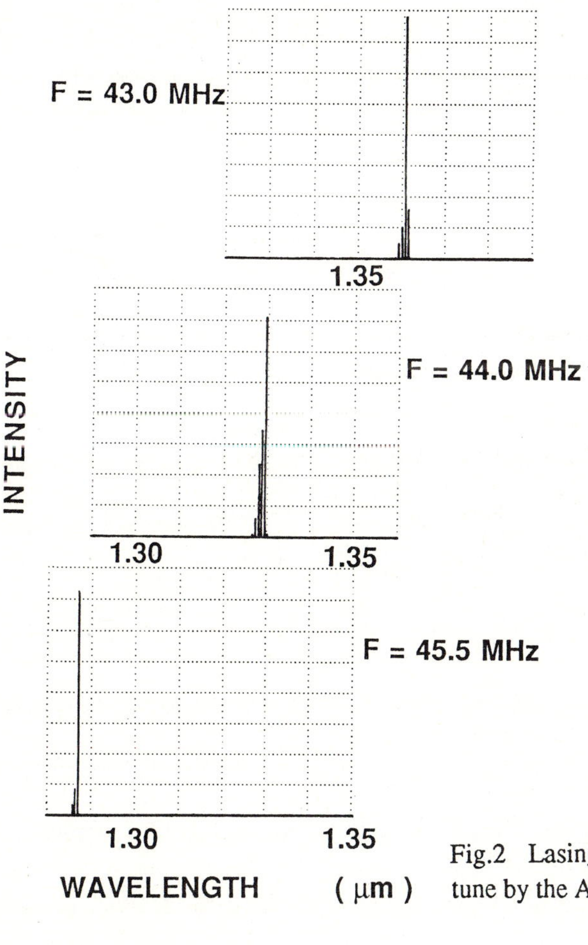

Fig.2 Lasing spectra of a fiber ring laser tune by the AO filter.

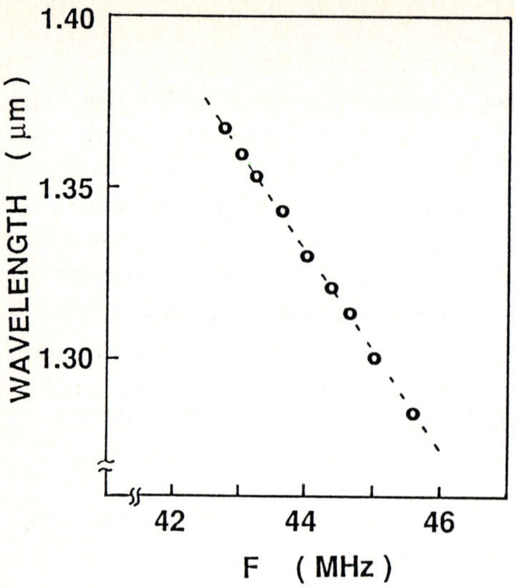

Fig.3 Tuning wavelength as a function of the acoustic drive frequency of the AO filter.

however, was observed up to this scanning time. The transit time depends on the length of the AO filters and the diameter of the beam, which indicates that the scanning speed would be improved by optimizing the structure of the AO filters.

4. Summary

We have demonstrated tunable fiber ring lasers with an AO filter which was successfully inserted in the cavity with a low loss of 5 dB. A wide tuning range of 80 nm (1285 to 1367 nm) was obtained by controlling the acoustic frequencies of the AO filters with the scanning time of 0.5 msec. The minimum spectral linewidth was as narrow as 15 kHz in this wide tuning range.

Acknowledgement

The authors would like to express their thanks to H. Horikawa, K. Nagai, and M. Kawahara for essential discussions.

References

[1] S. Oshiba, K. Nagai, M. Kawahara, and Y. Kawai, in Technical Digest of IOOC"89, Kobe, July, 1989, paper 19A2-5.
[2] S. Oshiba, K. Nagai, M. Kawahara, A. Watanabe, and Y. Kawai, Appl. Phys. Let., vol. 55, pp. 2383-2385, (1989).
[3] S. Oshiba, A. Matoba, M. Kawahara, and Y. Kawai, IEEE J. Quantum Electron., QE-23, 6, pp. 738-743, (1987).
[4] T. Yano and A. Watanabe, Appl. Opt., vol.15, pp.2250-2258, (1976).

Integrated Optical Switches for Wavelength-Division-Multiplex Communication Using Collinear Acousto-optic Interactions

Y. Kanayama, N. Goto, and Y. Miyazaki

Toyohashi University of Technology, Department of Information and Computer Sciences, Tempaku-cho, Toyohashi 441, Japan

Abstract. Integrated optical switches using collinear acousto-optic (A-O) interactions enable the switching of wavelength-division-multiplex (WDM) signals. To increase the WDM density, sidelobe suppression in filtering characteristics is investigated. A first method is to connect two switches, a second method is to modify an acoustic waveguide. The sidelobe can be suppressed under -18 dB by either method. Also, fundamental wavelength-selective switching is demonstrated.

1. Introduction

Wavelength-division-multiplexing (WDM) has been developed to increase transmission capacity and system flexibility [1,2]. Wavelength-selective switching devices [3] or wavelength filters [4] are required in WDM communication systems. From the view point of efficient use of optical power, the wavelength-selective switches are more suitable than filters in WDM switch systems. The authors have studied integrated optical wavelength-selective switches using collinear acousto-optic (A-O) interaction [5,6] and have proposed a switch matrix for WDM systems [7]. In the device, WDM optical signals are wavelength-selectively switched by frequency multiplexed surface acoustic waves (SAWs). In this paper, an improvement of the wavelength selectivity is discussed. A preliminary demonstration of wavelength-selective switching is also presented.

2. Wavelength-Selectivity of A-O Switch

A switch element is illustrated in Fig.1. Two optical channel waveguides whose propagation constants are slightly different are formed by Ti-indiffusion and Nb_2O_5 ridge on a Y-cut X-propagation $LiNbO_3$ substrate. A Ta_2O_5 SAW waveguide is deposited on the optical waveguides. An optical wave is switched between two waveguides due to TE_0-TE_1 mode conversion by collinear A-O interaction when SAW is applied under the Bragg condition, $K=k_1-k_0$, where K is a SAW propagation constant, k_1 and k_0 are optical propagation constants of TE_1 and TE_0 modes in the interaction region, respectively [5]. Since Y-branches act as mode splitters, incident and converted modes can be separated. A required SAW frequency for switching depends on the optical wavelength, and the A-O switching shows wavelength selectivity. Optical waves at multiple wavelengths are independently switched by frequency multiplexed SAWs. Fig.2 shows wavelength selective switching for WDM optical waves at λ_{i-1}, λ_i and λ_{i+1}. These optical waves are independently switched by SAWs at f_{i-1}, f_i and f_{i+1}, respectively. Since the minima of sidelobes do not appear periodically, the sidelobes cause crosstalk when the multiplex

 © Springer-Verlag Berlin, Heidelberg 1990

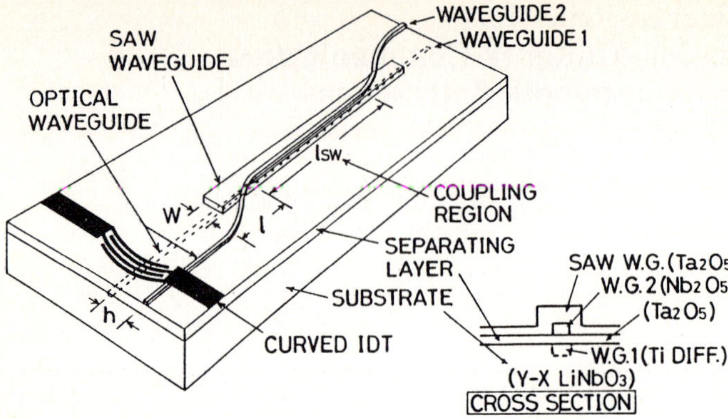

Fig.1 Collinear acousto-optic switch

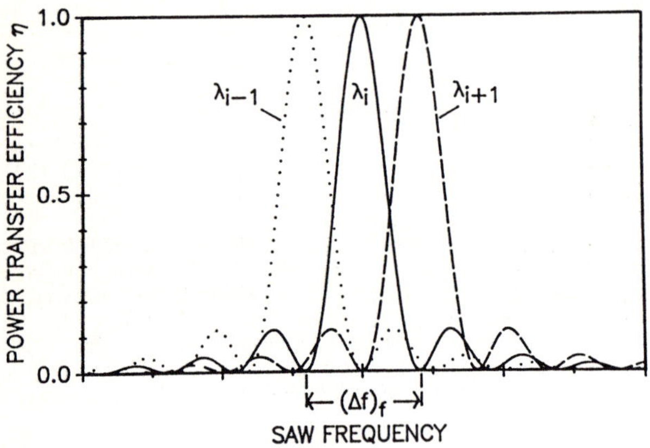

Fig.2 Frequency dependence of WDM switching

number increases. When the WDM number is 201, a crosstalk of -20 dB is resulted. The crosstalk can be suppressed by decreasing the switching ratio. However, the crosstalk must be suppressed by reducing the sidelobes to avoid optical insertion losses due to the decrease of the switching ratio.

Two methods to suppress the sidelobes are proposed. First, a cascade connection of two switch elements as illustrated in Fig.3(a) reduces the sidelobe from -9.3 dB to -18.6 dB as shown in Fig.4. An additional advantage of these cascaded switches is that a frequency shift due to the A-O interaction is compensated. Therefore, there is no difference in optical frequency between the input and output signals. Second, the sidelobes are also suppressed with a single switch element by modifying the A-O interaction strength along the interaction region. The A-O interaction strength is increased by increasing the SAW intensity. An arbitrary A-O interaction distribution along the interaction region is realized by introducing a modified SAW waveguide as shown in Fig.3(b). The SAW intensity can be increased by 10 percent when the SAW waveguide width is decreased from 100 μm to 50 μm while increasing the thickness to keep the SAW velocity constant. When the SAW waveguide is designed with a

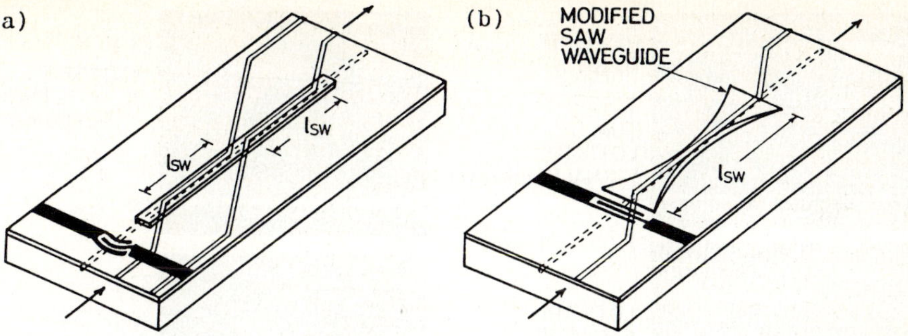

Fig.3 Suppression of sidelobes (a) by cascading two elements and (b)by modifying the SAW waveguide

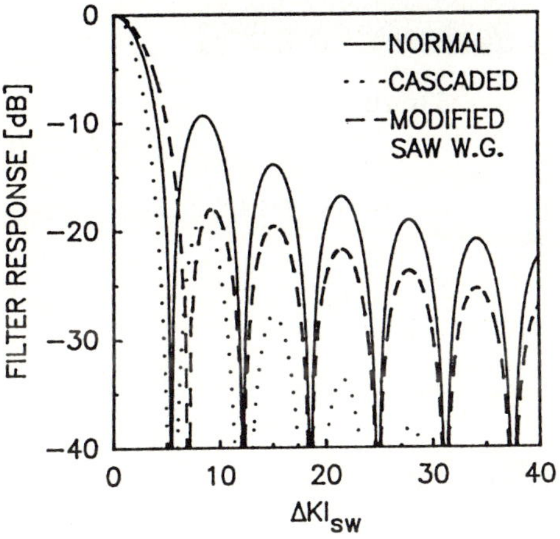

Fig.4 Filtering characteristics, where ΔK is a shift from Bragg condition. Dotted and dashed curves denote the filtering responses in two improved devices shown in Fig.3(a) and (b), respectively

width from 25 to 650 μm so that the SAW intensity varies as $I_a=I_0[1-0.4\cos(2\pi x/l_{sw})]$ in the propagation direction x, the first sidelobe reduces to -18.0 dB as shown in Fig.4.

3. Experiment of Wavelength-Selective Switching

A preliminary demonstration of a wavelength-selective switching is made with a slab-type waveguide device whose interaction length is 4 mm. Optical signals at two wavelengths, λ_1=755 nm and λ_2=785 nm, were independently switched by SAWs at f_1=72.2 MHz and f_2=81.9 MHz, respectively as shown in Fig.5. The full width of the filtering main lobe was about 5.04 nm. Theoretically, a wavelength interval of WDM can be as short as a half this full width.

4. Conclusion

We have investigated crosstalk suppression for wavelength-selective switching with collinear A-O switches. The switch elements are integrated laterally with an interval less than 500 μm and can be used to form a switch matrix for WDM systems.

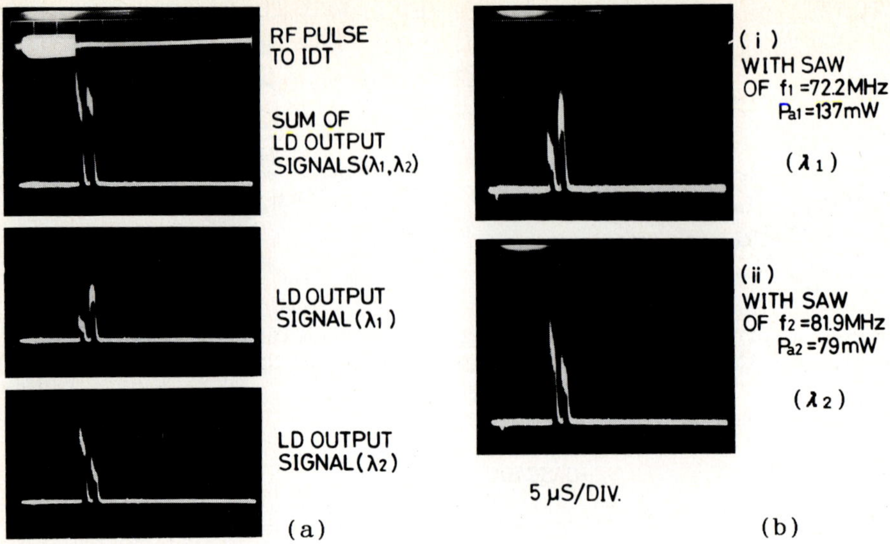

Fig.5 Demonstration of wavelength-selective switching; (a) input optical pulses and (b) switched outputs

References

[1]T. Yasui and H. Goto, IEEE Commun. Mag. **25**, 5, p.10 (1987).
[2]J.M. Senior and S.D. Cusworth, IEE Proc., **136 Pt.J**, 3, p.183 (1989).
[3]Y. Fujii, Opt. Lett., **14**, 18, p.1032 (1989).
[4]K.W. Cheung, D.A. Smith, J.E. Baran and B.L. Heffner, Electron. Lett., **25**, 6, p.375 (1989).
[5]N. Goto, Y. Miyazaki and Y. Akao, Trans. IECE Japan, **E66**, 1, p.21 (1983).
[6]N. Goto and Y. Miyazaki, Trans. IEICE Japan, E73, 1, p.116 (1990).
[7]N. Goto and Y. Miyazaki, Trans. IEICE Japan, **J71-C**, 5, p.678 (1988).

Part VII

Wavelength-Based Switching

Optical FDM Technologies for Future Lightwave Networks

K. Nosu and H. Toba

NTT Transmission Systems Laboratories, 808A,
1-2356 Take, Yokosuka-shi, Kanagawa-ken, 238-03, Japan

ABSTRACT. This paper sketches out perspectives on the potential role of optical frequency division multiplexing technologies in future telecommunication networks as well as their performance for developing related key optical devices.

1. Introduction

From the viewpoint of classical physics, light is an electromagnetic wave having an extremely high frequency bandwidth on the hundred THz order. Therefore, this inherent broad bandwidth potentially provides a large number of optical circuits and paths and is expected to realize not only simple point-to-point transmission, but also more sophisticated optical processing such as cross-connection, add/drop, cross-bar switching, etc,. This paper presents perspectives on the role of optical FDM Technologies in future telecommunication networks.

2. Applications of Optical FDM Technologies

The major functions of optical FDM in telecommunication networks are summarized in Table 1. Multi-channel large capacity long haul transmission is one of the most promising applications for the video communications era once high quality video terminals become widely used, just like telephones are today. The other functions, cross-connection, add/drop, and cross-bar switching, will realize more functional lightwave based telecommunications networks from subscriber loops, to intra-city networks to long haul trunks.

3. Key Technologies for Optical FDM Communication Systems and Related Device Technology

Optical channel number, optical frequency accuracy and lightwave tunability are major factors in designing optical FDM systems. The hatched area shown in Figure 1 indicates the present state of the art in laboratories and the open cube shows the future target. Optical device technology determines the performances shown in Fig.1. The technological trend for lightwave sources is shown in Figure 2. To realize tunable ranges more than 1 THz, optical frequency conversion using nonlinear optical effect seem to be indispensable. Due to the recent development of optical waveguide technology, periodic filters, ring resonators and other filters, whose structures are well known in microwave and millimeter wave fields, are promising candidates. Since open structure optical waveguides are weak against waveguide bending loss increases, higher refractive index difference waveguides are very important to develop more sophisticated waveguide filters, especially ones using ring resonators. Major waveguide requirements are illustrated in Figure 3. In addition to the devices described above,

Springer Series in Electronics and Photonics, Vol. 29 **Photonic Switching II**
Editors: K. Tada · H.S. Hinton © Springer-Verlag Berlin, Heidelberg 1990

Table 1. MAJOR FUNCTION OF OPTICAL FDM

	System Configuration	Items assigned to optical freq.	Applications	Absolute frequency accuracy	Key devices		
					filter	tunable LD	freq. conv.
Simple Multiplexing	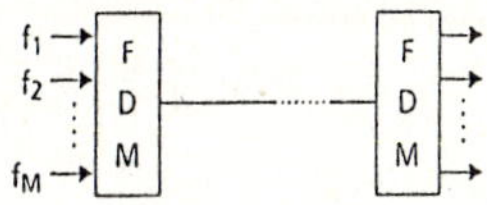	path or circuit	· trunk lines	moderate	○		
Broadcasting & Channel selection	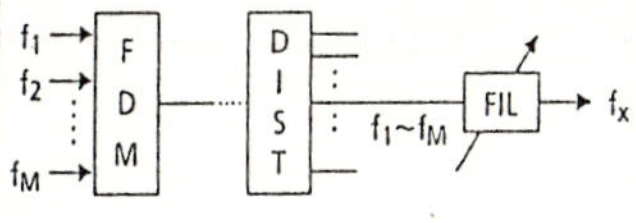	distributed circuit (program)	· CATV · LAN	moderate	○	○	
Add / Drop MUX (tapping)	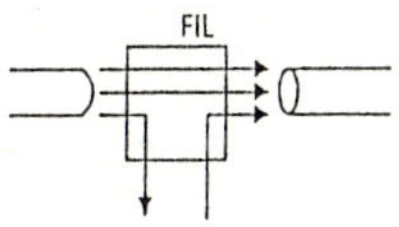	path	· Trunk network · LAN · inter-processor connection	severe	○	○	
Cross-Connect & Matrix Switch	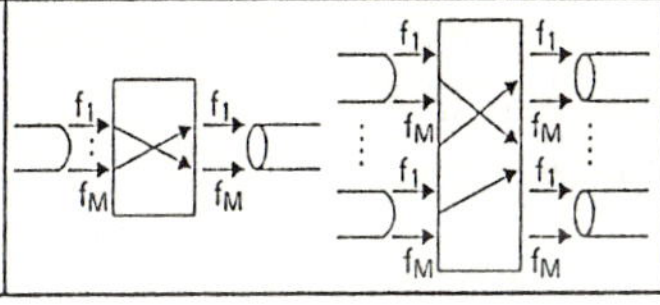	path circuit	cross-connect switch	severe	○	○	○

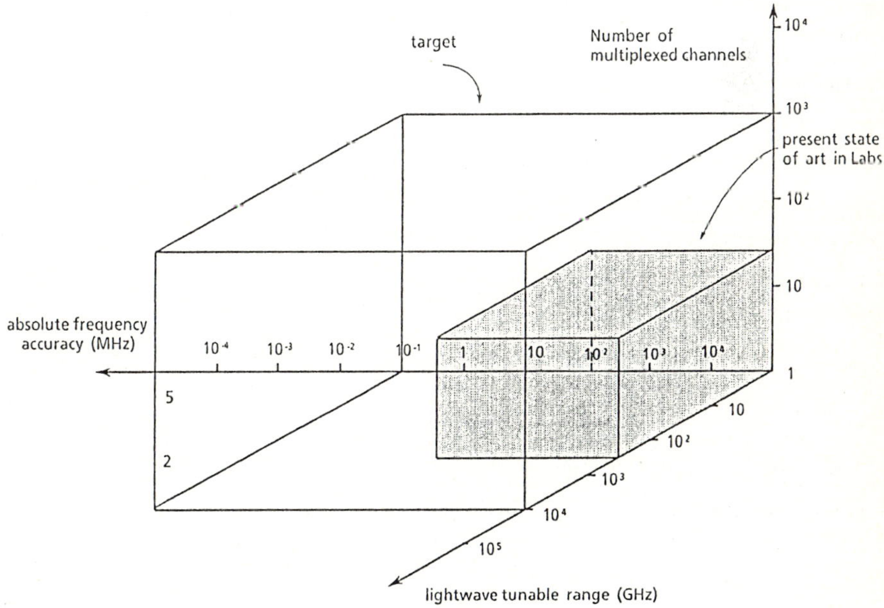

Fig. 1 OPTICAL FDM PERFORMANCE PARAMETERS

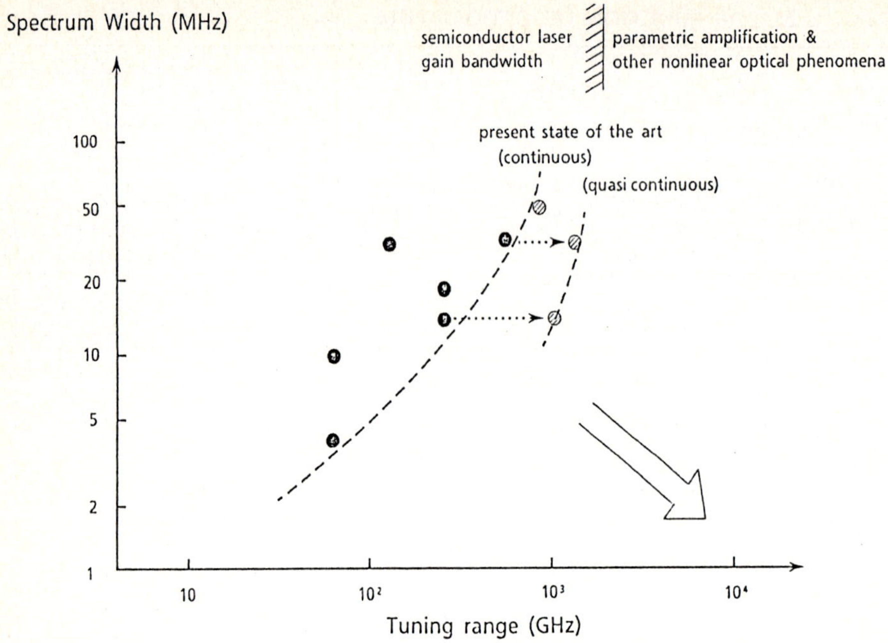

Fig. 2 TUNABLE LIGHT SOURCE PERFORMANCES

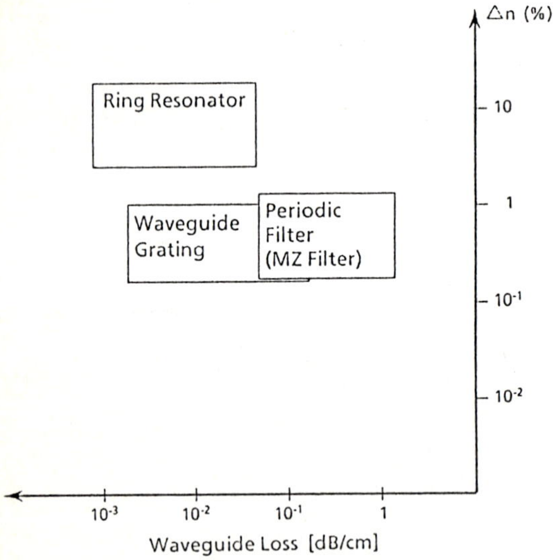

Fig. 3 REQUIREMENTS FOR OPTICAL WAVEGUIDES

common optical amplifiers, which can amplify a number of optical FDM signals simultaneously, are significantly improving the applicability of optical FDM technologies, from point-to-point transmission to multi-node networks.

4. Perspectives of Future Systems

A long term development strategy is shown in Fig. 4. By combining optical FDM channel section switches, drop/insertion switches, crossbar switches, and cross-connect switches among others, flexible and expandable networks can be realized. The potential applications for optical FDM in the near future will be found in STM based networks such as video distributions [4]. If more than one thousand optical FDM channels are available, STM based optical FDM or optical FDM + TDM, SDM networks can realize novel network architectures as alternatives and/or modifications of broadband ATM networks as we move towards the visual communication age.

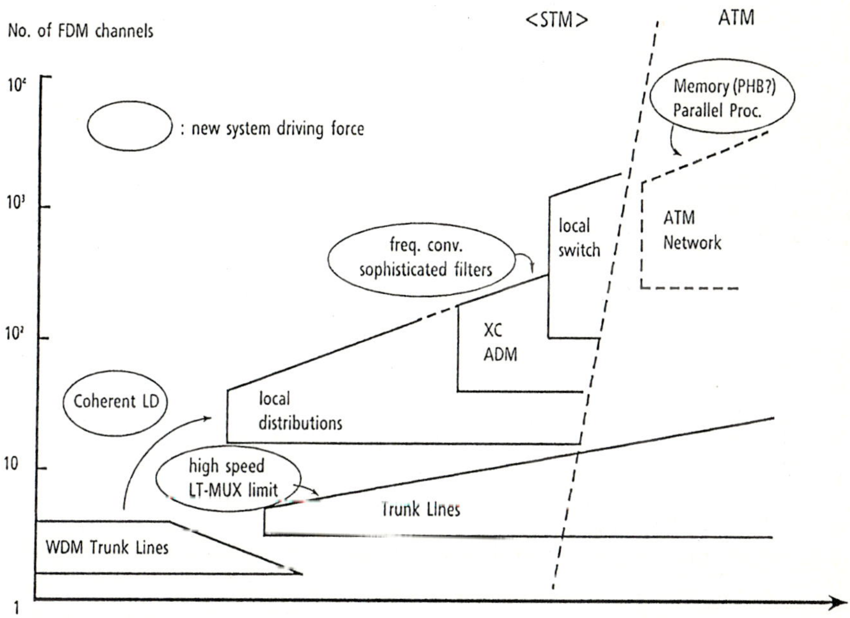

Fig. 4 OPTICAL FDM SYSTEM DEVELOPMENT STRATEGY

REFERENCES

1. K. Nosu and H. Toba. "Performance Consideration on High Capacity Optical FDM Networks," Globecom 88, 15.6, (1988)
2. E. J. Bachus, R. P. Braun, Ch. Caspar, H. M. Foisel, K. Heimes, N. Keil, B. Strebel, J. Vathke and M. Weickhman, "Coherent Optical Multicarrier Switching Node," OFC 89, PD13 (1989),
3. S.C. Liew and K. W. Cheung, "A Broadband Optical Local Network Based on Multiple Wavelengths and Multiple RF Subcarriers," ICC 89, 5.6 (1989)
4. H.Toba, K. Oda, K. Nakanishi, N. Shibata, K. Nosu, N. Takato and M. Fukuda., "100 Channel Optical FDM Transmission/Distribution at 622 Mb/s over 50 km," OFC '90, PD1 (1989)

Lightwave Frequency-Controlling Technologies for FDM Switching Star Network

S. Ohshima and M.S. Nakamura

Toshiba Corporation, 1 Komukai Toshiba-cho, Saiwai-ku, Kawasaki-shi 210, Japan

Abstract. We propose the technologies of LD frequency stabilization and broadband frequency detection for a FDM switching star network capable of multipoint-to-multipoint coherent transmission. Frequency stabilization of ±50MHz(0-50°C) and a detectable frequency range of 600GHz were demonstrated.

1. Introduction

Switching star networks employing lightwave FDM technologies have been proposed[1-4]. One paper reported multipoint-to-multipoint transmission networks[1]. In this network, each station is equipped with an optical sender and an optical receiver and has switching ability by itself. The stations are distributed in multiple locations. This network is quite useful and reliable for a future network because of the distributed switching control.

To realize multipoint-to-multipoint transmission networks, it is important to control lightwave frequencies. Several schemes to stabilize LD frequencies in CATV distribution networks, which are point-to-multipoint transmission networks, have been suggested[2-3]. These schemes cannot be applied to the multipoint-to-multipoint transmission networks.

In this paper, we propose frequency-controlling technologies for FDM switching star networks capable of multipoint-to-multipoint coherent transmission. It has been demonstrated that the LD frequency of a sender was stabilized to ±50MHz (0-50°C) and the frequency discrimination range of a local LD was 600GHz.

2. Lightwave switching star network

The switching star network[1] capable of multipoint-to-multipoint transmission is shown in Fig.1. Stations equipped with senders and receivers are connected with each other by a star coupler. Each sender transmits signals on a lightwave carrier of which the LD frequency is uniquely defined for each station. Each receiver can select any

Springer Series in Electronics and Photonics, Vol. 29 Photonic Switching II
Editors: K. Tada · H.S. Hinton © Springer-Verlag Berlin, Heidelberg 1990

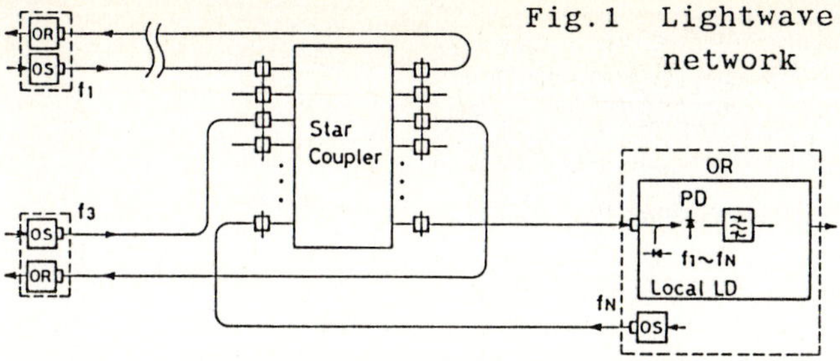

Fig.1 Lightwave switching star network

channel by tuning the local LD frequency. Thus lightwave signals are switched at each station.

Two technologies to control frequencies are necessary for realization of the multipoint-to-multipoint transmission network. One is LD frequency stabilization for a sender. Each LD frequency must be set to the unique frequency which has been allocated beforehand. The frequency accuracy should be less than a tenth of the channel frequency spacing. The other is the wide tuning technology for a receiver. All stations are not always working and the LD frequency would shift in the long term because of LD degradation. Therefore the local LD frequency should be detected to maintain the random access and the cold-start.

We suppose that the number of channels is 32, that the transmission scheme is 1.2Gbps CPFSK and that the channel frequency spacing is 8 times as large as the transmission rate in order to avoid crosstalk. In this case, the frequency accuracy for the sender should be better than 1GHz and the tunable range of the local oscillator should be wider than 300GHz.

3.Key Technologies

The LD frequency of the sender is stabilized by a Fabry-Perot resonator which can lock the LD frequency over a continuous spectral range. We adopt a Z-cut quartz etalon as the Fabry-Perot resonator[5]. The frequency temperature coefficient of the Z-cut quartz etalon is less than half that of fused silica. Because quartz exhibits the electro-optic effect, the cavity length of the quartz etalon can be directly modulated. By modulating the cavity length, the LD frequency is stabilized at the resonance frequency of the etalon. A sketch of the frequency stabilization scheme is shown in Fig.2. The maximum LD frequency deviation was ±50 MHz within the ambient temperature range of 0-50°C and the maximum capture range was ±30GHz. The module dimensions were 30x70x25mm.

Though a widely tunable LD has been developed[6], we have aimed at a broadband frequency detector for the local LD. A Mach-Zehnder interferometer is used to detect the

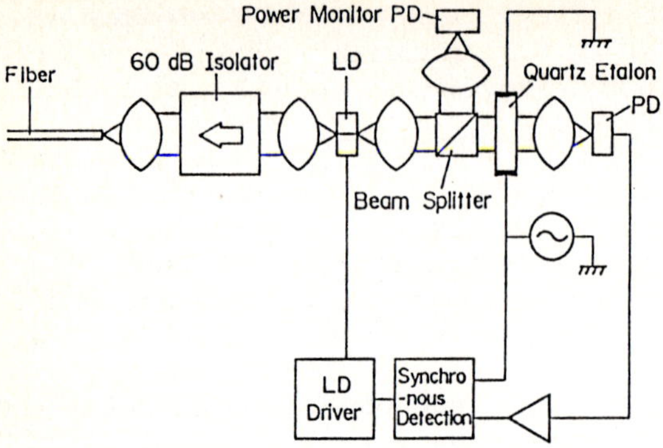

Fig.2 Sender frequency stabilization scheme using a quartz etalon

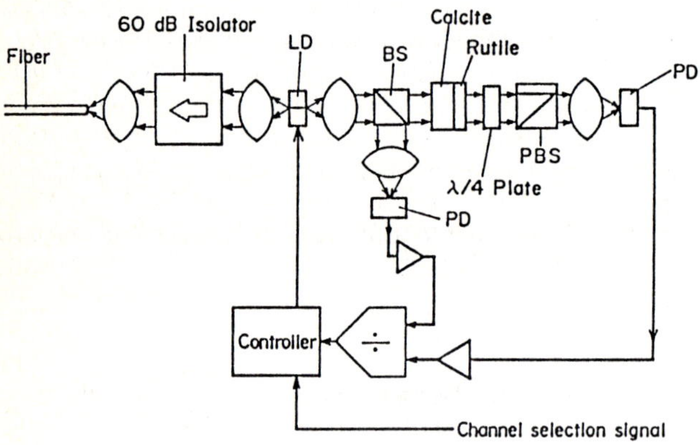

Fig.3 Frequency detection scheme using a birefringent Mach-Zehnder interferometer

frequency. A Mach-Zehnder interferometer has better linearity than a Fabry-Perot etalon. A sketch of the frequency detection scheme is shown in Fig.3. To make a path difference in the Mach-Zehnder interferometer without adjusting the wave surface, a birefringent crystal and a polarization beam splitter (PBS) are used. A calcite plate and a rutile plate are used as a birefringent crystal. Since the signs of the frequency temperature coefficients for the two materials are opposite, the temperature coefficient of the interferometer is small. The C axis of the birefrigent crystal is rotated by 45 degrees to the poralization direction of the LD light. The polarization state through the calcite and rutile plates is ellipsoidal. Though the ellipticity changes according to the frequency, the axes of the ellipsoid are fixed. Therefore a λ/4 plate can transform the ellipsoidal polarization into linear polarization.

Because the direction of the linear polarization rotates in accordance with the frequency, the frequency characteristic of the interferometer can be adjusted by rotating the PBS. The transmission ratio of a Mach-Zehnder interferometer changes sinusoidally as a function of the frequency. The maximum detectable frequency range, which is equal to a half period of the transmission, was 600GHz. The local LD module included this Mach-Zehnder interferometer and thermo-coolers. The maximum LD frequency deviation of this module was ±400MHz within the ambient temperature range of 0-50°C. The module dimensions were 30x80x25mm.

4.Conclusion

Two frequency controlling techniques for the lightwave FDM switching star networks have been proposed. One is LD frequency stabilization using a Z-cut quartz etalon. The frequency deviation within the ambient temperature range of 0-50°C was ±50MHz and the maximum capture range was ±30GHz. The other is the broadband frequency detector using the Mach-Zehnder interferometer, which is made of a calcite plate, a rutile plate and a PBS. The LD frequency deviation was ±400MHz and the maximum detectable frequency range was ±300GHz. By using these technologies, the switching star network could realize multipoint-to-multipoint transmission.

References

1)C.Baack,E.-J.Bachus,G.Heydt:Technical Digest,ECOC'87,79

2)K.Nosu,H.Toba,K.Iwashita:IEEE/OSA,J.Lightwave Technol. LT-5,1301(1987)

3)S.Yamazaki,M.Shibutani,N.Shimosaka,S.Murata,M.Shikada: Technical Digest,ECOC'88,86

4)E.Arthurs,J.M.Cooper,M.S.Goodman,H.Kobrinski,M.Tur,M.P.Vecci: Electron.Lett.24,119(1988)

5)S.Ohshima,M.S.Nakamura:Technical Digest,OFC'90,THE6

6)Y.Kotaki,S.Ogita,M.Matsuda,Y.Kuwahara,H.Ishikawa: Technical Digest,IOOC'89,19A2-4

Dynamic Wavelength Tuning for Broadband Optical Packet Switching

M.S. Goodman and H. Kobrinski

Bellcore, 445 South Street, Morristown, NJ 07960, USA

Abstract. This paper describes packet switching designs based on multiwavelength optical interconnection networks. The features of this approach are presented along with a review of multiwavelength experiments indicating the feasibility of the required fast tunable sources and receivers.

1. Introduction

In the broadband public telecommunication networks of the future, packet transport will be an important component of the total traffic carried since it provides a universal, multiple-bit-rate and application independent transport mechanism. Thus, necessary ingredients in these future networks will include very large capacity packet switches (> 50 Gbit/s throughput). The transition towards Gbit/s fiber optic trunks, which makes the use of transmission facilities more efficient and economical, would also imply switching systems with a smaller number of input/output ports but with higher data rates. Public network switches currently are based on electronic switching technology, however, the multi-Gbit/s fiber-optic transmission systems of the future may cause these switches to become bottlenecks in the broadband network.

The wavelength domain, combined with the space and time domains, can be used to design novel switching systems which may help alleviate this expected switching bottleneck. In this paper we discuss packet switching designs that are based on multiwavelength technology, and indicate some of the advantages of this approach.

Currently, electronic logic elements dominate the complex signal processing required by switching fabrics. Furthermore, buffer memories inherently are necessary components in packet switch fabrics to decouple the different characteristics of the incoming and outgoing signals, and to smooth their bursty statistics. Large, high-speed buffer memories that are currently available (along with those that are likely to be available in the near future) are based on purely electronic technologies. On the other hand, optical interconnection approaches have some unique advantages over purely electronic interconnection schemes. In particular, multiwavelength interconnection approaches offer data transparency, single stage control, non-blocking internal interconnection, multicast transport, and the potential for modular growth. Thus it seems that hybrid switching fabrics incorporating the power of electronics for logic and memory, and the flexibility of multiwavelength optical interconnection may offer the best of both approaches.

Figure 1 illustrates a general network architecture in which wavelength tuning, of either transmitters or receivers, provides single-stage routing. This scheme is referred to as 'broadcast and select' and can be implemented in several different ways[1]. One implementation [2], implied by Figure 1, includes tunable laser transmitters, a passive optical star coupler, and fixed, unique wavelength receivers. The star coupler broadcasts every input signal to all the output lines. An input-output connection at the physical level is accomplished by tuning the transmitter to the wavelength of the desired receiver.

In another implementation, also indicated in Figure 1, each input is uniquely characterized by a fixed wavelength transmitter. Multiple fixed receivers [3] or tunable receivers [4] at the output select the desired input signal through selection of the appropriate wavelength addresses. A unique feature of this optical implementation is the possibility for broadcasting or selective

Springer Series in Electronics and Photonics, Vol. 29 **Photonic Switching II**
Editors: K. Tada · H.S. Hinton © Springer-Verlag Berlin, Heidelberg 1990

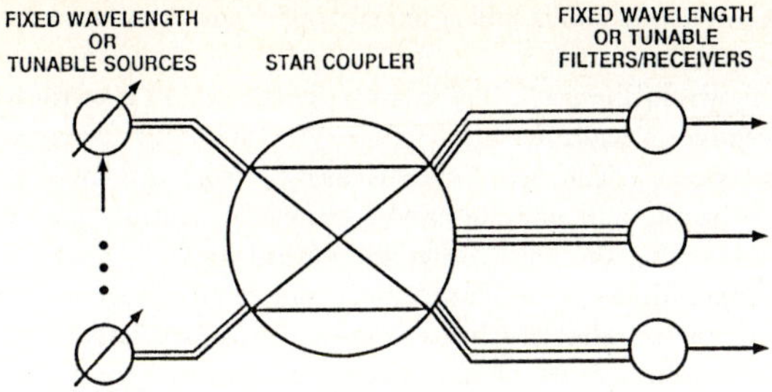

1. A Broadcast and Select Network

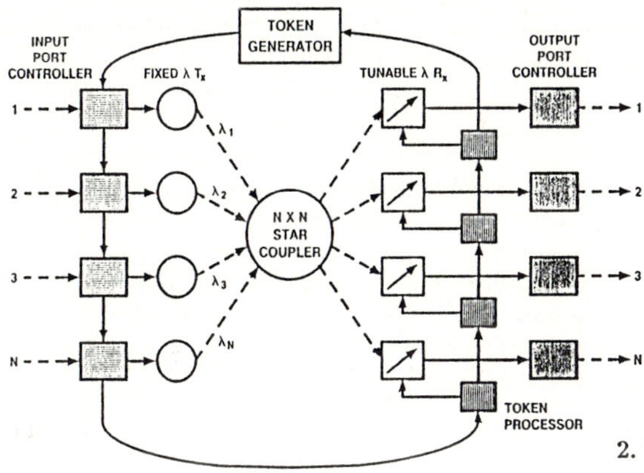

2. The Star-Track Switch Fabric

multicasting (more than one output can simultaneously receive signals from the same input). Other combinations where the tuning is performed at both the inputs and outputs, or with less tuning and more fixed-wavelength hardware, have also been considered[1]. Because of the independence of the wavelengths, architectures based on this 'broadcast and select' approach can support non-blocking connectivity between the inputs and outputs at Gbit/s transmission bit-rates.

Both the wavelength tuning control and the dynamics of the tuning are of paramount importance in switching applications [4,5]. Output port contention occurs when two input nodes try to communicate simultaneously with the same output. Switching control is required either to assure contentionless communication across the system or to resolve contention when it occurs. In order to illustrate the requirement of fast wavelength switching we assume typical parameters associated with packet switching systems. Assuming a 1000 bit packet size, and a 1 Gbit/s data rate, the packet transmission time is 1 μs. The implications for the tuning dynamics are: wavelength-switching times < several tens of nanoseconds, and several μs of residency time (i.e., the time needed by the tunable source or receiver to maintain a preset wavelength for transmission of the packet).

In the next section, we briefly describe a particular packet switch design - 'Star-Track'. We then discuss some of the technological requirements for the optical switch realization, and present some experimental results that demonstrate the feasibility of rapidly tunable laser transmitters and amplifiers.

2. An Example Multiwavelength Switch Fabric: The Star Track

A broadband optical multicast switch fabric, 'Star-Track', consists of a packet switch fabric with circuit switching capability and throughputs in the many Gbit/s range[4]. This switch fabric can support both prioritized demand access traffic and reserved traffic; and it can provide unicast and multicast switching in a unified framework.

An overview of the switch fabric is shown in Figure 2. The purpose of the switch is to route and transport packets (assumed to arrive on optical fiber input trunks) to their desired output ports for transmission on output fiber trunks. The Star-Track is formed from two internal networks, an optical 'star' transport network, and one (or more) electronic control 'tracks' surrounding the star and sequentially connecting the input ports and output ports.

The optical transport network is composed of optical transmitters and receivers connected through a wavelength insensitive optical star coupler. Each input port, i, has associated with it an optical transmitter with a unique wavelength λ_i. Each output port has one (or more) wavelength tunable receiver(s) capable of tuning over some portion of the spectrum of the input ports. The electronic control network is a 'major track' sequentially linking all the input and output ports. At the beginning (and end) of the track is a 'Token Generator' which generates control tokens and provides synchronization information to the fabric.

The essential operation of the switch may be described as follows: Packets arriving from input trunks are stored in buffers at the input ports. There are two control phases, which take place during a transmission cycle. Each input port *writes* information into the tokens indicating to which output port its packets are to be sent (reserving that output port) in the first control phase. The output port *reads* the tokens, and tunes its receiver to the appropriate input port wavelength during the second phase. As the token is emitted from the token generator (starting a new cycle), transmission of the packet is commenced at the input ports. The transmission and control cycles are overlapped in time, such that the control phase is being executed during the previous packet transmission cycle. Transmission occurs in parallel, from many input ports to available output ports, without collision, due to the multiwavelength transport.

Reserved packets should be assigned the highest priority. This priority class cannot be attained by 'promotion' from lower priority classes through the fairness mechanism. This is to insure that the reserved packet may be transported on the *very next* transmission cycle, without any interference.

Data packets arriving on optical fiber input trunks at the network transmission rate are converted first to an electronic bit-serial data stream, and then converted to parallel data streams incident on input port data buffers. (Additional input port processing, if required, can be provided at this stage.) A strobe signal is generated at the end of the control cycle by the token generator, initiating transmission within the switch. At this time, the input port converts the packet back to a serial data stream and transmits it to the output ports. The transmission within the switch is assumed to be bit-synchronous with transmission at rates > 1 Gbit/s. The output port, upon acceptance of a request registered in the token, and when available, determines its target input port, and fast-tunes (and locks) its receiver to the appropriate input port wavelength to receive the packet when transmission commences. The output port then converts the bit serial optical stream back to a parallel electronic digital data stream for input to the final output port buffer and retransmission across the network. The use of a wavelength selective element may be removed entirely by the use of fixed receiver arrays at each output port. Such a system could be implemented using either discrete or integrated receivers, with simple electronic logic 'enabling' a particular receiver, with all others at a given output port disabled. The resulting bit streams are then 'or-ed' together to produce the output.

Various devices have been recently demonstrated that can achieve the desired wavelength selectivity and tuning speed to implement the basic tunable multiwavelength architectures. Some of these demonstrations are discussed in the following section.

3. Experiments with Tunable Transmitters and Amplifiers

The central optical devices required for these multiwavelength optical interconnection networks are the wavelength tunable transmitters and receivers. In order to construct a switch with a significant increase in throughput beyond that available with conventional electronic switches, 32 or more ports will be required, with internal transmission rates of ≈ 1 Gbit/s. (Star-Track fabrics with at least 32 ports will not be limited by currently available fast electronics.) Optical power budget considerations may become the dominant limitation for switch sizes with several hundreds of ports. Therefore, the primary issue in the determination of the number of channels is the tuning range and the resolution of the wavelength-agile sources and receivers.

3.1 Tunable Transmitters

Several technologies exist which can be used for wavelength tunability; however, the fast switching requirement (in the nanosecond range as previously discussed) implies the use of semiconductor laser devices. The main candidates are multi-section, single longitudinal mode (SLM) lasers, i.e. Distributed FeedBack (DFB) and Distributed Bragg Reflector (DBR) lasers.

Wavelength tuning by current injection has been analyzed for both DFB and DBR structures [6,7,8]. Wavelength tuning with mode hopping (quasi-continuous) over 10 nm and continuously over 4 nm, has been demonstrated using a 3-section DBR laser [9]. A continuous tuning range of 2.1 nm has been demonstrated in a double-section DFB laser [10].

The response time for setting a given wavelength ultimately is dependent on the carrier lifetime of the tunable source. Typical recombination times for the excess carriers are in the range $0.3ns < \tau_r < 3$ ns. The longer times are associated with passive sections such as the tuning regions in a 3-section DBR laser, where the carrier lifetime is not decreased by the process of stimulated emission. Consequently, for high speed (>100's Mbit/s) tuning applications, as in frequency shift keying (FSK) transmission, DFB lasers are the usual choice.

Multi-section DBR lasers are somewhat more complicated to fabricate than DFB lasers, but they provide an additional degree of freedom in separately controlling the device gain, the Bragg condition and the phase matching. This results in increased tuning range available. In addition, the separate current controls allow us to switch wavelengths *and* modulate (either intensity or FSK) the data *simultaneously* on the same device[11,12]. The advantages of using a single device for both fast wavelength-switching and direct data modulation are significant: the elimination of external modulators improves both the simplicity of the implementation and the available power budget.

Figure 3 shows a DBR laser spectrum with wavelength tuning and direct intensity modulation at 200 Mbit/s, for 16 different wavelengths. The maximum variation in output power was ≈3 dB over the tuning range. The maximum continuous tuning range observed for this device was 2.2 nm. From crosstalk measurements, it was determined that in order to guarantee better than

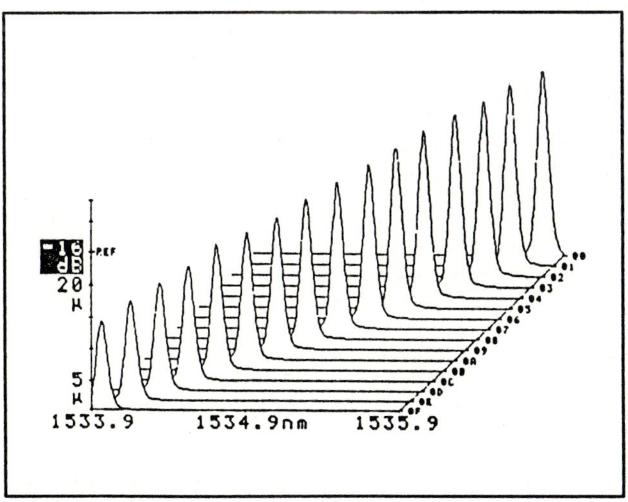

3. Spectrum of 16 wavelength tuned channels using a 3-section DBR laser

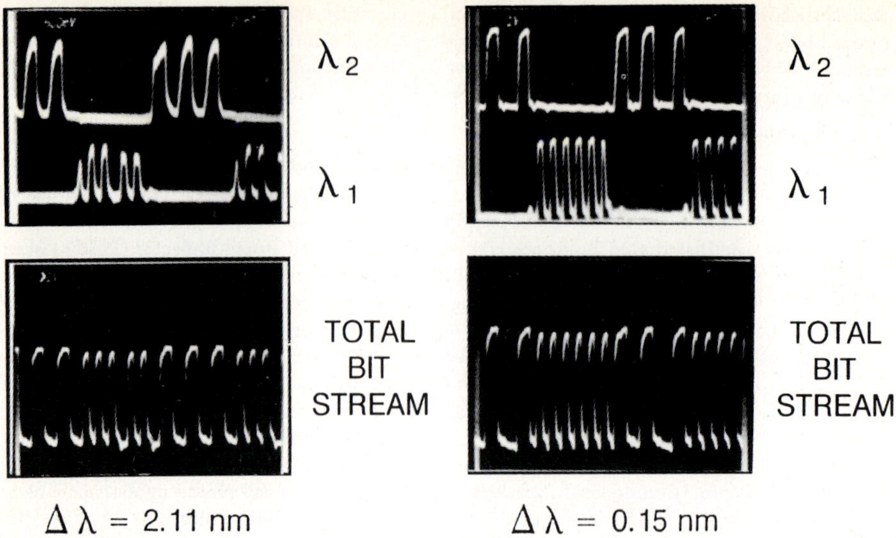

4. 2-Channel Packet Switching Simulation using a wavelength tunable transmitter and a fixed wavelength filter.

-10 dB crosstalk isolation (< 0.5 dB power penalty) with adequate extinction ratio, a minimum filter bandpass of 5 GHz and a channel spacing of 15 GHz ($\approx$0.12 nm) are required. Therefore, the total number of resolved wavelength channels is $\approx$ 20.

Figure 4 indicates the fast wavelength-switching characteristics of the device in a packet switching simulation at 200 Mbit/s. Wavelength-switching between two channels separated by $\Delta\lambda = 2.11$ nm is shown on the left. The two channels are resolved using a Fabry-Perot filter with a 6 GHz bandpass. Each packet contains 12 bits and has a duration of 60 ns. The first and last bits of each packet must be used as 'guard-bits' in order to achieve a bit-error rate, $BER < 10^{-9}$, indicating a transient interval of $\approx$ 10 ns when switching wavelengths.

Figure 4 on the right presents wavelength-switching for channels separated by $\Delta\lambda = 0.15$ nm. In this case, only one 'guard' bit between packets is necessary to insure $BER < 10^{-9}$, indicating a smaller transient interval of $\approx$ 5 ns. The tuning speed in these experiment was limited by the device capacity. A wavelength switching time of 1.8 ns was reported recently over a 1.25 nm tuning range [13].

Intensity data modulation produces a dynamic linewidth increase, or chirp, which determines the required optical filter bandpass at the receiving end, and hence, the minimum channel separation. This restriction can be alleviated by using FSK data format, modulating either the current into the active region, or superimposing the data signal onto the switching signal to modulate the passive section of the 3-section DBR laser. FSK modulation and direct detection with a single Fabry-Perot (FP) filter with resolution f_{FP}, implies a minimum channel spacing of $\approx 6f_{FP}$, for a power penalty <0.1 dB, provided that the laser linewidth is $< f_{FP}/10$ [14]. The linewidth of the 3-section DBR laser varies with the injected tuning current, and the laser used in this experiment had its largest linewidth of $\approx$ 80 MHz, and therefore the F-P filter resolution was set to $f_{FP} \approx 1$ GHz. The minimum channel spacing of 6 GHz and the total continuous tuning range of 2.2 nm, implied a maximum number of wavelength channels using these FSK conditions of $\approx$ 48.

The maximum FSK modulation bit-rate when modulating the passive sections of this device is limited by carrier lifetimes in these sections. External modulation of DBR lasers may combine adequate tuning ranges, switching speeds, and data rates for the applications considered in this paper.

3.2 Tunable Amplifiers

Tunable wavelength selection can be obtained by exploiting the resonant amplification in DFB structures biased below threshold. Measurements on these devices have shown continuing improvement in dc selectivity, gain, and tuning range [15,16]. A multisection phase-shift controllable DFB filter has been used to separate 8 channels over a 0.6 nm tuning range [17]. More recently, a 0.95 nm tuning range with a constant 24.5 dB optical gain, has been demonstrated in a similar device, corresponding to selection of 18 wavelength-channels[18]. The dynamical properties under both fast wavelength-switching and high bit-rate data streams have also been investigated [19]. In this paper we present experimental results for the fast-tuning properties of distributed-feedback amplifiers (DFB-OA), including BER measurements at >1 Gbit/s for packets switched between two wavelengths to simulate the high-speed transient conditions of a packet switching system. Two high-bit-rate wavelength-multiplexed signals provided the input in a simulation of the transient conditions of a packet switching system using a uniform (single-section) DFB-OA for fast wavelength-selection. The two 1.2 Gbit/s input waveforms and the 100 Mhz switching signal to the amplifier, all driven by the same clock, are shown in Figure 5. The wavelength of the DFB-OA, biased at $I_a = 0.95 I_{th}$, was switched between λ_1 and λ_2, separated by $\Delta\lambda = 0.23$ nm. A BER$< 10^{-9}$ was measured for the total detected pattern of packets at λ_1 and λ_2 of Figure 5a. Next, the BER was measured for one signal channel while another channel at the other wavelength was directly overlapping it in time to simulate the crosstalk conditions in a system like the Star-Track. The resultant waveforms are shown in Figures 5b and 5c. These measurements of BER under bursty transmission at 1.2 Gbit/s data rate indicate that no bits are lost during the fast transitions from one wavelength to another, in other words, switching in less than 1 ns was attained.

Tunable receivers can also be implemented using tunable transmitters as local oscillators in coherent receiver schemes; or through the combination of parallel wavelength demultiplexers, receiver arrays, and electronic selection circuitry (preferably in an OEIC form).

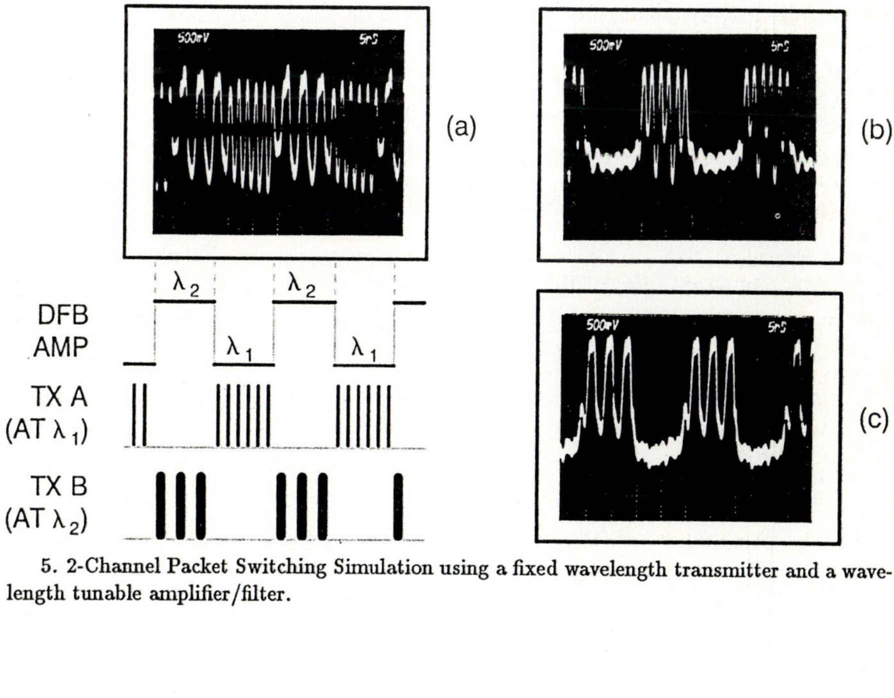

5. 2-Channel Packet Switching Simulation using a fixed wavelength transmitter and a wavelength tunable amplifier/filter.

4. Conclusion A multiwavelength optical interconnect fabric for packet switching has been described that features large throughput and incorporates both unicast and multicast packet transport. System requirements for this and similar multiwavelength architectures indicate the need for 32 to 128 individually selectable channels with Gbit/s transmission rates. Random access times between these channels are required to be less than $\approx$ 10% of a packet time; implying access times typically less than $\approx$ 50 ns. In addition, residency times of up to several μs, during which the wavelength must be stable are required. Tunable laser transmitters and tunable filters have been studied under conditions simulating packet transmission. The number of accessible channels has been demonstrated to be $>$ 50 using DBR lasers with several nm continuous tuning range. Switching with $<$ 15 ns access time has been demonstrated, meeting the switching time requirements stated above as well.

Wavelength selective filters based on DFB optical amplifiers have been shown to be capable of selecting Gbit/s data packets with nanosecond switching times and with BER $< 10^{-9}$ satisfying the timing requirements. However, currently available tuning ranges imply $<$ 20 wavelength channels. Additional studies of the noise characteristics of the DFB-OA's and their effect on the overall power budget are also needed.

With the anticipated further development of wavelength tunable sources and receivers as well as OEIC technologies, multiwavelength optical networks may continue to stimulate new approaches to the challenges of emerging broadband systems.

References

1. M.S. Goodman, 'Multiwavelength Networks and New Approaches to Packet Switching', *IEEE Commun. Mag.*, p25-37, 1989.

2. E. Arthurs, M.S. Goodman, H. Kobrinski, and M.P. Vecchi, 'HYPASS: An Optoelectronic Hybrid Packet Switching System', *IEEE J. on Sel. Areas. Commun.*, 6:1500-1510, 1988.

3. H. Kobrinski, R.M. Bulley, M.S. Goodman, M.P. Vecchi, C.A. Brackett, L. Curtis, and J.L. Gimlett, 'Demonstration of High Capacity in the Lambdanet Architecture: A Multiwavelength Optical Network', *Elect. Lett.*, 23:824, 1987.

4. T.T. Lee, M.S. Goodman, and E. Arthurs, 'A Broadband Optical Multicast Switch', *To appear: Proc. Int. Switch. Symposium, ISS'90*, Stockholm, 1990.

5. M.P. Vecchi, 'Component Technologies for Multiwavelength Broadband Packet Switching Systems', *Proc. Integrated Optics and Optical Fiber Commun., IOOC'89*, pp21D2-4, Kobe, Japan, 1989.

6. M. Kuznetsov, 'Theory of Wavelength Tuning in Two-segment Distributed Feedback Lasers', *IEEE J. of Quant. Elect.*, QE-24:1837-1844, 1988.

7. K. Kobayashi and I. Mito, 'Single Frequency and Tunable Laser Diodes', *IEEE J. Lightwave Tech.*, 6:1623-1633, 1988.

8. L.A. Coldren and S.W. Corzine, 'Continuously-tunable Single-frequency Semiconductor Lasers', *IEEE J. of Quant. Elect.*, QE-23:903-907, 1987.

9. S. Murata, I. Mito, and K. Kobayashi, 'Tuning Ranges for 1.5 μm Wavelength-tunable DBR Lasers', *Electron. Lett.*, 24:577-579, 1988.

10. Y. Yoshikuni, K. Oe, G. Motosugi, and T. Matsuoka, 'Broad Wavelength-tuning Under Single Mode Oscillation with a Multi-electrode Distributed Feedback Laser', *Electron. Lett.*, 22:1153-1154, 1986.

11. H. Kobrinski, M.P. Vecchi, T. Chapuran, J.B. Georges, C.E. Zah, C. Caneau, S.G. Menocal, P.S.D. Lin, A.S. Gozdz, and F.J. Favire, 'Simultaneous Fast Wavelength-switching and Intensity Modulation using a Tunable DBR Laser', *Photon. Tech. Lett.*, 2:139-142, 1990.

12. H. Kobrinski, M.P. Vecchi, T. Chapuran, J.B. Georges, C.E. Zah, C. Caneau, S.G. Menocal, P.S.D. Lin, A.S. Gozdz, and F.J. Favire, 'Fast Wavelength-switching and Simultaneous FSK Modulation using a Tunable DBR Laser', *Elect. Lett.*, 26:308-310, 1990.

13. N. Shimosaka, M. Fujiwara, S. Murata, N. Henmi, K. Emura, and S. Suzuki, 'Photonic Wavelength-division and Time-division Hybrid Switching System Utilizing Coherent Optical Detection', *Proc. 15th European Conf. on Comm. ECOC'89*, p268, Gothenburg, 1989.

14. I.P. Kaminow, P.P. Iannone, J. Stone, and L.W. Stulz, 'FDMA-FSK Star Network with a Tunable Optical Filter Demultiplexer', *IEEE J. Lightwave Tech.*, 6:1406-1414, 1988.

15. H. Kawaguchi, K. Magari, K. Oe, Y. Noguchi, Y. Nakano, and G. Motosugi, 'Optical Frequency-selective Amplification in a Distributed Feedback-type Semiconductor Laser Amplifier', *Appl. Phys. Lett.*, 50:66-67, 1987.

16. K. Magari, H. Kawaguchi, K. Oe, Y. Nakano, and M. Fukuda, 'Optical Signal Selection with a Constant Gain and a Gain Bandwidth by a Multi-electrode Distributed Feedback Laser Amplifier', *Appl. Phys. Lett.*, 51-52:1974-1976, 1987.

17. M. Nishio, T. Numai, S. Suzuki, M. Fujiwara, M. Itoh, and S. Murata, 'Eight Channel Wavelength Division Switching Experiment using Wide Tuning Range DFB LD Filters', *Proc. ECOC'88*, Vol. 2, p49, Brighton, 1988.

18. T. Numai, S. Murata, and I. Mito, '1.5 μm Tunable Wavelength Filter using a Phase Shift Controlled DFB Laser Diode with a Wide Tuning Range and a High Constant Gain', *Appl., Phys. Lett.*, 54:1859-1860, 1989.

19. H. Kobrinski, M.P. Vecchi, E.L. Goldstein, and R.M. Bulley, 'Wavelength Selection with Nanosecond Switching Times using DFB Laser Amplifiers', *Electron. Lett.*, 24:969-971, 1988.

Dense FDM Coherent Optical Switching System

B. Glance and O. Scaramucci

AT&T Bell Laboratories, Crawford Hill Laboratory,
Holmdel, NJ 07733, USA

We report the results obtained with a dense FDM coherent optical star network operated as a computer-controlled high capacity switch. The system provides switching connectivity between the input signals modulating the transmitting lasers and the output signals demodulated by optical receivers. It consists of eight FSK-modulated channels multiplexed by a 16×16 optical star coupler. Six of these channels are generated by individual digitally tuned optical frequency synthesizers. The frequency of these six channels can be randomly selected from a set of equi-spaced frequencies (2.2 GHz). Two of the sixteen multi-channel signals, coming out of the star coupler, are demultiplexed by their individual digitally tuned heterodyne receivers. Each receiver can randomly select a desired channel by computer command. Switching is performed by rerouting, from the computer keyboard, the transmitted channels between the receivers. The receivers can identify missing channels. More importantly, the digital tuning of the frequency synthesizers and of the receivers is insensitive to variation of the frequency-tuning current relationship of the lasers.

1. Introduction

Optical star networks using dense Frequency Division Multiplexing (FDM) and coherent detection techniques provide the possibility to build a high capacity transmission system interconnecting a large number of users.[1,3] Such a system can also be used as a high capacity switch providing rerouting connectivity between the input signals modulating transmitting lasers and the output signals demodulated by optical receivers. Two system configurations can be used to achieve this result. In one configuration, the incoming signals are each transmitted at a different fixed optical frequency. They are multiplexed by a star coupler. Demultiplexing is achieved by means of a tunable optical receiver (Fig. 1).

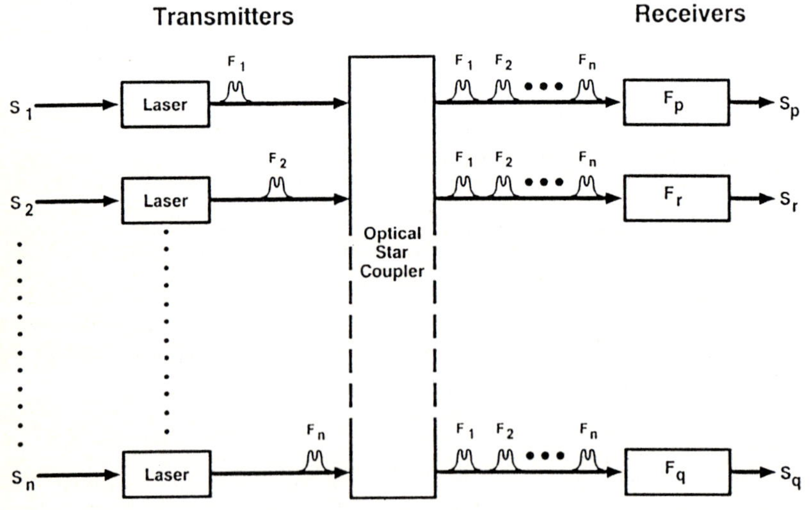

Fig. 1 Schematic of the Switch Circuit

Springer Series in Electronics and Photonics, Vol. 29 **Photonic Switching II**
Editors: K. Tada · H.S. Hinton © Springer-Verlag Berlin, Heidelberg 1990

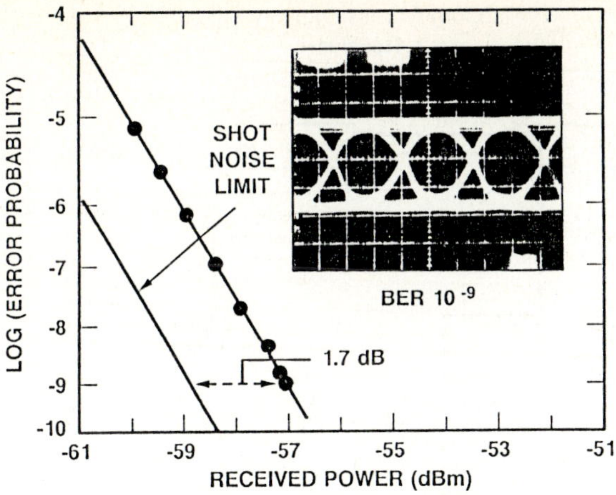

Fig. 2 Receiver Sensitivity Obtained with a FSK-Modulated Signal at 200 Mb/s

The same result can also be obtained by switching the optical frequencies of the incoming signals and with receivers tuned at a fixed optical frequency. The use of dense FDM maximizes the number of incoming signals that can be multiplexed in the limited tuning range of presently available lasers. The use of optical coherent receivers provides the high selectivity needed to retrieve the desired signal from the incoming signals multiplexed at close frequency intervals [4]. Coherent receivers also provide the high sensitivity needed to detect the low level signals caused by the attenuation of the star coupler.

Heterodyne detection is the most commonly used technique for coherent optical reception. This technique yields the best compromise between receiver sensitivity and laser linewidth requirement. Such a receiver uses a Local Oscillator (LO) laser to transpose the received optical signal to the radio frequency domain where signal selection can be performed with an electrical filter.

Recently available three-contact semiconductor lasers [5,8] fulfill many of the requirements needed by dense FDM optical systems. These sources generate a single-mode signal tunable over 1 THz near a wavelength of 1.53 μm. A receiver sensitivity of 74 photons/bit was demonstrated in a multi-channel heterodyne system using three-contact tunable lasers [9] to generate and receive signals FSK-modulated at 200 Mb/s and spaced at 2.2 GHz frequency intervals (Fig. 2).

The narrow linewidth (4 MHz) signal needed to achieve this sensitivity was only obtained over a small fraction of the 1-THz tuning range of the laser. However, recently reported results indicate that generation of a narrow linewidth signal tunable over a large tuning range is feasible [10]. One can therefore envision switching systems based on FDM and coherent detection techniques capable to provide interconnectivity between 400 users at a data rate of 200 Mb/s. Realization of even larger switching systems may be possible if the tuning range of the lasers can be increased.

Switching in these systems will be computer-controlled. Connection between users will be made by selecting the optical frequencies of the transmitted signal and of the received channel by computer commands. We present the first demonstration of a multi-channel heterodyne optical star network providing this capability. The result has been obtained using random access digitally tuned optical frequency synthesizers [11] for generating the transmission signals and random access digitally tuned optical heterodyne receivers [12] for selecting the desired channels. The entire optical system is computer-controlled. The computer activates/deactivates the system, calibrates the digital tuning of the individual lasers, protects the calibrations against frequency drifts of the lasers and allows the user to select the frequency of the transmission signals and the frequency of the received signals. It also identifies missing channels when an absent channel is selected.

2. System Description

The system consists of eight transmitted FSK-modulated signals multiplexed by a 16×16 optical star coupler (Fig. 3). Six of these channels are provided by individual optical frequency synthesizers. Their optical signal is generated by a three-contact tunable Multiple Quantum Well Distributed Bragg

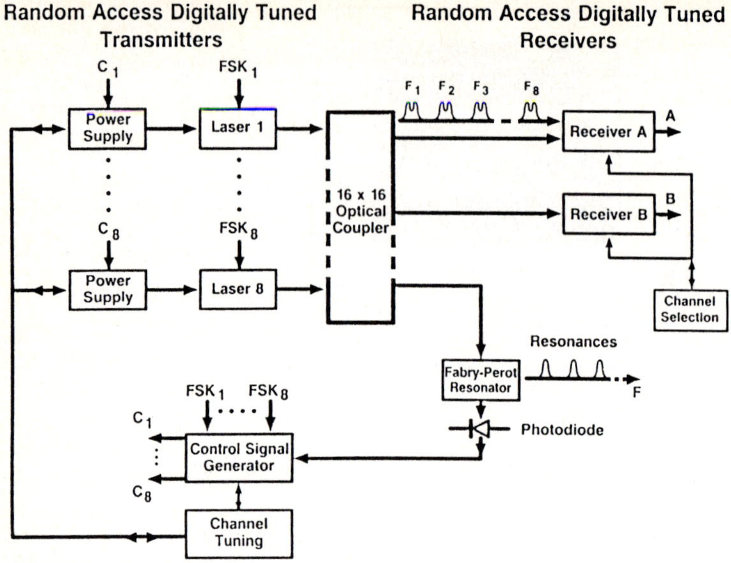

Fig. 3 Schematic of the Computer-Controlled Optical Star Network

Reflector (MQWDBR) laser [7,8] operating around 1.53 μm. The carrier frequency of these six channels can be randomly selected from a set of equi-spaced frequencies (2.2 GHz) by command from the keyboard of a Personal Computer (PC). The comb of frequencies is the same for each optical frequency synthesizer. It is defined by a set of resonances of a Fiber Fabry-Perot (FFP) cavity [13] having a free spectral range of 2.2 GHz. Each optical frequency synthesizer is calibrated by storing in the PC memory the tuning current values locking its laser frequency to the resonances of the FFP. The number of available frequencies provided by each optical frequency synthesizer varies from 34 to 69 depending on the continuous tuning range of the first DBR mode of the laser.

The tuning bandwidth used in this experiment is a small fraction of the 1-THz tuning capability of the laser. The bandwidth is limited because we used a single variable current resistively divided between the Bragg and the phase sections of the laser to tune the frequency. This tuning arrangement provides a continuous frequency tuning over the first DBR mode which facilitates greatly the tuning calibration of the system. The 1-THz tuning range of the three-contact laser is distributed over twelve overlapping DBR modes. The frequency discontinuities occurring at the DBR mode overlaps complicate considerably the calibration process. We are presently resolving this problem. Preliminary results indicate that the digital random access capability of the optical frequency synthesizer can be extended to the 1-THz tuning range of the laser. The two remaining channels are generated by monolithic two-contact MQWDBR lasers. Since these lasers are barely tunable, they are operated at fixed frequencies each locked to a separate resonance of the FFP.

Each channel is locked to one of the resonances of the FFP by an independent error signal [14]. The individual error signals are obtained by photodetecting the output of the FFP fed by the FDM signal (Fig. 4). The amplified photocurrent is correlated with the eight different random modulating FSK signals. The result yields the separate error signals needed for independently locking each channel. To obtain this result, each channel is FSK-modulated at 200 Mb/s by an independent modulator.

The digital tuning calibration depends on the stability of the relationship between the laser frequency and its tuning current. The digital tuning calibration is protected against variations of the above relationship by a process that automatically corrects the set of stored current values when such a variation occurs. The result is achieved by a servo-control circuit [11,12] which detects the drift of the laser free-running frequency relative to the tuning current. This information is then used by the PC to reevaluate the set of stored current values of the frequency-drifting laser. The result yields an absolutely stable correspondence between the "current addresses" and the corresponding frequencies selected during the calibration process.

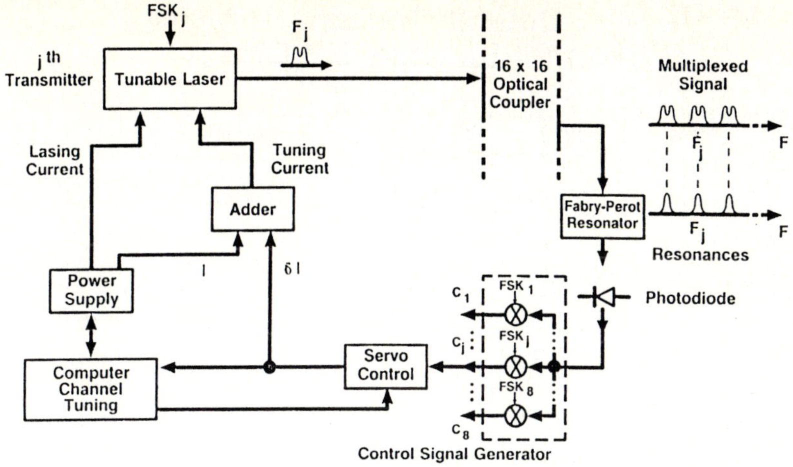

Fig. 4 Schematic of the Random Access Digitally Tuned Optical Frequency Synthesizer

Two of the sixteen output multi-channel signals, coming out of the star coupler, are demultiplexed by their individual optical heterodyne receiver. Both receivers operate identically. Heterodyning is performed by a balanced-photodiode mixer using a three-contact tunable MQWDBR lasers as an LO (Fig. 5). The IF signal, centered at 700 MHz, is demodulated by a delay discriminator. Each receiver can randomly access each of the eight channels. The channel selection process is performed by means of a computer that controls the frequency of the LO laser through its biasing current. The channels are randomly accessed by selecting from the computer memory the biasing current value that tunes the LO frequency near the desired channel. A control circuit is then activated to frequency-lock the LO signal at one IF frequency interval below the desired channel. The receiver digital tuning is initially calibrated in a manner similar to the optical frequency synthesizers except that the comb of FFP resonances is replaced by the comb of the received channels and that the error signal is generated by the delay-discriminator used for demodulation. The digital tuning calibration is protected against drifts of the LO free-running frequency versus tuning current by automatic reevaluation of the stored current values as previously described for the optical frequency synthesizer. The effectiveness of this protection was tested by verifying that the digital tuning calibration remains unchanged in the presence of a 20 GHz drift of the laser free-running frequency induced by a temperature change of 1°C.

The power spectra of the eight channels and that of one LO observed by means of a high resolution scanning Fabry-Perot interferometer are displayed in Fig. 6. The channels are locked to successive

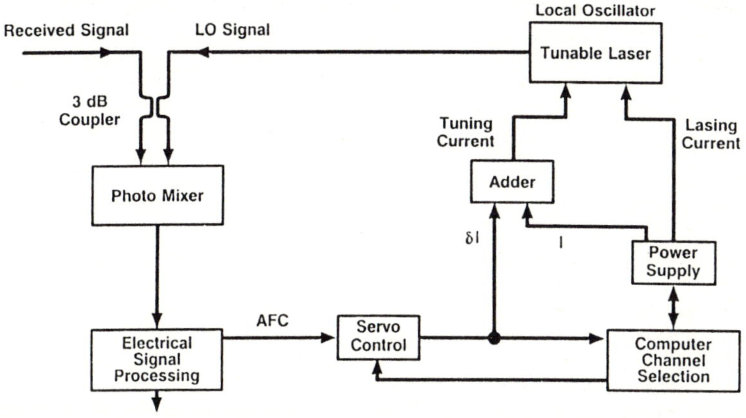

Fig. 5 Schematic of the Random Access Digitally Tuned Optical Heterodyne Receiver

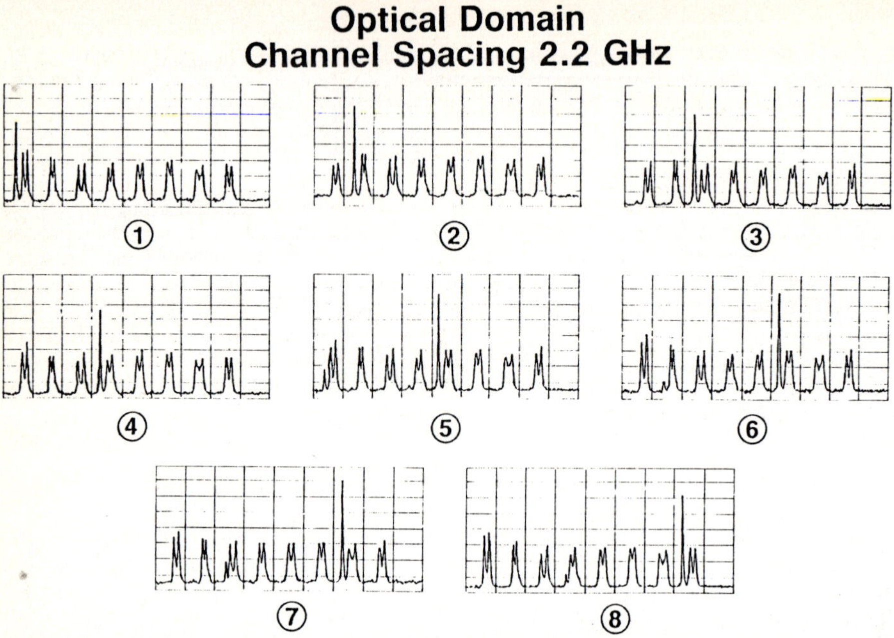

Fig. 6 Power Spectra of the Eight Channels and of the LO Signal Observed in the Optical Domain

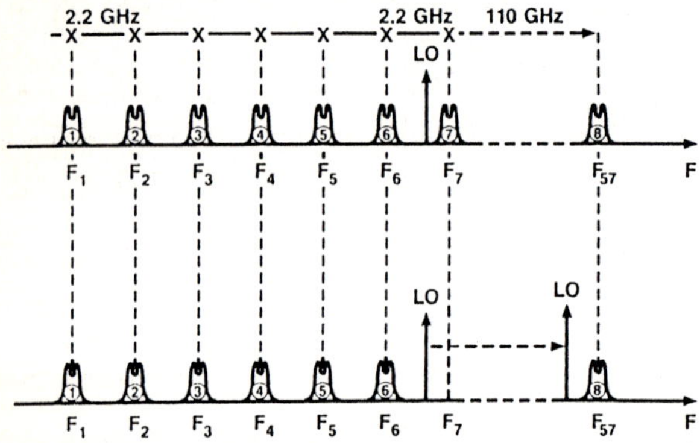

Computer Display

Channel		
Selected	Received	Status
7	7	Present

Channel		
Selected	Received	Status
7	57	7 TO 56 Missing

Fig. 7 Computer Display of Channel Status

resonances (spaced by 2.2 GHz) of the FFP resonator shown in Fig 4. The LO signal is tuned to receive successively each of the eight channels (one IF frequency interval of 700 MHz below the selected channel).

The computer displays the channel selected by a receiver and identifies the reception of the channel. The computer can also detect absent channels. When the receiver is tuned to an absent channel, the LO locks to the next available channel. The computer then displays the absence of the selected channel and which channel is received instead. If the two above channels are separated by more than one channel interval (2.2 GHz) the computer then displays all the channels missing in the empty frequency interval. Thanks to the large capture range of the receiver, this capability is provided even when the next available channel is located 100 GHz away, as illustrated in the example shown Fig. 7. Channel traffic information is thus provided by the PC.

3. Switching Operation

The star network was operated as a switch by re-routing the modulating signals applied to the transmitting lasers between the outputs of the two receivers. The result can be obtained by operating the transmitters at fixed optical frequencies and by tuning the optical receivers to select the desired channels. The result can also be obtained by using fixed frequency tuned receivers and transmitters switched at different optical frequencies. Figure 8 illustrates an example of channels transmitted at fixed frequencies and receivers tuned to select desired channels.

Figure 9 displays several examples of channel exchange between the two receivers observed in the optical domain. These examples show six optical channels spaced at 2.2 GHz frequency intervals and the position of the two LO signals set at one IF interval below the channel selected by their respective receivers (the LO signals are identified by their different amplitudes). Channel selection is activated by the computer, which is programmed to tune the LO lasers of each receiver simultaneously. The switching is initiated by entering a two-digit number on the keyboard of the computer; the first digit selects the channel received by receiver A, the second digit selects the channel for receiver B. Channel exchange between the receivers is obtained by inverting the two-digit command. The switching time is a few milliseconds. Its duration is limited by the time response of the AFC circuit controlling the receiver IF locking.

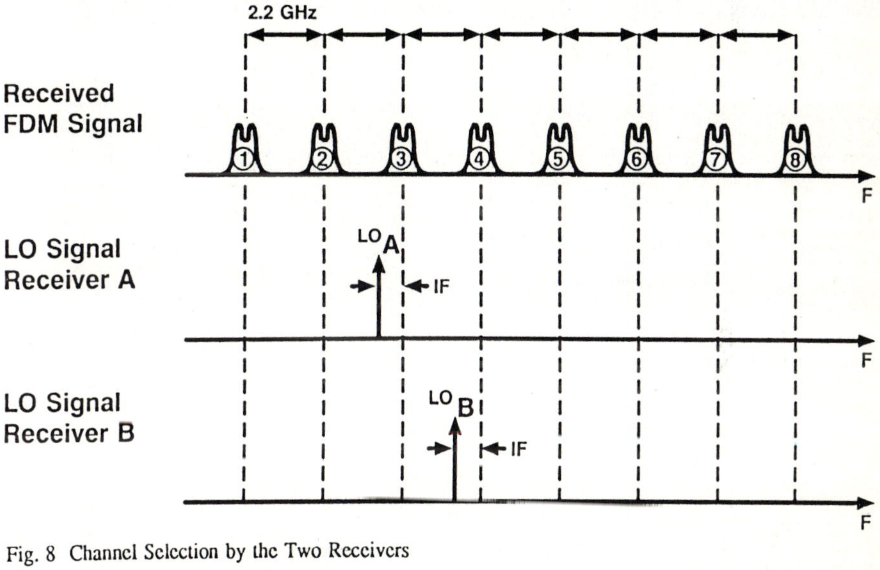

Fig. 8 Channel Selection by the Two Receivers

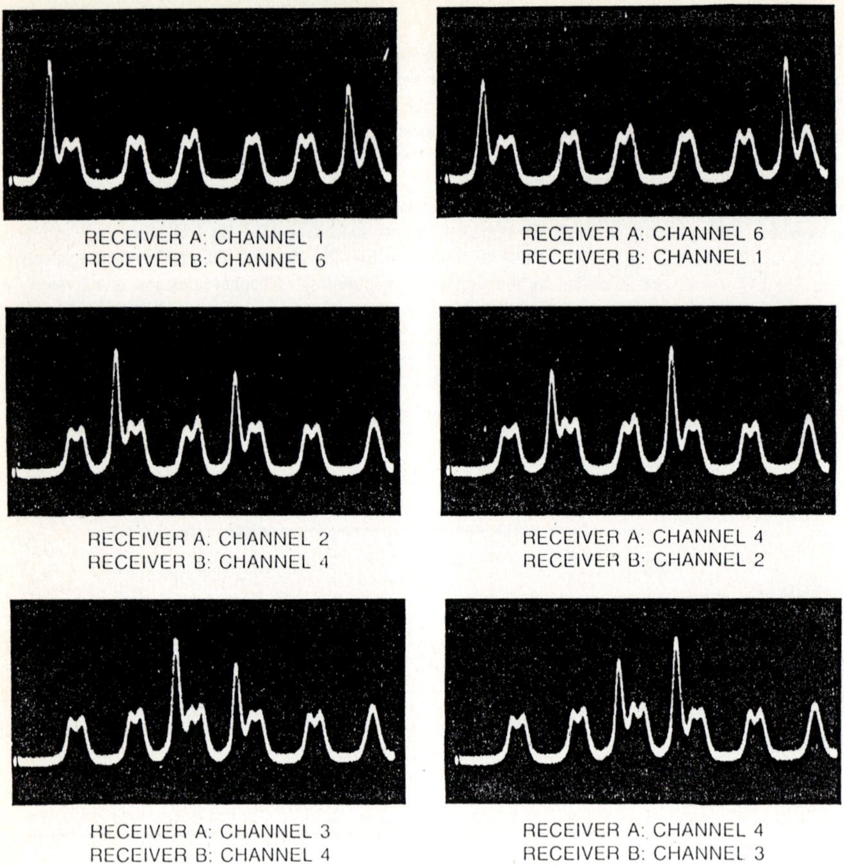

Fig. 9 Channel Exchange Observed in the Optical Domain

Conclusion

We have demonstrated the use of an FDM coherent optical star network as an optoelectronic switch. The experimental system consists of a 16×16 star coupler fed by eight optical channels FSK modulated at 200 Mb/s and spaced by 2.2 GHz. Two of the multichannel signals exiting the star coupler were demultiplexed by their respective heterodyne receivers. Each receiver could randomly access any channel by addressing the keyboard of a computer controlling both receivers. Such a switch can potentially interconnect a large number of users communicating at a high bit rate.

Acknowledgment

The authors are grateful to T. L. Koch, U. Koren, R. P. Gnall and C. A. Burrus for supplying the lasers, to L. D. Tzeng for supplying the high sensitivity photoreceiver, to J. Stone and L. W. Stulz for supplying the fiber Fabry-Perot and to K. T. Gayliard-Veguilla and P. D. Magill for providing the laser circuit mount.

References

1. M. Fujiwara, N. Shimosaka, M. Nishio, S. Susuki, S. Yamazaki, S. Murata and K. Kaede, "A Coherent Photonic Wavelength - Division Switching Systems for Broadband Networks", ECOC'88 11-15 Sept. 1988, Brighton Center, Brighton, U.K.

2. N. Nishio, T. Numai, S. Suzuki, M. Fujiwara, M. Itoh and S. Murata, "Eight-Channel Wavelength-Division Switching Experiment Using Wide-Tuning-Range DFB LD Filters", OFC'89 6-9 Feb. 1989, Houston, Texas.

3. K. Y. Eng and M. Santoro, "Multi-Gb/s Optical Cross-connect Switch Architectures: TDM versus FDM Techniques", Proceeding IEEE Global Telecommunications Conference, pp 999-1003 Dallas, Texas November 27-30, 1989.

4. B. Glance, L. D. Tzeng, T. L. Koch, O. Scaramucci, K. C. Reichmann and U. Koren, "High-Performance Coherent Optical Communication System" Elect. Letters, Vol. 25, No. 14 pp. 932-933, July 1989.

5. S. Murata, I. Mito and K. Kobayashi, "Over 720 GHz (5·8 nm) Frequency Tuning by a 1·5 μm DBR Laser With Phase and Bragg Wavelength Control Regions", Electron. Lett., 1987, 23, pp. 403-405.

6. Y. Kotaki, M. Matsuda, M. Yano, H. Ishikawa and H. Imai, "1·55 μm Wavelength Tunable FBH-DBR Laser", ibid., 1987, 23 pp. 325-327.

7. T. L. Koch, U. Koren, R. P. Gnall, C. A. Burrus and B. I. Miller, "Continuously Tunable 1.5 μm Multiple-Quantum-Well InGaAs/InGaAsP Distributed-Bragg-Reflector Lasers," Electron. Lett. 24, 1431 (1988).

8. T. L. Koch, U. Koren and B. I. Miller, "High Performance Tunable 1.5 μm InGaAs/InGaAsP Multiple Quantum Well Distributed Bragg Reflector Lasers," Appl. Phys. Lett. 53, 1036 (1988).

9. B. Glance, T. L. Koch, O. Scaramucci, K. C. Reichmann, L. D. Tzeng, U. Koren and C. A. Burrus, "Densely Spaced DFM Coherent System With Near Quantum-Limited Sensitivity and Computer - Controlled Random Access Channel Selection", Elect. Letters, Vol. 25, No. 14, pp. 883-885, July 1989.

10. M. Okai, N. Chinone and T. Harada "Narrow Linewidth Tunable Multiple Quantum Well x/y Shifted Distributed Feedback Laser", Technical Digest OFC'90, San Francisco, 22-26 Jan. 1990.

11. B. Glance, O. Scaramucci, T. L. Koch and J. Stone, "Optical Frequency Synthesizer", Elect. Letters, Vol. 25, No. 17, Aug. 1989.

12. B. Glance and O. Scaramucci, "Optical Heterodyne Receiver Providing Random Channel Selection", Elect. Letters, Vol. 25, No. 19, Sept. 1989.

13. J. Stone, "Optical-fibre Fabry-Perot Interferometer With Finesse of 300", Elect. Letters, Vol. 21, No. 11, May 1985.

14. B. Glance, J. Stone, K. J. Pollock, P. J. Fitzgerald, C. A. Burrus, B. L. Kasper and L. W. Stulz, "Densely Spaced FDM Coherent Star Network With Optical Signals Confined to Equally Spaced Frequencies", Journal of Lightwave Techn., Vol. 16, No. 11, November 1988.

An Optical Phase-Locked Loop Using an Acousto-optic Frequency Shifter

J. Nishikido and A. Himeno

NTT Communication Switching Laboratories, 3-9-11 Midori-cho, Musashino-shi, Tokyo 180, Japan

Abstract. A new optical phase-locked loop for coherent optical signal processing is presented. Experimental phase locking is successfully demonstrated for periods of over 10 minutes.

1. Introduction

The optical phase-locked loop (PLL) is the most important component for future fiber-optic communications systems. Application of these PLLs focuses on homodyne receivers [1, 2] in coherent optical transmission systems. Other significant applications may also include optical frequency synchronization for dense frequency division multiplexing networks and phase locking circuits in the field of coherent optical signal processing such as discrete Fourier transform [3]. For these other applications it is indispensable to lock the relative optical phases between modulated lights. In this paper an optical PLL using an external optical frequency shifter is proposed for that purpose.

2. Configuration

The proposed PLL consists of an external optical frequency shifter, a voltage controlled oscillator (VCO), an optical coupler, and a photodetector, as shown in Fig. 1. To lock the relative optical phases between more than two modulated lights, unlike homodyne receivers, the external optical frequency shifter is used instead of a local light source. In a case where light 2 is generated by a local light source, the PLL is equivalent to the conventional one. The VCO integrates phase error and

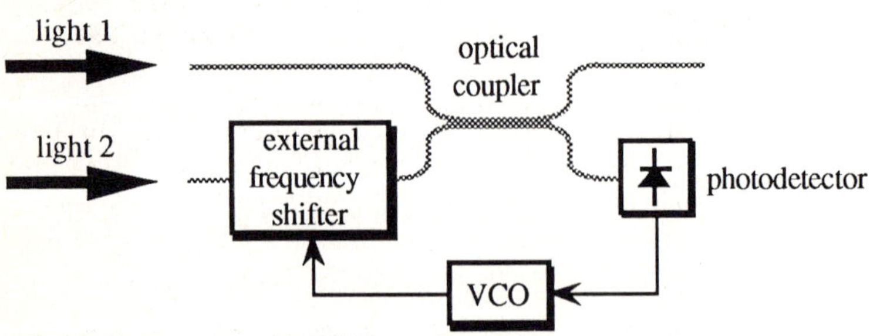

Fig. 1. Proposed optical PLL configuration

Springer Series in Electronics and Photonics, Vol. 29 **Photonic Switching II**
Editors: K. Tada · H.S. Hinton © Springer-Verlag Berlin, Heidelberg 1990

drives the external frequency shifter. This phase error is then fed back to the external frequency shifter via the VCO.

3. Experimental results

The experimental setup is shown in Fig. 2. Diode pumped solid state ring YAG lasers oscillating at a 1.319 μm wavelength were used for the frequency-stabilized light sources. The oscillation frequency of the YAG lasers stabilized within ± 2 MHz. An acoustooptic (AO) modulator composed of PbMoO4 AO medium, optical beam collimators, and pigtail optical fibers was used as the external frequency shifter. The minimum insertion loss of the AO frequency shifter, which was driven by 1 W of acoustic power, was 5 dB at a driving frequency of 93.5 MHz, as shown in Fig. 3. The optical 3 dB bandwidth of the frequency deviation was more than 12 MHz. The tunable range of the VCO was 3 MHz wide at the center frequency of 93.5 MHz. For stable operation, optical polarization maintaining fibers were used for the

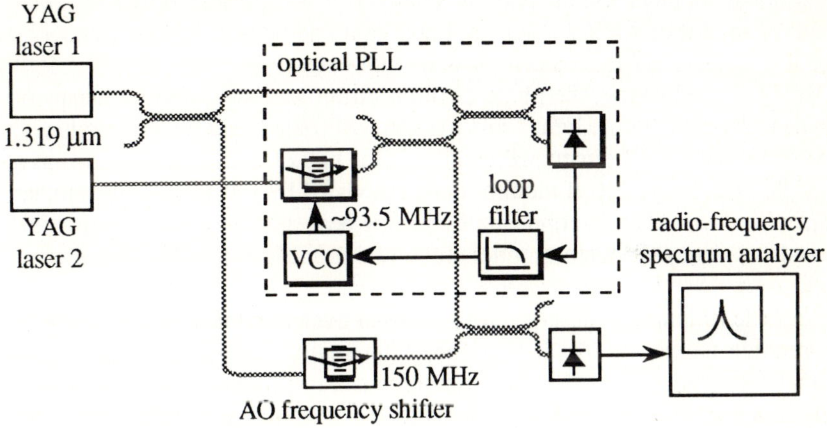

Fig. 2. Experimental setup

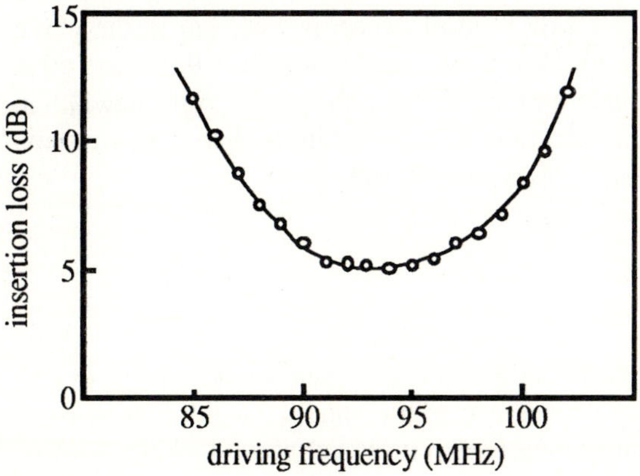

Fig. 3. Insertion loss characteristics of AO frequency shifter

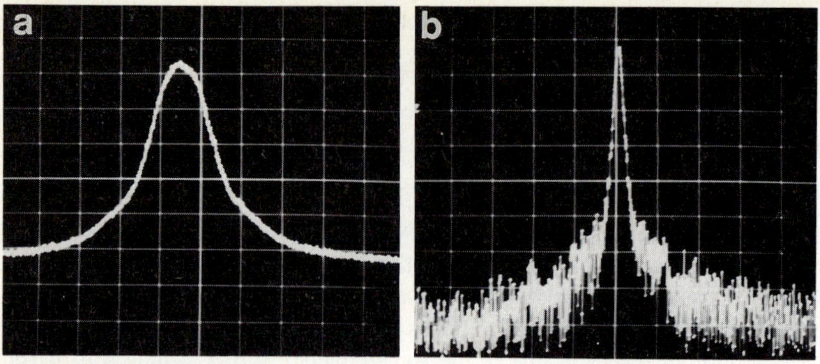

(horizontal: 10 kHz/div., vertical: 10 dB/div.)

(horizontal: 1 kHz/div., vertical: 10 dB/div.)

Fig. 4. a Frequency-deviated YAG laser 2 spectrum and **b** phase error spectrum

3 dB directional couplers and the pigtails. Phase error was processed by a lag filter with a time constant of $\tau = 9.9 \times 10^{-5}$ s. Loop gain and natural angular frequency were set at $K = 4.2 \times 10^7$ rad/s and $\omega_n = 6.5 \times 10^5$ rad/s.

To verify phase locking, the phase error spectrum was measured by means of heterodyne technique. Before light 1 and frequency-deviated light 2 were combined, they were split using 3 dB directional couplers. One half of each light was used as a monitor light. The frequency of monitor light 1 deviated by 150 MHz using another AO frequency shifter. The intermediate frequency beat spectrum, between monitor light 2 and frequency-deviated monitor light 1, was measured as the phase error spectrum.

Stable phase locking was observed for periods of over 10 minutes and was broken only by relative frequency drift between the light sources. A pull-in range of 300 kHz and a lock range of 2.8 MHz were measured and respectively agreed well with the pull-in range of 400 kHz and the lock range of 3 MHz that were calculated from the loop parameters and VCO characteristics. The frequency-deviated light 2 spectrum was measured by the delayed self-heterodyne method [4], as shown in Fig. 4a. The spectrum was equivalent to the phase error spectrum in the free running state. The full width at half maximum was 7.5 kHz. The phase error spectrum in the phase locking state is also shown in Fig. 4b. The measured linewidth of the phase error spectrum was less than the measurable limit of 100 Hz. The phase error linewidth in the phase locking state narrowed to less than 1/75th that in the free running state. This result confirms that the optical phases were well locked.

4. Conclusion

An optical phase-locked loop using an external optical frequency tuner was proposed. Phase locking was demonstrated by the stable operation and linewidth narrowing of the phase error spectrum. This configuration allows optical phase locking between modulated lights and is useful for coherent optical signal processing and optical frequency synchronization.

Acknowledgment. The authors wish to thank Dr. Ikutaro Kobayashi and Ken-ichi Yukimatsu for their helpful discussions and encouragements.

References

1. D. J. Malyon, D. W. Smith, R. Wyatt: Electron. Lett. **22**, 421-422 (1986)
2. J. M. Kahn, B. L. Kasper, K. J. Pollock: Electron. Lett. **25**, 626-628 (1989)
3. M. E. Marhic: Opt. Lett. **12**, 63-65 (1987)
4. T. Okoshi, K. Kikuchi, A. Nakayama: Electron. Lett. **16**, 630-631 (1980)

A 2.5 Gb/s Optical FDM Switch Experiment with a 15,000 GHz Tuning Range

M.A. Santoro[1], *K.Y. Eng*[1], *J. Stone*[2], *and T.L. Koch*[1]

[1]AT&T Bell Laboratories, Crawfords Corner Road, Holmdel, NJ 07733, USA
[2]AT&T Bell Laboratories, Crawford Hill Laboratory, Holmdel, NJ 07733, USA

Abstract. We describe a 2.5 Gb/s optical FDM switch using frequency-locked monolithic tunable lasers, a star coupler, and a computer-controlled tunable two-stage fiber Fabry-Perot filter.

1. Introduction

A star-based optical FDM approach to designing a switch offers the possibility of large sizes (1000×1000) and high data rates (>1 Gb/s). FDM techniques have been successfully demonstrated in experiments on star-based local network applications [1,2]. However, a star network has many practical challenges due to the geographic separation of the users, e.g. synchronization, control of the FDM comb, polarization maintenance, channel reservation and so on. Most of these difficulties can be minimized in an optical switch environment where the components are confined to a "box." In an FDM optical cross-connect switch, high-speed data from various inputs modulate different lasers at different frequencies (FDM comb). The laser signals are then combined in a star coupler, and a frequency tunable receiver is used at each output port to retrieve an input signal of choice. This arrangement of fixed transmitter/tunable receiver facilitates broadcasting and multicasting in the switch. The tunable receiver in an FDM switch can be implemented using a coherent approach with a tunable laser LO [3], or using a direct-detection approach with a tunable optical filter. We describe here the latest results of an on-going experiment using monolithic tunable lasers and a tunable two-stage fiber Fabry-Perot (FFP) filter covering 15,000 GHz. Our laboratory demonstration is believed to be the first such system achieving the widest tuning range (15,000 GHz) at 2.5 Gb/s.

2. Experiment and Results

The experimental set-up is depicted in Fig. 1. The four transmitting lasers are all two-section MQW DBR lasers, each tunable over ≈1,000 GHz in the 1.5 μm region [4]. Each laser is coupled to the PM fiber pigtail of a 2×2 $Ti{:}LiNbO_3$ waveguide switch element through an optical isolator

Springer Series in Electronics and Photonics, Vol. 29 **Photonic Switching II**
Editors: K. Tada · H.S. Hinton © Springer-Verlag Berlin, Heidelberg 1990

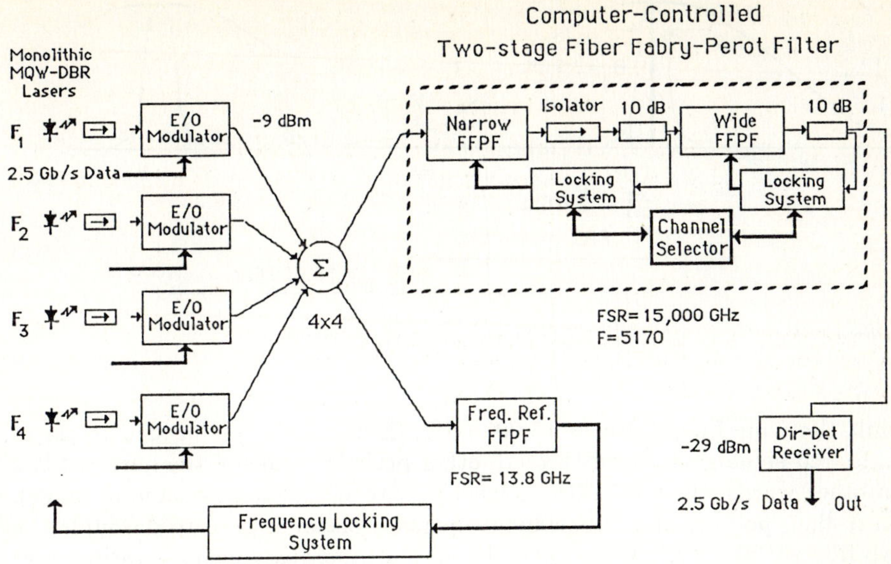

Figure 1: Experimental Set-up

(isolation > 70 dB). The waveguide switch was used as an external modulator providing on/off modulation at 2.5 Gb/s. The four transmitting laser signals are summed in a 4×4 wavelength-independent star coupler (excess loss = 0.5 dB). The power coupled into the star was -9 dBm. One of the star coupler outputs is connected to a fixed FFP filter (FSR = 13.8 GHz, finesse = 25) providing a reference frequency comb for stabilizing the input laser frequencies [1,5]. The transmitting laser frequencies are each dithered with a slightly different signal for the purpose of channel identification and locking. The frequency locking circuits can resolve dithering signals as close as 10 Hz apart, and the frequency locking system can support at least 1,000 channels. The reference FFP filter itself was found to be sufficiently stable in the MHz range under normal laboratory room temperature condition without special stabilization control. In our laboratory demonstration, two lasers (F_1 and F_4) were locked to channels separated by ≈600 GHz; F_3 was midway between F_1 and F_4, and F_2 was positioned just one channel (13.8 GHz) away from F_1.

A star coupler output is connected to a tunable direct-detection receiver which can receive any of the four transmitting signals. The tunable receiver consists of a tunable two-stage (i.e., two in tandem) FFP filter followed by a direct-detection receiver. The two-stage approach for increasing the overall filter finesse was first used by Kaminow et al.[6] in a Vernier configuration. In contrast, our design employs a narrow filter followed by a wide filter. Each of the FFP filters is computer controlled for automatic tuning, and their characteristics are superimposed in Fig. 2. The narrow filter (FSR = 700 GHz, finesse = 240) provides the desired filter bandwidth (2.9 GHz)

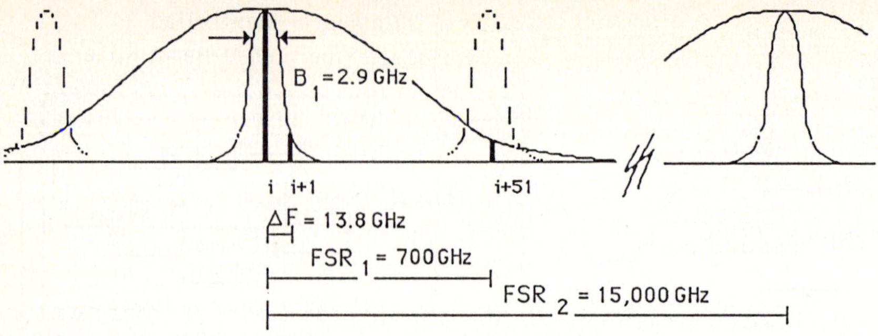

Figure 2: Narrow and Wide Filter Characteristics

while the wide filter (FSR = 15,000 GHz, finesse = 110) is needed to reject undesired signals admitted through other periodic peaks of the narrow filter (in other words, the wide filter is used to extend the effective tuning range). An in-line, polarization-independent optical isolator is used to decouple the two filters. The FFP filters currently in our laboratory demonstration were not optimized to yield minimum loss (total loss = 11 dB). Efforts are being pursued to fabricate lower-loss filters (total loss reduced to 5 dB perhaps). Tuning of the filters is done by ramping the voltage applied to the PZT element of the filters until the correct channel signature is detected. The narrow filter remains locked to the desired channel dithering signal, while the wide filter is locked to yield maximum power. The overall finesse of the combined filters was measured to be $\approx$5170. The direct-detection receiver consists of an InGaAsP APD/preamp module followed by a two-stage delay-line equalizer and amplifier. At 2.5 Gb/s, it provides a sensitivity of -29 dBm (BER of 10^{-10}). To assess the interference effects due to other carriers, the FFP filters were removed, and one desired channel was set above other interfering carriers. A 0.5 dB power penalty was measured when the total carrier-to-interference ratio (CIR) was 10 dB. In the present filter design, the expected CIR is $\approx$ 14 dB.

3. Discussion and Conclusions

We have described a 2.5 Gb/s FDM optical switch experiment using a tunable two-stage FFP filter. The FFP filter offers the advantage of being a compact component potentially suitable for future commercial applications. Its tuning range in our two-stage design is large enough to support a 1000$\times$1000 multi-Gb/s cross-connect optical switch. Furthermore, the FFP filters can be made polarization insensitive, making them ideal to use in an FDM-star switch fabric. However, power budget considerations requires an increase in launched laser power into each input of the star coupler proportional to the number of input lines implemented. For a 1000$\times$1000 switch, a 10 dBm launched power into the star coupler is needed if the coupler excess and FFP filter losses can be reduced to < 10 dB. One option is to use an optical amplifier which is rapidly becoming a viable technology. In any case, the extraordinary tuning range of a two-stage FFP filter design is demonstrated in this experiment.

REFERENCES

[1] B. S. Glance et al., "Densely Spaced FDM Coherent Star Network With Optical Signals Confined to Equally Spaced Frequencies," *J. Lightwave Technol.*, November 1988.

[2] L. G. Kazovsky and R. E. Wagner, "Multichannel Coherent Lightwave Technology," *ICC '88.*

[3] K. Y. Eng et al., "An FDM Coherent Optical Switch Experiment With Monolithic Tunable Lasers Covering a 1,000 GHz Range," *Photonic Switching Topical Meeting*, Salt Lake City, Utah, March 1-3, 1989.

[4] T. L. Koch et al., "High Performance Tunable 1.5 μm InGaAs/InGaAsP Multiple-Quantum-Well Distributed-Bragg-Reflector Lasers," *Appl. Phys. Lett.*, September 1988.

[5] J. Stone and L. W. Stulz, "Pigtailed High-Finesse Tunable Fibre Fabry-Perot Interferometers With Large, Medium and Small Free Spectral Ranges," *Electronics Letters*, 16 July 1987.

[6] I. P. Kaminow et al., "A Tunable Vernier Fiber Fabry-Perot Filter for FDM Demultiplexing and Detection," *IEEE Photonics Tech. Lett.*, January 1989.

Application of Optical FSK Signals to Wavelength-Division Switching Systems

M. Nishio [1], *S. Suzuki* [1], *T. Numai* [2], *and M. Fujiwara* [2]

[1]C & C Systems Research Laboratories, NEC Corporation,
4-1-1 Miyazaki, Miyamae-ku, Kawasaki 213, Japan
[2]Opto-electronics Research Laboratories, NEC Corporation,
4-1-1 Miyazaki, Miyamae-ku, Kawasaki 213, Japan

Abstract. A wavelength-division switching network, using optical FSK signals, has been proposed. It has low crosstalk characteristics and can achieve greater wavelength-division multiplexity than that using intensity-modulated signals.

1. Introduction

A photonic wavelength-division (WD) switching system is expected to provide large-line-capacity for broadband signals[1]. The possibility of 8-WD multiplexity in the switching network, using phase-shift-controlled DFB LD tunable wavelength filters[2], has been reported[3]. With the aim of increasing WD multiplexity, much effort has been put into expanding the wavelength tuning range for the tunable filter. However, even if a wide tuning range is achieved, the multiplexity in the conventional WD switching network, using Intensity-Modulated (IM) signals, will be limited by the crosstalk characteristics. This paper proposes the application of optical Frequency-Shift-Keying (FSK) signals to WD switching systems. It enables the switching system to improve WD multiplexity and also to combine with the FSK wavelength-division multiplexed (FSK-WDM) transmission system[4]. Optical amplifiers can be easily introduced to both systems, due to low cross modulation for the FSK-WDM signal[5]. To confirm the advantage of the multiplexity improvement in the proposed WD switching network, the WD multiplexity, taking crosstalk characteristics into account, is studied and the results compared with those for the switching network using IM signals.

2. WD switching network using FSK signals

Figure 1 shows a new wavelength switch (λ-switch), using FSK signals. An input FSK-WDM signal is split and transmitted to individual tunable wavelength filters. Each tunable filter selects a wavelength corresponding to either signal logic level "1" or "0" for a specific FSK signal, so that the FSK signal can be converted to an IM signal. The IM signal from the tunable filter is applied to an FSK modulator with optical input, and is

Springer Series in Electronics and Photonics, Vol. 29 Photonic Switching II
Editors: K. Tada · H.S. Hinton © Springer-Verlag Berlin, Heidelberg 1990

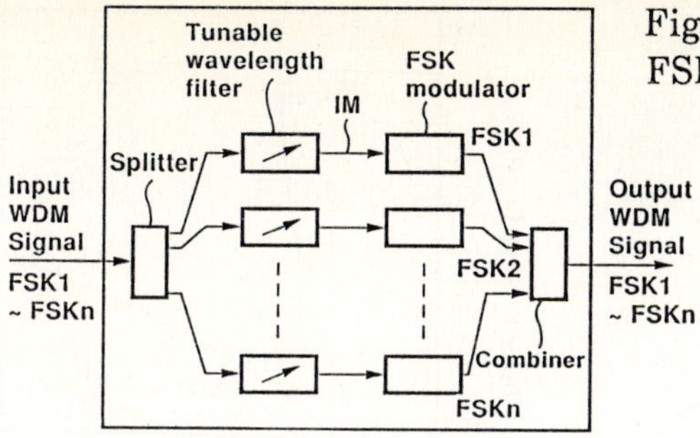

Fig. 1 Wavelength switch using FSK signals.

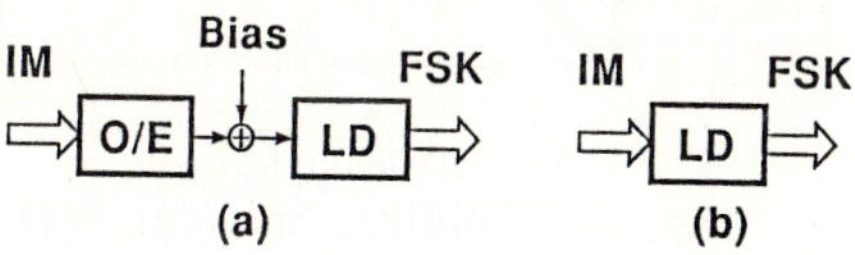

Fig. 2 FSK modulators with optical input.

FSK modulated onto a pre-assigned-wavelength light carrier. As a result, the wavelength can be interchanged between FSK signals.

There are two kinds of FSK modulators with optical input. One is the FSK modulator, using an O/E (opto-electronic) converter and an LD (laser diode) combination, as shown in Fig. 2(a). The other uses an IM signal injection into the DFB LD[6], as shown in Fig. 2(b). When an external IM signal is injected into the DFB LD, the electron carrier density changes, the refractive index also changes, and, the output frequency for the DFB LD shifts, according to the injected IM signal intensity. As a result, FSK modulation can be achieved by IM signal injection into the DFB LD. The FSK modulator in Fig. 2(b) indicates the possibility of fabricating an all-optical WD switching network.

In this λ-switch, crosstalk is the main cause of the FSK modulator input signal extinction ratio degradation. Thus, a switching network using FSK signals has better crosstalk characteristics than that using IM signals. To confirm this, the following section shows results of calculating WD multiplexity n for the λ-switch, using the FSK modulator shown in Fig. 2(a).

3. WD multiplexity comparison between FSK and IM signals

Figure 3 shows a phase-shift-controlled DFB LD tunable filter transmission spectrum used for the calculation. It is assumed that the pass-band shape is Lorentzian with 1-Å 10-dB down bandwidth , and that there are 5-Å spaced 1-dB Lorentzian sidelobes in the out-band with X= 16-dB attenuation from the peak gain. Figure 4 shows the relation between WD multiplexity n and power penalty (P.P.) due to crosstalk at a 10^{-10} error rate in the O/E converter, where channel separation $\Delta\lambda c$ is 1 Å and the wavelength deviation $\Delta\lambda d$ for the FSK signals is 0.25 Å. It can be

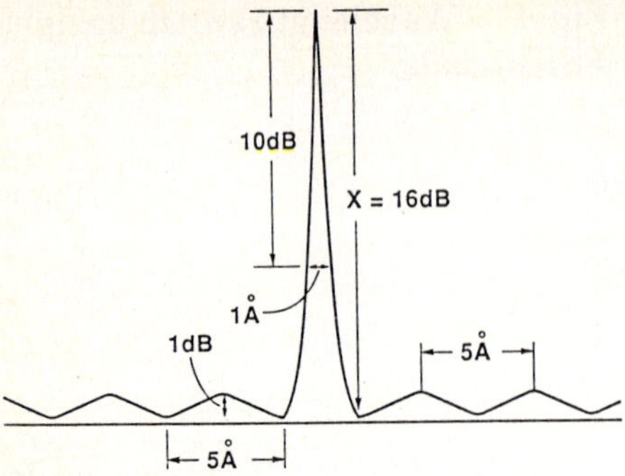

Fig. 3 Simplified tunable filter transmission spectrum.

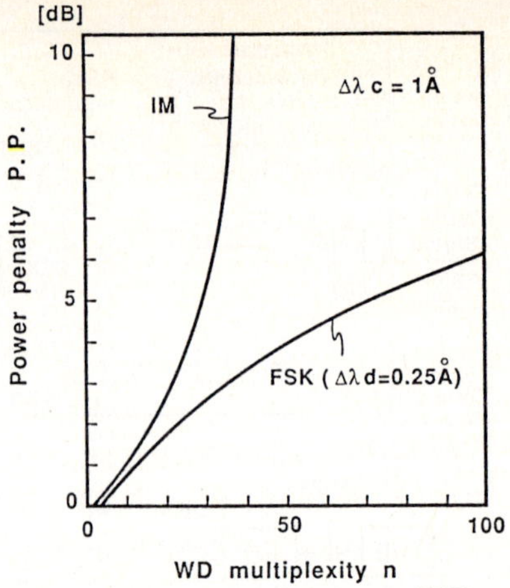

Fig. 4 Relation between WD multiplexity and power penalty.

concluded that the n value in the switching network, using FSK signals, is much larger than that using IM signals, if the large P.P. can be permitted. Figures 5 and 6 show required channel separation $\Delta\lambda c$ and wavelength tuning range W, as function of the n value for 1-dB and 4-dB P.P., respectively. When permitting 1 dB for P.P., the n values are 11 and 16 for IM signals and FSK signals, respectively. When allowing 4 dB for P.P., the n value for FSK signals can reach 50, if 50 Å can be obtained for W, while the n value for IM signals cannot reach more than 25. It is therefore

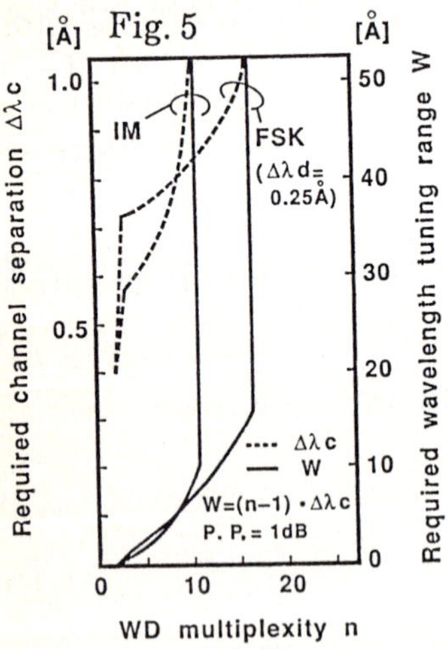

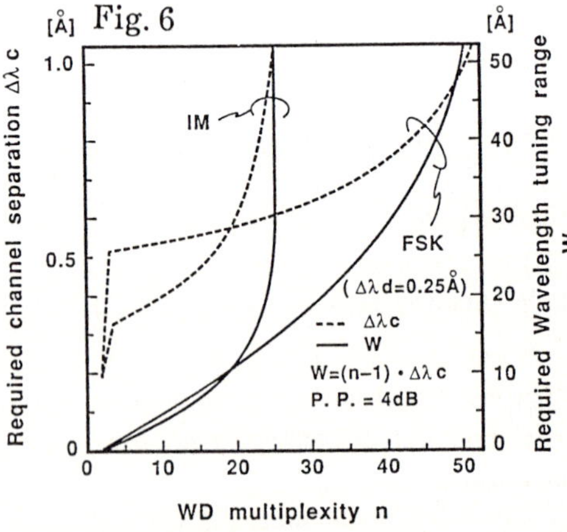

Fig. 5 Required channel separation and wavelength tuning range as a function of WD multiplexity for 1-dB power penalty.

Fig. 6 Required channel separation and wavelength tuning range as a function of WD multiplexity for 4-dB power penalty.

concluded that the crosstalk characteristics in the WD switching network using FSK signals is better than that using IM signals. The same advantage, as mentioned above, can be expected for the λ-switch using the FSK modulator shown in Fig. 2(b), because the crosstalk characteristics in the input IM signal for the FSK modulator are common to that for the FSK modulator of Fig. 2(a).

4. Conclusion

Application of FSK signals to WD switching systems has been proposed to improve WD multiplexity. Taking crosstalk characteristics into account, it has been shown that greater WD multiplexity in the switching network, using FSK signals, can be achieved, compared to that using IM signals. The proposed WD switching network is suitable for fabricating large-line-capacity WD switching systems in the broadband communications network with FSK-WDM transmission systems.

References

1. S. Suzuki and K. Nagashima, "Optical broadband communications network architecture utilizing wavelength-division switching technology," Topical Meeting on Photonic Switching, Salt Lake City, U.S.A., ThA2, pp. 21-23, March, 1987
2. T. Numai, S. Murata and I. Mito, "1.5μm tunable wavelength filter using a phase-shift-controlled distributed feedback laser diode with a wide tuning range and a high constant gain," Appl. Phys. Lett., 54, pp. 1859-1860,1989
3. M. Nishio, T. Numai, S. Suzuki, M. Fujiwara M. Itoh and S. Murata, "Eight-channel wavelength-division switching experiment using wide-tuning-range DFB LD filters," European Conference on Optical Communication, Conference Publication Number 292-Part 2, pp.49-52, 1988
4. I. P. Kaminow, P. P. Iannone, J. Stone and L. W. Stulz, "FDMA-FSK star network with a tunable optical filter demultiplexer," IEEE J. Lightwave Technol., 6, pp. 1406-1414, 1988
5. T. Saitoh and T. Mukai, "Recent progress in semiconductor laser amplifiers," IEEE J. Lightwave Technol., 6, pp.1656-1664, 1988
6. K. Inoue and N. Takato, "Wavelength conversion for FM light using light injection induced frequency shift in DFB-LD," Electron. Lett., 25, pp.1360-1362, 1989

Photonic Wavelength-Division Switching Network Using a Parallel λ-Switch

M. Nishio and S. Suzuki

C&C Systems Research Laboratories, NEC Corporation,
4-1-1 Miyazaki, Miyamae-ku, Kawasaki 213, Japan

Abstract. A parallel λ-switch, which offers both space and wavelength switch functions, can reduce number of wavelength filters and converters. It has a growable structure and non-blocking one-to-many connections capability.

1. Introduction

A photonic wavelength-division (WD) switching system is suitable for fabricating broadband networks[1]. For fabrication of large-line capacity switching networks, a unit switch with large multiplexity is necessary. This paper proposes a new WD unit switch, a parallel λ-switch, which has large multiplexity and can reduce optical devices. This paper also reports that the numbers of required wavelength filters and converters for the parallel λ-switch are much fewer than those for the conventional WD switching networks. The modular growth and one-to-many connections capability without internal blocking, for the proposed switch, are also studied. Moreover, a multi-stage switching network, using parallel λ-switches, is shown.

2. WD unit switch

Figure 1 shows a conventional unit switch, a crosspoint matrix switch[2]. In this structure, wavelength switches (λ-switches), which consist of tunable wavelength filters and wavelength converters, are arranged at every crosspoint. The multiplexity for this unit switch is m·n, where m and n are the numbers of input WD multiplexed (WDM) signals and WD multiplexity, respectively. The greatest drawback of this crosspoint matrix switch is that it requires large numbers of tunable wavelength filters and wavelength converters.

To overcome this problem, a new unit switch, a parallel λ-switch, is proposed, as shown in Fig. 2. The parallel λ-switch consists of multi-input sub-switches, arranged in parallel with each optical WDM output signal. All WDM input signals are split into every sub-switch. In the sub-switch, an mx1 optical switch transmits one of the m WDM input signals to a

Springer Series in Electronics and Photonics, Vol. 29 **Photonic Switching II**
Editors: K. Tada · H.S. Hinton © Springer-Verlag Berlin, Heidelberg 1990

Fig. 1 Crosspoint matrix switch

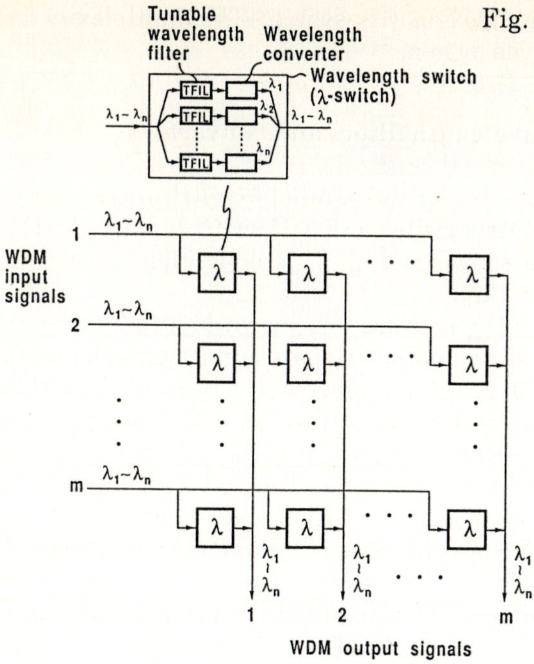

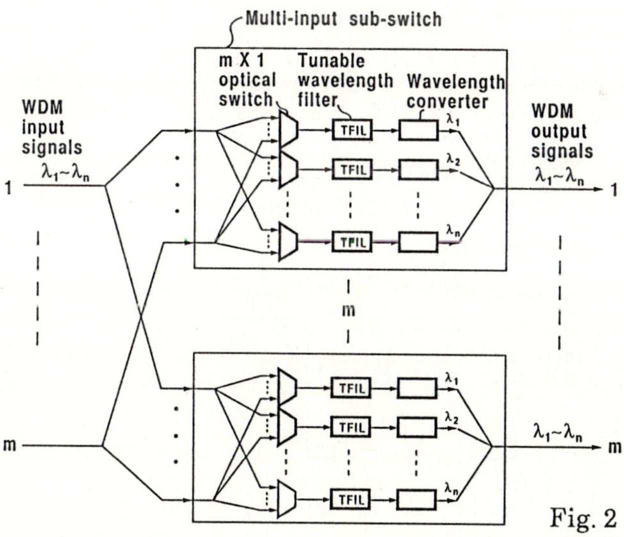

Fig. 2 Parallel λ-switch

tunable filter, which selects a specific wavelength signal. The tunable filter output signal is applied to a wavelength converter. Therefore, an arbitrary wavelength signal multiplexed on an arbitrary optical WDM input signal can be converted to a pre-assigned wavelength signal. Wavelength converter output signals are multiplexed to form individual WDM output signals. As a result, an equivalent space-division switching network for the parallel λ-switch is a matrix switch. The multiplexity for the parallel λ-switch can be given by m·n.

3. Numbers of required wavelength filters and converters

Numbers of required optical devices for the parallel λ-switch are compared with that for the crosspoint matrix switch and a λ^3 switching network[1]. The λ^3 switching network, as shown in Fig. 3, is equivalent to a non-blocking three-stage Clos network.

The results for each switching network are shown in Table.1, where numbers of optical WDM input and output signals are set to be equal to WD multiplexity n. Therefore, total line-capacity N is n^2. In this comparison, a wavelength demultiplexer (DMUX) and a multiplexer (MUX) in the λ^3 switching network are assumed, consisting of n fixed wavelength filters and a combiner, respectively. The mx1 optical switch in the parallel λ-switch is composed of m gate switches and a combiner.

Device structures of wavelength filters and converters are generally more complex than those of optical gate switches. Therefore, it is important to compare the total amount of required wavelength filters and converters, as a function of line-capacity N. The results are shown in Fig. 4. It can be concluded that the number of these devices required for the parallel λ-switch is much fewer than those for other switching networks.

As for modularity, the parallel λ-switch has an advantage that multi-input sub-switches can be set in proportion to the required line-capacity.

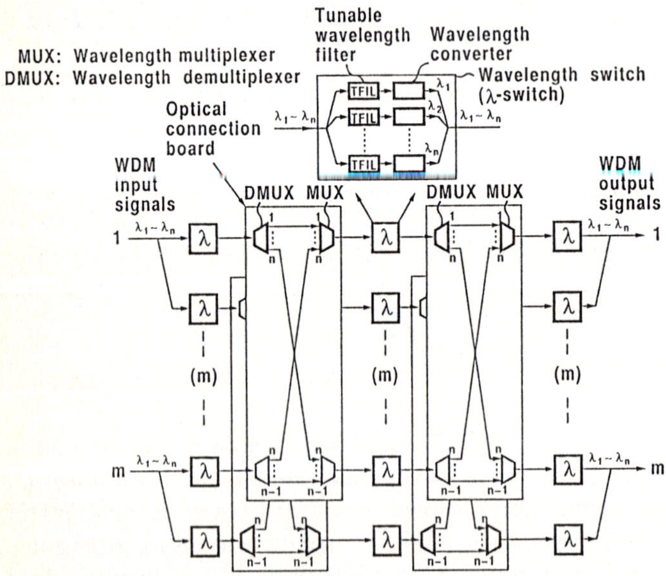

Fig. 3 λ^3 switching network

Table 1. Number of required optical devices

	Line capacity N	Wavelength filter		Wavelength converter	Gate switch
		Tunable	Fixed		
Parallel λ-switch	n^2	n^2	—	n^2	n^3
Crosspoint matrix	n^2	n^3	—	n^3	—
λ^3	n^2	$6n^2$	$\simeq 4n^2$	$6n^2$	—

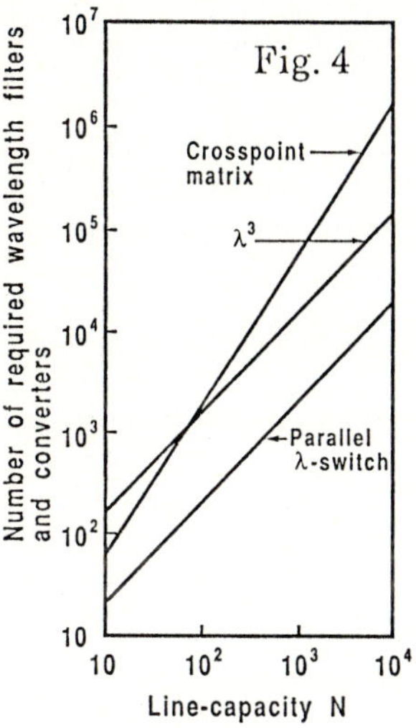

Fig. 5

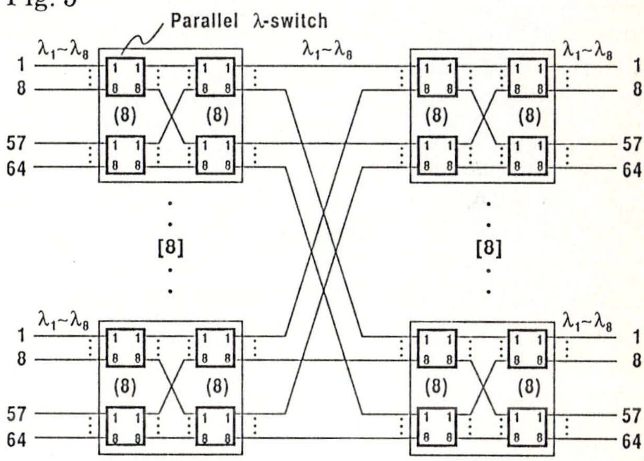

Fig. 4 Number of required wavelength filters and converters as a function of line-capacity N

Fig. 5 4,096-line four-stage junctor network using parallel λ-switches

Moreover, from Fig. 2, it is evident that the parallel λ-switch can make one-to-many connections without internal blocking.

While the WD multiplexity will remain small in the near future, large line capacity switching networks can be constructed using multi-stage switching networks with the parallel λ-switches. Figure 5 shows a 4,096-line four-stage junctor switching network, using 8×8 parallel λ-switches as unit switches, which have only 8 WD multiplexity.

4. Conclusion

A new wavelength-division switching network using the parallel λ-switch has been proposed. This proposed switch requires much fewer wavelength filters and converters than conventional switching networks. The parallel λ-switch can provide high modularity and make one-to-many connections without internal blocking. Moreover, it has been shown that a 4,096-line four-stage junctor switching network can be constructed using the parallel λ-switches as unit switches.

References

1. S. Suzuki, and K. Nagashima, Topical Meeting on Photonic Switching, ThA2, pp. 21-23, March, 1987
2. B. Strebel, E. J. Bachus, and J. Vathke, IEEE Global Telecommunications Conference & Exhibition, Conference Record Vol. 1, pp. 32-36, 1989

Part VIII

Self-Routing Switching

Ultrafast Photonic Packet Switch with Optical Output Buffer

Y. Shimazu, M. Tsukada, and S. Kikuchi

NTT Communication Switching Laboratories, 3-9-11 Midori-cho, Musashino-shi, Tokyo 180, Japan

ABSTRACT A new ultrafast photonic packet switch is based on a time-division multiplexing distribution network with novel fiber optic output buffers for throughput control.

1. Introduction

The asynchronous transfer mode (ATM) is considered to be a key multiplexing mode for supporting various bit-rates of services in future communication networks. Space division packet switches for the ATM have been extensively studied. They are classified into input-buffer switches [1], crosspoint-buffer switches, and output-buffer switches [2]. The output-buffer switch has the greatest potential because it has simple control logic and excellent traffic characteristics [1].

This paper presents an ultrafast photonic ATM (ULPHA) output-buffer switch based on a time-division multiplexing (TDM) network [3]. In this switch, the output buffering is realized simply by optical delay lines and switches.

2. ULPHA Switch Description

The ULPHA switch is an n-input, n-output packet switch based on an optical time division "broadcast and select" network (Fig.1), in which the bit-rate of electrical input and output is V and that of optical TD highways is n x V. The principle of the ULPHA switch is as follows. The packet coder converts the electrical input packets with a length of T into ultrafast optical packets with a length of T/n by modulating the optical pulses (width: t) from the gain-switched LD and compressing their periods. The packet coder at the same time generates slow address packets wavelength-multiplexed with the data packets. The star coupler time multiplexes all the packets from the packet coders on the TD highways. The filter passes certain packets only when it detects address matching. The buffer makes the interval of each arriving packet longer than T to adjust its throughput to V. The packet decoder converts the packet lengh back to T and oututs it as an electrical signal. The ULPHA switch realizes an ultra-high throughput with its simple structure. It can realize a 64 x 64 switch with 1-Gb/s inputs/outputs, since 10-20 ps optical pulses are obtained from LD [4]. Switch components are proposed in the following.

Springer Series in Electronics and Photonics, Vol. 29 **Photonic Switching II**
Editors: K. Tada · H.S. Hinton © Springer-Verlag Berlin, Heidelberg 1990

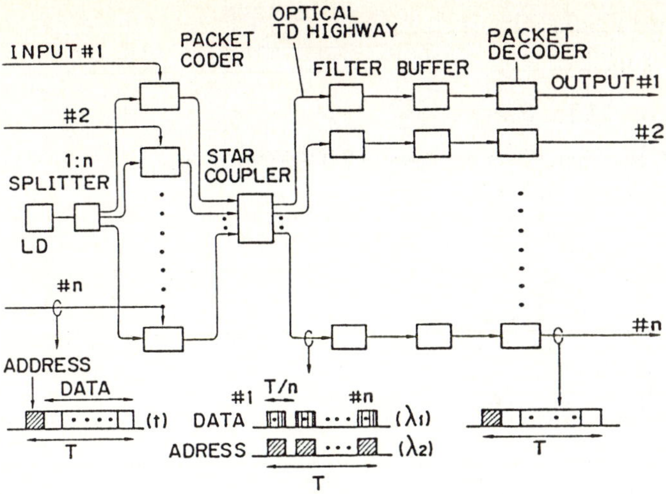

Fig. 1 Block diagram of ULPHA switch based on an optical time division network.

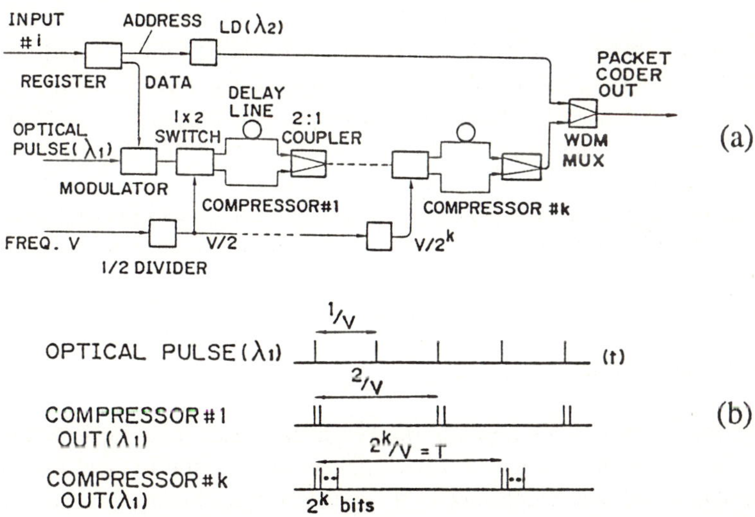

Fig. 2 (a) Configuration of packet coder. (b) Timing diagram for packet coder. Electrical RF signals with frequencies V/2 to $V/2^k$ drive each 1 x 2 optical switch.

At the packet coder (Fig. 2), a packet is stored in a register, and its data and address are separately read. The data of the packet is converted into modulated optical pulses with a width t, a bit-rate V and a wavelength λ_1. Compressor #1 sends the modulated pulses alternately to port 1 and port 2 of its 1 x 2 switch and combines them again at a separation of t. Compressors #2 to #k ($k=\log_2 L$) compress the periods of pulses further until 2^k optical pulses with a separation of t are obtained in every $2^k/V$ period. The address of the packet is converted into a slow optical signal

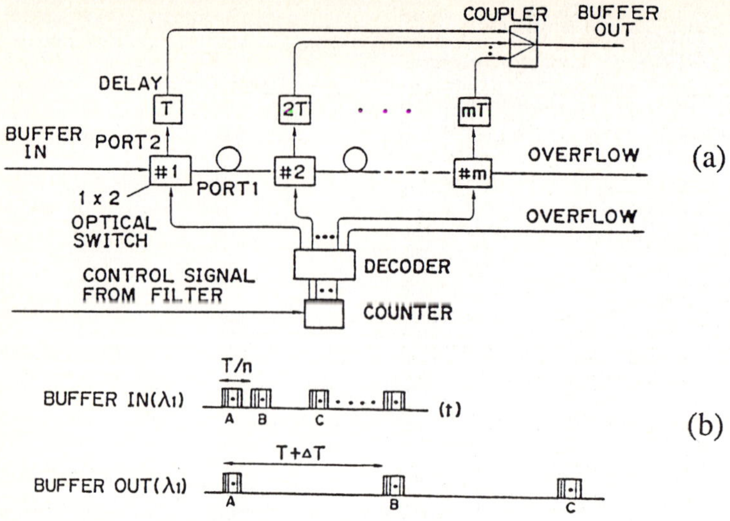

Fig. 3 (a) Buffer configuration. (b) Timing diagram for buffer.

with a length up to T/n and a wavelength λ_2 by an LD. The data and address packets are sent to the star coupler and multiplexed on the optical TD highway.

At the filter, the data and address packets from the optical TD highways are wavelength demultiplexed. The address is converted into an electrical signal and compared with the address of the filter. The optical data packet passes through the filter only when there is address matching. The electrical control signal created by the address matching is relayed to the buffer.

At the buffer (Fig. 3), the arriving data packets enter k stages of the 1 x 2 optical switch, pass through each port 1, and exit port 2 of switch #i leading to the iT delay line when a switching signal is applied to the switch #i. This switching signal comes from the electrical decoder with the counter. The counter increments at every packet arrival by a control signal from the filter, and decrements every period of T after the last packet arrival. Then, each packet comes out from the buffer output with a time interval longer than T. This buffer is like a FIFO memory in traffic control.

At the packet decoder, the packet is first divided into parallel highways. Then, optical pulses are superimposed on each bit on the parallel highways, detected by a photodetector, and stored in a register. The data in the register is read out every cycle of T with a packet bit-rate of V.

3. Feasibility Analysis and Experiment

For the hardware implementation, the packet coder requires eight stages of 1 x 2 switches with 1:2 couplers/splitters for the target 64 x 64 switch in Section 2. The total required length of the buffer is calculated using the M/D/1 model [2] as 20, 30 and 50 respectively for 10^{-3}, 10^{-6} and 10^{-9} packet loss rates with 80% switch load.

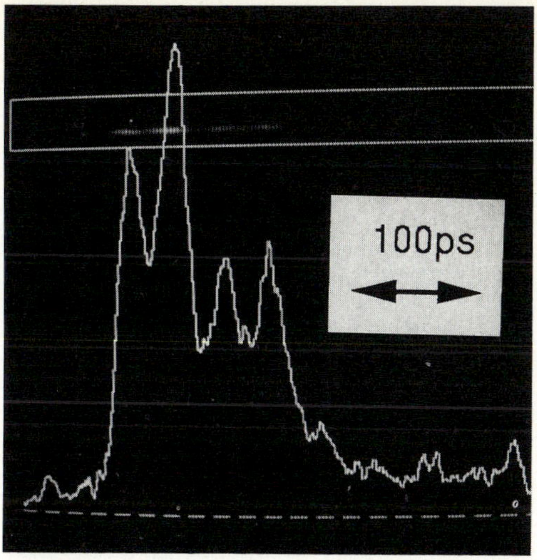

Fig. 4 A 25 Gb/s 4-bit packet monitored with a streak camera having 10-ps time resolution (all high level).

The packet coder and the buffer above can be realized, since an 8 x 8 optical switch with 64 crosspoints is available as a commercial product.

The ULPHA switch has a large power loss. The packet coder, star coupler, buffer and packet decoder have respective splitter/coupler losses of 24 dB, 18 dB, 18 dB, and 9 dB to achieve the 64 x 64 switch. This requirement in power can be satisfied by inserting erbium-doped fiber ampiifiers along the packet's path.

An experiment on the compressor of the packet coder was conducted. In this experiment, 44-ps optical pulses generated from the gain-switched DFB-LD (1.3 μm) with 1-GHz repetition rate were put into dispersion-shifted single-mode fiber, and transformed into 22-ps pulses. Using two stages of compressor made from Ti:$LiNbO_3$ waveguide optical switches and 2 x 1 couplers, a 4-bit packet signal of 25 Gb/s was generated and monitored with a synchronous streak camera having 10-ps time resolution as shown in Fig. 4.

With its high throughput and simple structure, the ULPHA switch has a great potential as a future photonic ATM switch.

References

[1] M.J. Karol, et al., " Input vs. output queueing on a space-division packet switch," Globecom '86 Conf. Rec., November 1986, pp.659-665.

[2] Y.S. Yeh, et al., "The Knockout Switch: A simple, modular architecture for high-performance packet switching," ISS '87, March 1987, pp.801-808.

[3] P.R.Prucnal, et al., "Ultrafast all-optical synchronous multiple access fiber networks," IEEE J. Select. Areas Commun., Vol.SAC-4, No.9, Dec. 1986

[4] A.Takada, et al., "High-speed picosecond optical pulse compression from gain-switched 1.3-μm DFB-LD through highly dispersive single-mode fiber," IEEE J. Lightwave Tech., Vol.LT-5, No.10, October 1987

Photonic ATM Switching Network

S. Kuroyanagi, T. Shimoe, and K. Murakami

Fujitsu Laboratories Ltd., 1015 Kamikodanaka, Nakahara-ku, Kawasaki 211, Japan

Abstract. This paper proposes a photonic ATM switching network using multi-input optical buffer memory based on wavelength division switching.

1. Introduction

Broadband ISDN (B-ISDN) is expected to offer high-speed data and full-motion video services. The Asynchronous Transfer Mode (ATM) is considered to be the most appropriate for B-ISDN because of its bandwidth flexibility and time transparency. The switching speed of ATM switches using n inputs and 1 output buffer memory is generally n times as fast as the link speed [1]. This paper proposes a photonic ATM switch using multi-input optical buffer memory based on wavelength division switching. The switching speed of this photonic ATM switch is equal to the link speed.

2. Switch Configuration

2.1 Switching Principle

Figure 1 shows the configuration of the n x n photonic self-routing module (PSRM). The PSRM consists of cell selectors and optical buffer memories. Each input interface module (IIM) identifies cells arriving at the inlet by the virtual channel identifier (VCI) in the cell header. The IIM converts the wavelength of each cell to the wavelength for the appropriate outlet decided in the call set-up phase. Using the cell

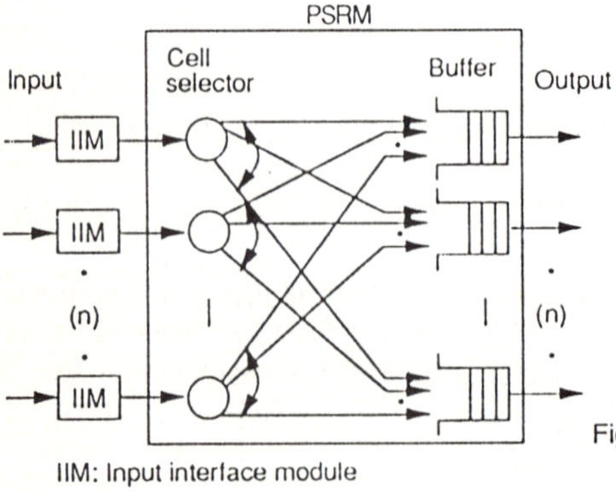

Fig.1 Photonic self-routing module (PSRM)

Springer Series in Electronics and Photonics, Vol. 29 Photonic Switching II
Editors: K. Tada · H.S. Hinton © Springer-Verlag Berlin, Heidelberg 1990

wavelength as routing information, the cell selectors route each cell arriving at the inlets to one of the optical buffer memories at the outlets. Buffer memories store the cells temporarily to prevent cell collision at the outlet.

2.2 Cell Selector Switch

The cell selector consists of a splitter and n fixed-wavelength filters as shown in Fig. 2. Cells are distributed by the splitter and are led to fixed-wavelength filters. Each filter extracts only a specific wavelength cell. For example, fixed-wavelength filter Fi extracts only the wavelength λi cell.

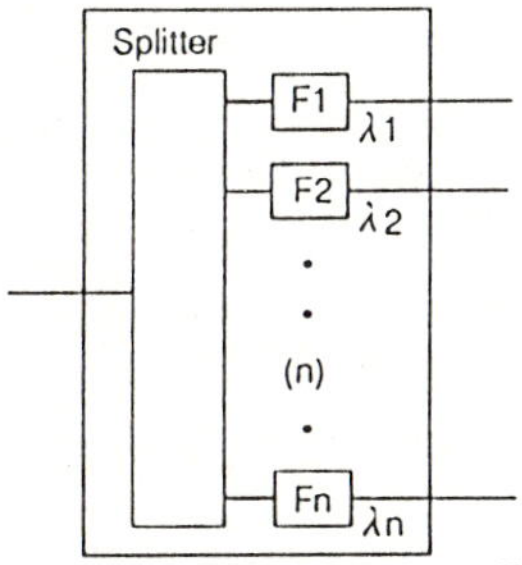

Fig.2 Cell selector switch

2.3 Optical Buffer Memory

Figure 3 shows two types of optical buffer memory. The buffer memory consists of m-linked optical memory with a capacity equal to one cell length and an n x m switch based on wavelength division switching. In type 1, the n x m switch is constructed of n wavelength conversion lasers and m fixed-wavelength filters. Type 2 consists of n fixed-wavelength conversion lasers and m wavelength tunable filters. In type 1, the wavelength conversion lasers convert the wavelength of each cell to an arbitrary wavelength on a cell-by-cell basis, and the cells are

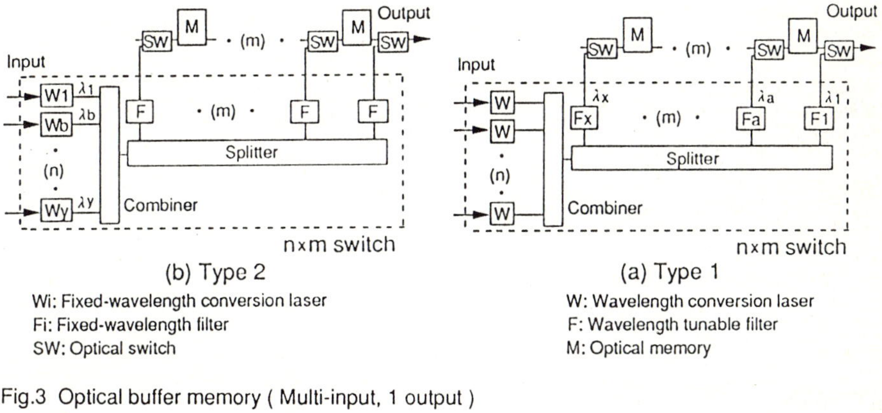

Fig.3 Optical buffer memory (Multi-input, 1 output)

distributed to the fixed-wavelength filters. Each filter extracts only a specific wavelength cell. In type 2, each fixed-wavelength conversion laser converts the wavelength of the cell to a specific wavelength. The wavelength tunable filters extract an arbitrary wavelength cell on a cell-by-cell basis. Finally, for both types, the cells are directed into one of the optical memories. For example, optical memory is constructed of fiber delay line. In this way, it is possible to construct a multi-input optical buffer memory, because n cells can be stored at the same time[2]. A photonic packet switch using only one fiber delay line has been proposed[3].

The buffer capacity at the outlet of the PSRM in the M/D/1 model is determined by the highway traffic load and the required cell-loss rate. If we select a cell-loss rate of 10^{-10} and a highway traffic load of 0.8, the buffer capacity is about 50 cells[4].

2.4 Wavelength Control Device Characteristics

The worst case for crosstalk noise in the type 1 buffer memory occurs when n cells from n inlets arrive at the same outlet at the same time. Crosstalk noise can be attenuated by suitable arrangement of the tunable wavelength of each filter. Figure 4a shows an example. The wavelength separation between the adjacent fixed-wavelength filters is more than 3 times as wide as the minimum wavelength separation. In the type 2 buffer memory, crosstalk noise is the same as in type 1 if the wavelength separation between the each fixed-wavelength conversion laser is more than 3 times as wide as the minimum wavelength separation (Fig. 4b). Figures 5a and b show the required wavelength separation and wavelength tuning range for both buffer memories. The wavelength separation of type 2 is r times as wide as that

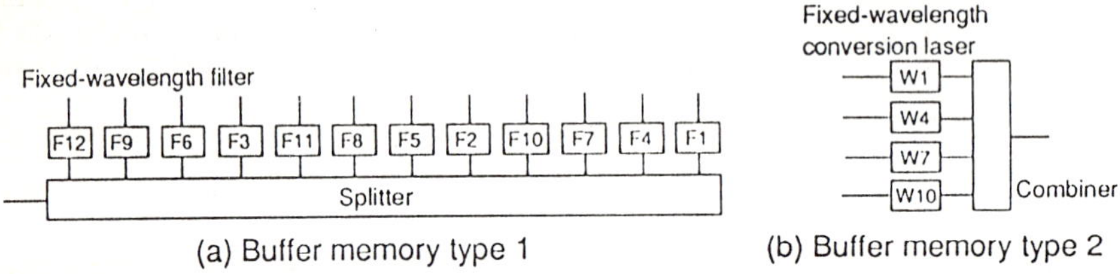

Fig.4 Wavelength assignment (n = 4, m = 12)

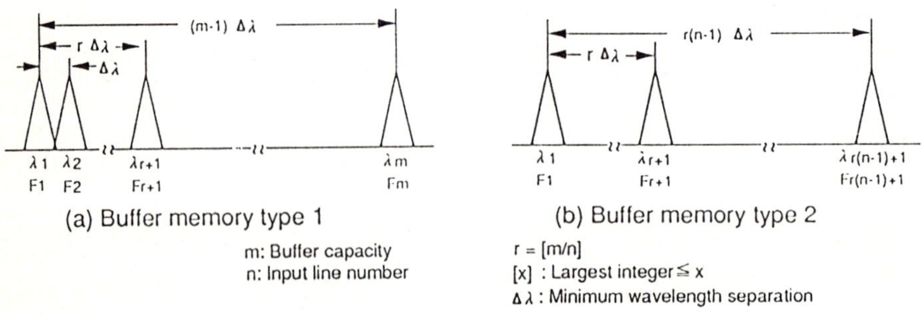

Fig.5 Wavelength filter and conversion laser characteristics

of type 1. The wavelength tuning range is $(m-1)\Delta\lambda$ for type 1, and $r(n-1)\Delta\lambda$ for type 2.

The type 1 wavelength conversion laser and the type 2 wavelength tunable filter are required for high-speed switching as the highway speed increases.

3. Multi-stage Switching Network

A large switching network is constructed by connecting multiple PSRM stages. Figure 6 shows a three-stage example. In this configuration, there are multiple paths between a first-stage PSRM and a third-stage PSRM. This allows the traffic flow to be routed efficiently between the input highway and the output highway[4]. The output interface module (OIM) replaces the VCI at the input highway with the VCI for the output highway.

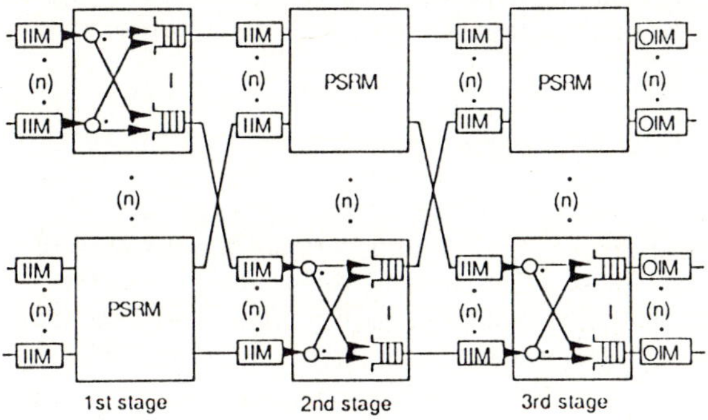

OIM: Output interface module

Fig.6 3-Stage Switching Network

4. Conclusion

This paper has proposed a photonic ATM switching network using multi-input optical buffer memory based on wavelength division switching. The switching speed is equal to the link speed.

Acknowledgment

The authors thank Dr. H. Takanashi for his encouragement and guidance. The authors also thank the members of the FUJITSU LABORATORIES Communication and Space Division.

References

[1] M. J. Karol, M. G. Hluchyj, and S. P. Morgan,"Input vs. Output Queueing on A Space-division Packet Switch," GLOBECOM'86 19.4, 1986

[2] Y. S. Yeh, M. G. Hluchyj, and A. S. Acampora, "The Knockout Switch: A Simple, Modular Architecture for High-performance Packet Switching," ISS'87 B10.2, 1987

[3] S. Suzuki, H. Suzuki, K. Kasahara, M. Fujiwara, and T. Takeuchi, "Photonic Packet Switch Using Optical Buffer Memories," Topical meeting on Photonic switching FA3, 1989

[4] K. Hajikano, K. Murakami, E. Iwabuchi, O. Isono, and T. Kobayashi, "Asynchronous Transfer Mode Switching Architecture for Broadband ISDN," ICC'88 29.3, 1988

ATM Photonic Switching Node Architecture Based on Frequency Switching Techniques

A. de Bosio, P. Cinato, B. Costa, A. Daniele, and E. Vezzoni

CSELT – Centro Studi E Laboratori Telecomunicazioni,
Via G. Reiss Romoli 274, I-10148 Torino, Italy

Abstract. An opto-electronic switching node architecture based on ATM and frequency division switching techniques is proposed. The node architecture, capable of handling frequency multiplexed packet input flows, is based on optical memories realized with fiber rings and a multiconnection network with tunable filters and frequency converters.

1. Introduction

The rapid improvements in optical devices and components are accelerating the introduction of optical technology in the telecommunication network. As far as optical switching is concerned, it is reasonable to expect that it will be first introduced in Local Area Networks (LAN), Metropolitan Area Networks (MAN), as well as Optical Digital Cross-connect Systems (ODCS); further, when the technology is mature, it will find economical applications also in the local and trunk switching nodes as well as in the Remote Switching Units (RSU).

When projecting a switching node model for the public telecommunication network, the following requirements have to be taken into consideration: the architecture must be easily integrated in the future Broadband Integrated Services Digital Network (B-ISDN) [1], [2]; the switching technique must exploit the potential of the optical components [3]; the control of the system must be capable of re-using most of the software for signalling and services, as it constitutes the major cost of the present switching systems.

Along this line an optical switching node architecture is proposed, based on the combined utilization of ATM (Asynchronous Transfer Mode) and optical frequency division switching techniques. A modular structure has been chosen, which allows an easy cooperation with existing electrical switching nodes according to an add-on philosophy.

The node has also been given the capability of switching optical frequency multiplexed packet input flows: the structure of the internal subsystems has been defined following the basic idea that, in order to exploit the wide-band of the optical components, it is necessary to operate directly on the frequency multiplexed flows.

2. ATM Opto-electronic Switching Node Architecture

An ATM opto-electronic switching node for the B-ISDN can be modelled as in Fig.1: the Exchange Termination (ET) block provides the line termination

Springer Series in Electronics and Photonics, Vol. 29 **Photonic Switching II**
Editors: K. Tada · H.S. Hinton © Springer-Verlag Berlin, Heidelberg 1990

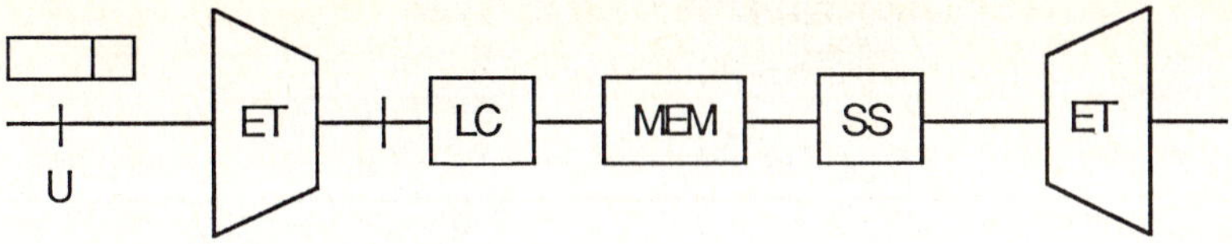

U: User-Network Interface
ET: Exchange Termination
LC: Label Converter
MEM: Memory block
SS: Space Switch

Fig. 1: ATM switching node block diagram

functions, the channel concentration, and multi-demultiplexing, as well as the synchronization functions; the Label Converter (LC) block performs the label translation, which is necessary to establish a virtual connection avoiding the use of labels unique over the whole network; the Memory (MEM) block realizes packet storage for dynamic packet contention resolution; the Space Switch (SS) block performs packet switching from one link to another.

It is reasonable to expect that different strategies of frequency introduction in the B-ISDN would mainly impact on the switching node architecture at the Exchange Termination level. In other words, there exists a switching system nucleus, made up of LC, MEM and SS blocks, which is quite general, and maintains its validity for different levels of frequency utilization at the main B-ISDN interfaces.

In this section, the architecture of an opto-electronic switching node nucleus will be investigated further with reference, in particular, to the model shown in figure 2 (patent pending), which is made up of four subsystems: the first for packet label conversion (LC), the second for packet storage (MA-OFLM), the third for space and frequency switching (SS), and the fourth for switching node control (SNC).

The model is characterized by a set of N input/output links; on each input/output link a set of n input/output optical frequencies is multiplexed; each input/output frequency is carrying a flow of packets, each identified by a label.

Every input packet is then characterized by an input triple made up of an input link, an input frequency and an input label; the switching function is performed by transforming these triples into previously defined (during the signalling phase) output triples, and associating them with the output packets.

The label conversion in the input packet is done in the Label Converter subsystem (LC): it is realized by reading a translation memory (electrical), where an output label has been written during the call setup phase and modulating the new header. This function is managed by a local control, which also performs the signalling splitting and insertion, the empty packet recognition as well as some other peripheral functions. A small fraction of the optical power is tapped out by an optical splitter, detected, amplified and sent to the control unit, that on recognizing the packet label, provides for the label replacement by driving a local laser source, to modulate and insert the new label bit sequence, through an optical switching element.

After label conversion, the packets enter the Multiple Access Optical Fiber Loop Memory (MA-OFLM) subsystem, which is based on a Frequency Conversion (FC) input stage (exploiting for instance tunable multi-electrode DFB laser diode or

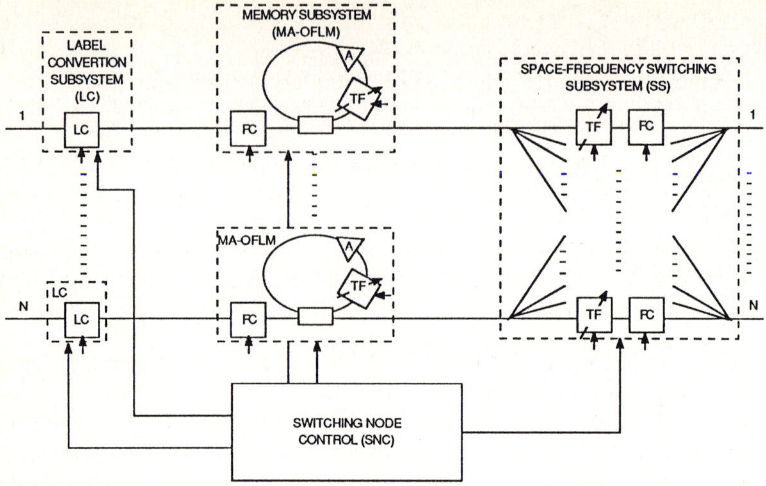

MA-OFLM: Multiple Access Optical Fibre Loop Memory
FC: set of Frequency Converters
TF: set of Tunable Filters
A: wideband amplifier

N: number of links
n: number of channels frequency multiplexed over a link
N*n: total number of channels

Fig. 2: Model of the opto-electronic ATM switching node

nonlinear effects in optical amplifiers), an optical fibre loop, and a frequency selective output stage located in the Space Switching (SS) block. In MA-OFLM every packet is assigned a frequency for storage purposes in the fiber loop, being the number of memory positions equal to the number of frequencies available at the output of the frequency converter. In particular, the memory address is identified by the frequency utilized to multiplex the packets in the loop.

After frequency conversion and storage, the packets enter a multiconnection network (the SS subsystem), based on passive 1xM/Mx1 splitters/combiners; on each internal connection, a couple made up of a tunable filter (made for instance of a tunable multi-electrode DFB filter/amplifier) and a frequency converter performs the packet selection and the frequency switching. In order to extract packets from the loop memory, the Switching Node Control (SNC) tunes a number of tunable filters in the multiconnection network to the proper frequency while, at the same time, in order to free the previously used frequencies in the memory loop, the corresponding filters in the loop are tuned on a garbage frequency.

The Switching Node Control (SNC) subsystem realizes the switching function by means of a routing algorithm similar to the one utilized in the Coprin matrix [4] based on Electrical Routing Queues (ERQ), each associated with an output link-output frequency couple, where the information related to the memory position (memory loop storage frequency) of the input packets is written. The control, after reading the ERQs sets the tunable filters and the frequency converters in SS as well as in MA-OFLM subsystems for packet extraction and routing.

The system works synchronously: time is slotted into network cycles; during each network cycle, all the packets are extracted from MA-OFLM and sent to SS, where the tunable filters select the right packets, while the frequency converters perform the frequency switching. After that, the filters in the loop memory are properly set to erase the packets already switched.

It must be noted that the proposed memory system, based on fiber rings associated with tunable filters allows the simultaneous reading of many packets, which would be difficult to realize using an electrical memory.

A preliminary evaluation of the node dimensioning has been made, considering a node capable of serving 100,000 users, each generating a mean traffic volume of 0.15 Mbps, with a packet length of 424 bit (48 bytes body and 5 bytes header), a channel bitrate of 600 Mbps and a link utilization of 0.4; the number of input/output channels must be 64. If 8 channels are frequency multiplexed on each input/output link, a total of 24 frequency positions in the memory loops has been evaluated to be necessary.

In order to better understand the application field as well as the limits of the proposed model, a 2x2 demonstrator is planned to be implemented.

3. Conclusion

The model of an opto-electronic switching node, based on the combined use of fast packet and frequency switching techniques has been presented. This model has been defined following a modular approach based on the definition of four subsystems with precise interfaces and functionalities that can improve independently as the optical technology matures.

The proposed model is capable of handling frequency-multiplexed streams of input packets of some Gbps per link; the lack of multiple frequency converters still forces the demultiplexing of the information flows for the frequency switching function; the optical storage is accomplished by using optical fibre rings associated with tunable filters and frequency converters.

Some technological improvements in opto-electronic components would strongly reduce the cost of a practical implementation. In fact, to avoid either the space-frequency switching matrix duplication or speed-up techniques, frequency converters and filters capable of switching in less than 100 ps should be devised. Moreover, as the cost of the model is strongly dependent on the number of frequency converters, efforts should be made to develop less expensive optical frequency converters as well as fast multiple frequency converter devices.

References

[1] A. de Bosio, F. Melindo, "Deterministic and statistic circuit assignment architectures for optical switching systems", Proc. Top. Meet. on Photonic Switching, Lake Tahoe, Mar. 1987.

[2] P. Cinato, A. de Bosio, "Optical technology applications to fast packet switching", Proc. Top. Meet. on Photonic Switching, Salt Lake City, Mar. 1989.

[3] S. Suzuki, et al., "Photonic packet switch using optical buffer memories", Proc. Top. Meet. on Photonic Switching, Salt Lake City, Mar. 1989.

[4] M. Devault, et al., "The Prelude ATD experiment: assessments and future prospects", Journal on Selected Areas in Communications, V.6, N. 9, Dec. 1988.

A New Timing Architecture for Optical ATM Switching Systems

Y. Takahashi and E. Amada

Central Research Laboratory, Hitachi, Ltd., Kokubunji, Tokyo 185, Japan

Abstract. This paper describes a new optical signal format and a new timing scheme for optical ATM (Asynchronous Transfer Mode) switching systems. The scheme achieves bit and cell synchronization utilizing this format.

1. Introduction

Optical switching systems using space division optical switches have very promising practical applications [1], [2]. However, some optical signals may be lost when the optical switch is activated. This is due to a finite switching time and the lack of a memory function in such optical switches. Therefore, it is difficult to extract a timing signal from the switched optical signal at the output port, because signal loss causes irregular operation of common retiming circuits.

A special optical receiver may solve this retiming problem, but it may make the system very complicated. One practical method is direct clock distribution. However, this method can not compensate for gate delay time variation caused by temperature change. Therefore, it is difficult to incorporate it in high speed systems. Cell synchronization is another problem in high bitrate data streams, because it requires very high speed operation of logic circuits. Thus, a simple method of cell synchronization is preferable.

Parallel transmission of clock and cell synchronization signals with different wavelengths is a good solution, but it takes more wavelengths and makes the system complicated.

We propose a new optical signal format that includes guard time for switching action. We also describe a retiming and cell synchronization scheme utilizing the proposed optical signal format in optical ATM (Asynchronous Transfer Mode) switching systems.

2. Signal Format

The proposed optical ATM switching system uses electrical buffer memories for input and output ports and an optical space

Springer Series in Electronics and Photonics, Vol. 29 **Photonic Switching II**
Editors: K. Tada · H.S. Hinton © Springer-Verlag Berlin, Heidelberg 1990

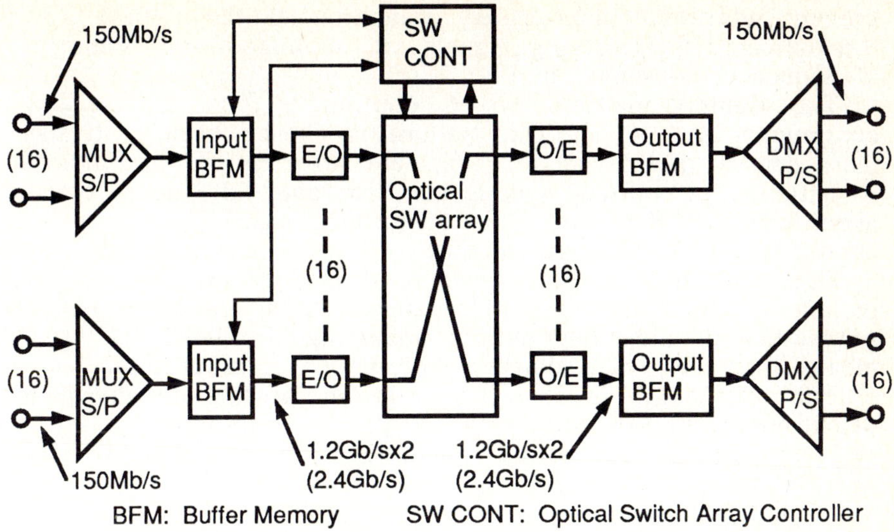

Fig. 1 Schematic view of proposed ATM switching system

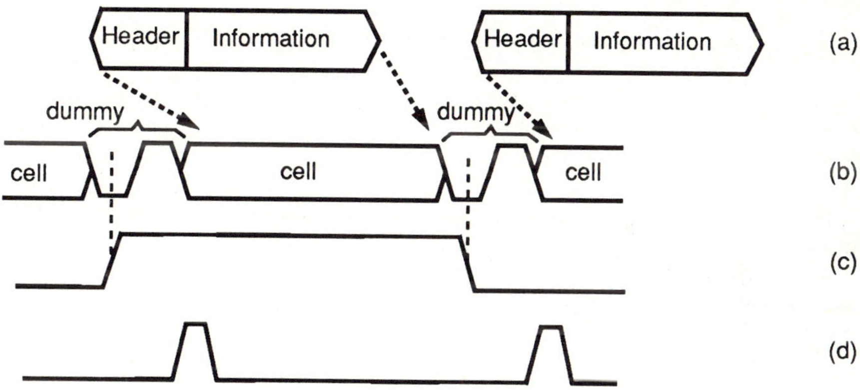

Fig. 2 Optical signal format and pulse shapes

division switch for switching the optical signal [1]. Its schematic structure is shown in Fig. 1. The optical switch changes its paths cell by cell. At optical switch transition, a few bits of the optical signal may be lost. Therefore, some guard time is required to prevent signal loss. A dummy signal pattern is added to the input signal in the guard time interval.

Each input cell is 53 bytes and has a header and data, as shown in Fig. 2 (a). The 16 input signals (150 Mb/s) are scrambled, have dummy patterns added to them and are multiplexed to two high speed signals (two 1.2 Gb/s). Each high speed signal, together with a dummy data pattern, Fig. 2 (b), is transformed into an optical signal and transmitted at a different wavelength. To prevent additional unnecessary transition at the optical signal, the optical switch driving signals are applied at the center of a '0' sequence, as shown in Fig. 2 (c).

The dummy pattern has 6 continuous bits of '0' and 6 continuous bits of '1'. This pattern is chosen because it has a mark density of 0.5 and 6 '0' bits are sufficient for optical switch transition. The optical switch has rise and fall times, both of about 3 ns [3]. Therefore, 6 bits of 1.2 Gb/s signal (≈ 5 ns) should be sufficient for the guard time.

The input data is scrambled, so it is very rare for the same pattern as the dummy pattern to appear in the optical signal. The probability of the same pattern appearing is $0.00024(1/2^{12})$. This means that the added dummy pattern can form a unique word.

The cell length is 53 bytes. Addition of this dummy pattern decreases switching efficiency to 95 %. Improvements in rise and fall times of optical switches can increase switching efficiency.

3. Timing Scheme

3.1 Retiming Circuit at Output Port

The structure of an optical receiver at an output port is shown in Fig. 3. There are neither reductions in data nor additional transients, because the switching action is done within a sequence of the '0' pattern. Therefore, a common retiming circuit using a PLL or a tank circuit can extract a timing signal from the received optical signal. Using the clock signal from the retiming circuit, the decision circuit can regenerate the data, as in Fig. 2 (b).

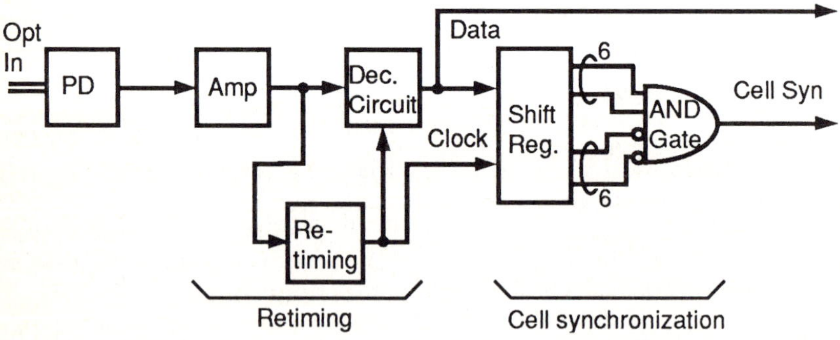

Fig. 3 Retiming and cell synchronization scheme at output port

3.2 Cell Synchronization

Because the dummy pattern (a sequence of 6 bits of '0' and 6 bits of '1') forms a unique word, we can use it to generate a cell synchronization signal. A 12-bit shift register and an AND gate, as shown in Fig. 3, can be used for cell synchronization. The AND gate, which has 6 positive inputs and 6 negative inputs, outputs a cell synchronization pulse, as shown in Fig. 2 (d). An output buffer memory (Output BFM) writes the received data using this cell synchronization pulse to mark the cell starting point. This process also rejects the dummy pattern.

Naturally, a pattern the same as the dummy pattern can appear. Some protection mechanism for preventing incorrect synchronization will be needed. A protection method in digital transmission systems can be applied.

4. Conclusions

The authors have proposed a new optical signal format and described a retiming and cell synchronization scheme in optical ATM switching systems. The proposed optical signal format has dummy data in the optical signals for preventing loss of optical signal during optical switching actions.

This makes it possible for a common optical receiver to extract a timing signal from the optical signal at an output port. Simple logic circuits can produce a cell synchronization signal from the dummy data. We believe this new scheme will have extensive applications in optical space division switching systems.

References

[1] E. Amada, Y. Takahashi, and H. Inoue: "A Photonic Approach to ATM Switching," GLOBECOM'89, No. 50.2, Dallas, Texas, November, 1989

[2] M. Esenberg, N. Mehravari: "Feedback Scheme for Frame Synchronisation in Photonic Network of Time-Multiplexed Space Switches", IEE Electron. Lett. vol. 25, No. 11, pp. 710-712, May 1989

[3] Y. Takahashi, H. Inoue, T. Kato, and E. Amada: "High-Speed Switching Characteristics of Integrated 4x4 InP Optical Switch Array", IEE Electron. Lett. vol. 25, No. 15, pp. 964-965, July 1989

All-Optical Self-Routing Crossbar Switch

M. Hashimoto, M. Fukui, and K. Kitayama

NTT Transmission Systems Laboratories, 1-2356 Take, Yokosuka-shi, Kanagawa-ken 238-03, Japan

Abstract: An all-optical self-routing crossbar switch utilizing the wavelength as routing information and an optically addressable spatial light modulator is proposed. Preliminary experimental results are also shown.

1. Introduction

In parallel computing architecture, optical interconnection can offer efficient processor-to-processor and/or processor-to-memory communication. The crossbar network is the most powerful non-blocking network although the number of communication lines is limited. The crossbar network is also a basis for multistage interconnection networks. In an asynchronous packet switching environment, whose control is totally distributed, self-routing is a promising method and has been studied thoroughly(1). Several optical self-routing systems have been demonstrated. One is the optical spread spectrum switching technique(2), in which each bit is encoded with a particular code sequence representing the destination address.

In this paper, an optical crossbar switch employing self-routing optical control with wavelength-encoded TAG is proposed, and a proof-of-concept experiment has been successfully conducted.

In the two-dimensional $N \times N$ optical crossbar, each incoming signal is evenly split into N signals and distributed to a vertical column of the two-dimensional switching matrix as shown in Fig. 1 (a) (3). The output lights on each horizontal row are then combined to produce N outgoing signals. By turning on the appropriate switches in the switch matrix, one can achieve arbitrary interconnections between the incoming and

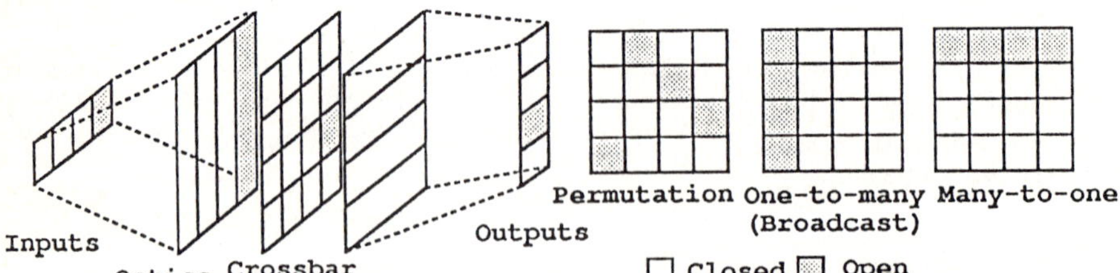

(a) Optical configuration (b) Basic mask patterns
Fig. 1 Optical crossbar switch

Springer Series in Electronics and Photonics, Vol. 29 **Photonic Switching II**
Editors: K. Tada · H.S. Hinton © Springer-Verlag Berlin, Heidelberg 1990

outgoing ports. One desirable aspect of the switching system is its outstanding flexibility in that it can create various combinations of the switching patterns. The three basic schemes: one to one, one to many, and many to one, are illustrated in Fig. 1 (b). Other switching architectures may not be capable of broadcast-type distribution or may require all connections to be changed to satisfy a single new demand for a connection.

2. Principle of Switching and Self-Routing

Figure 2 shows a photonic switch controlled by control light input. As described below, the optically addressable spatial light modulator (SLM) acts as a matrix of photonic switches with high parallelism. Figure 3 shows schematically the architecture of the proposed self-routing optical crossbar switch. The key components are a special class of optical bandpass filters and an optically addressable SLM. The bandpass filter consists of several horizontal wavelength filters whose center wavelengths are the destination addresses. Therefore, the number of center wavelengths corresponds to the number of destination addresses. A cell consists of the TAG pulse with wavelength λ_i preceding the information field with wavelength λ_0. Here, λ_i (i=1, 2, ---, n) corresponds to the destination address, port i. In this switch, SLM is the reflection-type and modulates the polarization of the readout light according to the TAG information. The incident light is expanded over a vertical column of the two-dimensional switching matrix of the SLM through an anamorphic lens. The TAG light passing through the wavelength filter λ_i from incoming port j activates the (i, j) element of the switching matrix SLM, which modulates the polarization state of the information field at 90° . As the

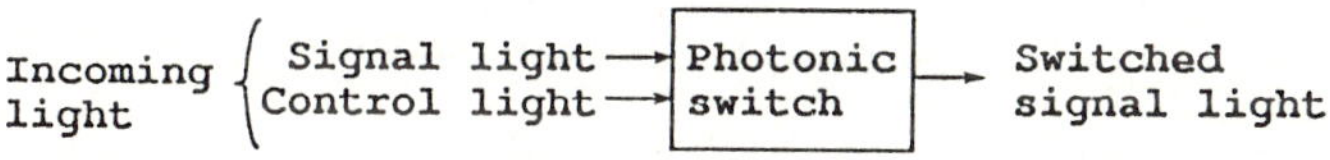

Fig. 2 Photonic switching element

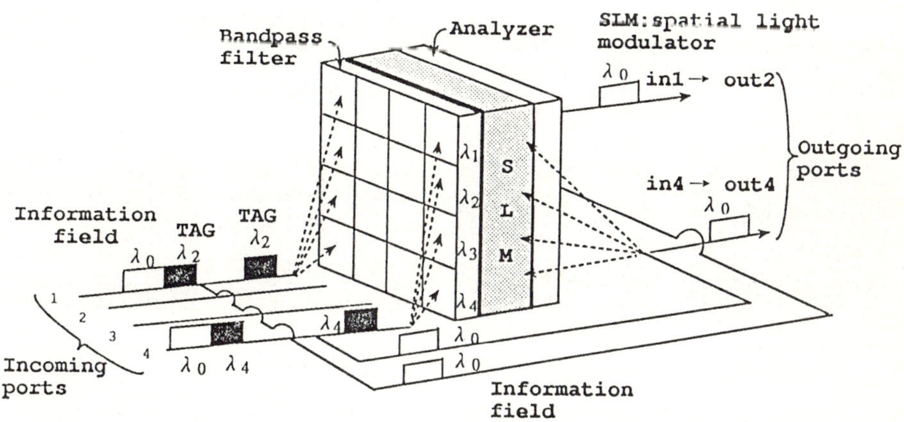

Fig. 3 Principle of self-routing with optical crossbar switch

analyzer is designed to allow only the modulated light to pass, the modulated light from the SLM can pass through the analyzer. In this way, one cell can be switched from port j to port i.

3. Preliminary Experimental Results

A preliminary experiment was conducted with the experimental setup shown in Fig. 4. For simplicity, just 1×2 crossbar switching was demonstrated. A He-Ne laser (λ_0=633nm) was used to generate the information field. An argon laser was used as the TAG light source. The TAG wavelengths of λ_1=488nm and λ_2=515nm, were assigned to the outgoing ports, port 1 and port 2, respectively. The center wavelength of the bandpass filter in the right half part was tuned to 488nm, while in the left half it was 515nm. BSO($Bi_{12}SiO_{20}$) (4) was used for the photoconductive and photorefractive material of the optically addressable SLM. Its photorefractivity is sensitive from 360nm to 550nm but is relatively insensitive over 600nm. Minor changes from the basic configuration in Fig. 3 were made for experimental convenience; the information field and TAG were combined after the TAG lights passed through the wavelength filters. The analyzer was designed to pass only unmodulated light.

Figure 5 shows the experimental results. The incoming patterns of TAG lights λ_1 and λ_2 after passing the bandpass filters are shown in Figs. 5(a) and (b) in order, λ_1=488nm and λ_2=515nm. In this case, the bar state of port 1 to 1 is obtained with the TAG light λ_1, while the cross state of port 1 to 2 is realized with the tag light λ_2.

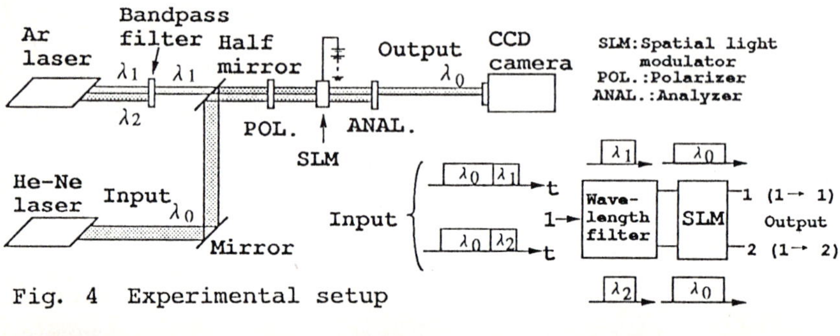

Fig. 4 Experimental setup

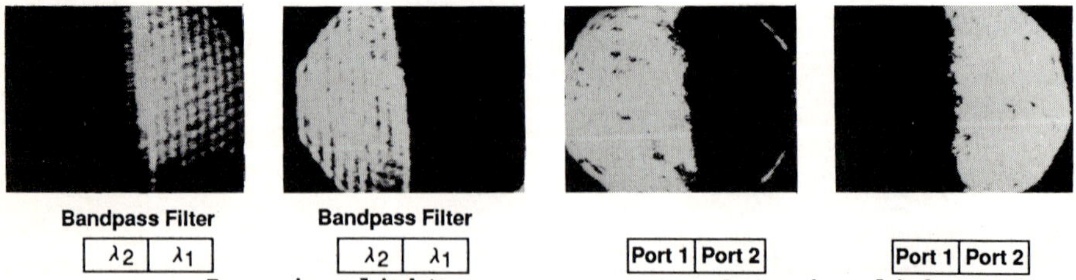

(a) TAG λ_1=488nm (b) TAG λ_2=515nm (c) λ_0=633nm (d) λ_0=633nm

Fig. 5 Experimental result

The readout lights with TAG wavelengths λ_1 and λ_2 are shown in Figs. 5 (c) and (d), respectively. In the case of (c), it is obvious that the information field was switched to port 1, and in (d), it was switched to port 2.

4. Conclusions

In conclusion, a new all-optical self-routing method that uses a wavelength and optically addressable SLM together with wavelength filters has been proposed. The TAG wavelength corresponds to a destination address. 1×2 crossbar switching has been demonstrated experimentally. In the experiment, BSO was used in an optically addressable SLM. An extension to a multi-channel system using narrower bandwidth wavelength filters(5), (6) with laser diodes (LD) as the light sources is being made.

Various types of optically addressable SLMs have been developed(7). Using an optically addressable SLM that is addressed by LD and can switch LD light of $1.3\sim 1.55\,\mu m$ for transmission systems, the proposed optical crossbar switch will have a variety of applications not only in optical processing or computing but also in optical transmission or switching systems.

Acknowledgement The authors would like to thank Dr. S. Shimada, Dr. H. Kimura, Dr. H. Ishio and Dr. T. Matsumoto for their encouragement.

References:

1 For example, J. Turner, IEEE, COM, 36, 734 (1988).
2 P. R. Prucnal, D. J. Blumenthal, P. A. Perrier, IEEE, COMMs. MAGAZINES, 25, 50 (1987).
3 A. A. Sawchuk, B. K. Jenkins, and C. S. Raghavengra, COMPUTER 20, 50 (1987).
4 P. Aubourg, J. P. Huignard, M. Hareng, R. A. Mullen. Appl. Opt., 21, 3706 (1982)
5 J. Minowa and Y. Fujii, Appl. Opt. 27, 1385 (1988).
6 K. Kikushima, K. Nawata, and M. Koga, IEEE JQE., QE-25, 163 (1989).
7 For example, Appl. Opt., 28, 4939 (1989).

Photonic Lattice Type Self-Routing Switching Network

N. Ogino and M. Fujioka

KDD Kamifukuoka R&D Laboratories, 2-1-15 Ohara,
Kamifukuoka-shi, Saitama 356, Japan

Abstract. A photonic lattice type self-routing switching network for ATM switching systems is proposed. This network has a small cell transfer delay time.

1. Introduction

Application of self-routing switching networks is a promising approach to construct the ATM switching systems that will be used in broadband ISDN. As to the type of self-routing network, most studies have centered on the Banyan network, its improvement [1], and the Manhattan street network [2].

In this paper, a lattice type self-routing network is proposed. This proposed network is characterized by its high reliability. By application of a photonic switch to this network, a small cell transfer delay time can be obtained. In this paper, the structure and action of the photonic lattice type self-routing network are described.

2. Lattice type self-routing network

Fig.1 shows the structure of a 2-dimensional lattice type self-routing network. A switch element (SE) is bidirectionally connected to a cell handling unit (CHU) which accommodates inlet and outlet circuits. Each SE is bidirectionally linked to the adjacent SEs up and down, and right and left. Bidirectional loops composed of the links connecting two SEs are therefore formed vertically and horizontally.

Each SE has a row address and a column address. Each SE transfers cells based on the row address and the column address of destination SE which are written in the cell label. Here, we consider 2^3 routes from an origin SE to a destination SE. A cell is first transferred vertically to make the cell's horizontal position identical with that of the destination SE, or first transferred horizontally to make the cell's vertical position identical to that of the destination SE. In the case where the cell is transferred vertically, two transfer directions can be considered. Also in the case

Springer Series in Electronics and Photonics, Vol. 29 Photonic Switching II
Editors: K. Tada · H.S. Hinton © Springer-Verlag Berlin, Heidelberg 1990

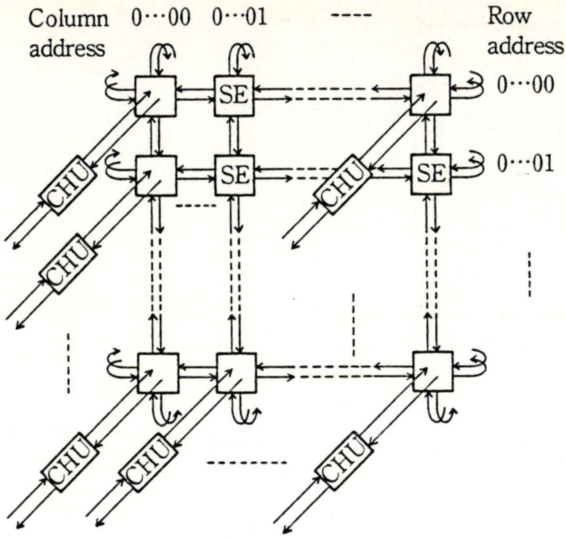

Fig.1 Structure of 2-dimensional lattice type self-routing network

where the cell is transferred horizontally, two transfer directions can be considered.

3. Application of bistable directional coupler type of photonic switch

Fig.2 shows the structure and principle of the mechanism of the bistable directional coupler type of photonic switch [3]. This switch achieves hysteresis characteristics by feedback of light intensity in one output port to the switching drive voltage. When there is no input light, the cross-state of the switch is maintained. If both bias light and address light are given to the switch, the switch enters the bar-state, following which the bar-state is maintained until the bias light is removed. If only the bias light is given to the switch, the cross-state is preserved. The switch can therefore be controlled by the presence or absence of the address light.

The cell is composed of a start flag, a label and an information field. In the start flag, no light is transmitted to reset the photonic switch. In the label, routing lights and address lights each having different wavelength are transmitted. The values of three bits to identify 2^3 routes are expressed by the presence or absence of three routing lights. The value of each bit indicating the address of the destination SE is expressed by the presence or absence of the corresponding address light.

Fig.3 shows a part of the switch circuit in the SE. This circuit has the function of transferring cells from the left SE to the right SE, the up SE, the down SE or the CHU. The switch A and B are controlled by the

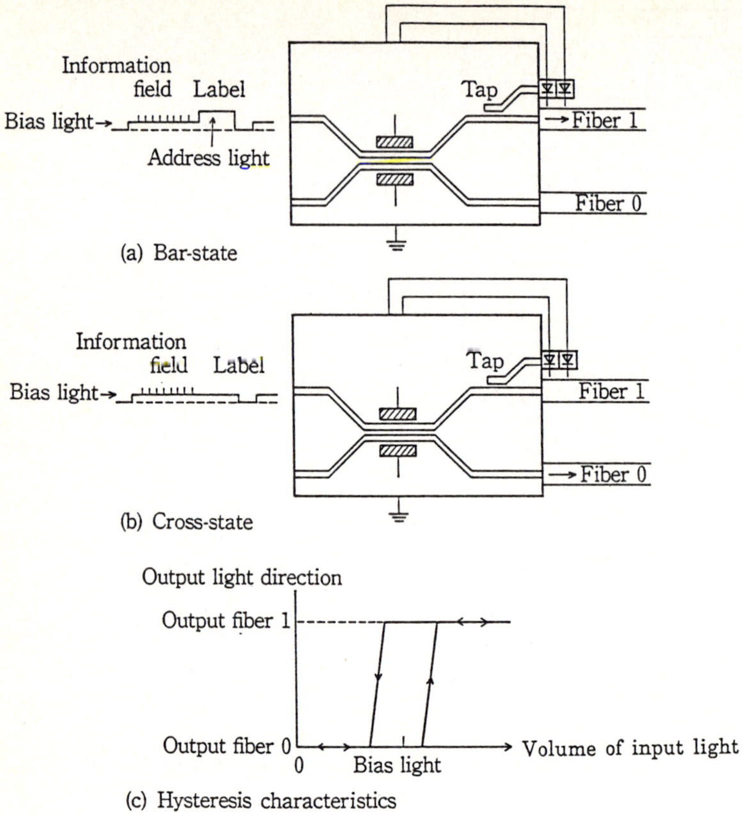

Fig.2 Bistable directional coupler type of photonic switch

address lights, and include an AND gate in their feedback circuits. The outputs of these AND gates become "1" only when the row or column address of the SE including the AND gates is identical to the row or column address of the destination SE. The switch C which determines the transfer direction of cells is controlled by a routing light.

4. Reduction of delay time by the photonic switch

In the case of the photonic switch, unlike that of the electronic switch, all the address bits in the cell label can be analyzed simultaneously by wavelength division multiplexing. For this reason, the length of the label can be reduced and the delay time to send the label decreases. In the photonic switch, the path is switched automatically while the label passes the switch, and the delay time for analyzing the label becomes zero. When the propagation delay can be disregarded by integration of the switching circuit, therefore, cell synchronization is needed only at the entrance of

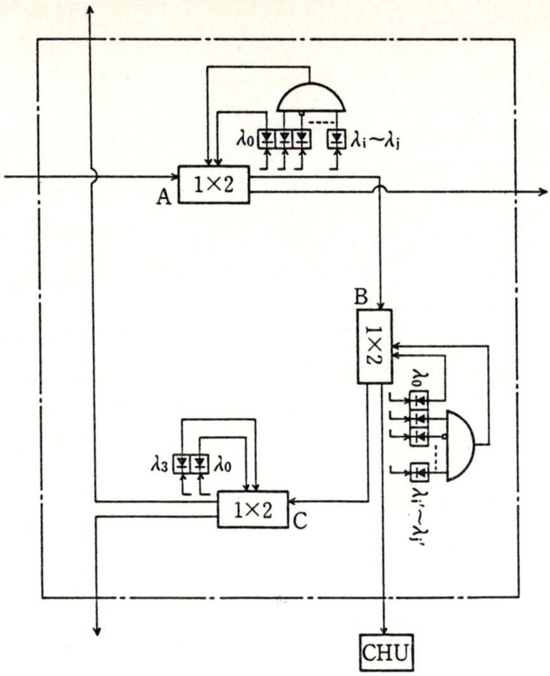

Fig.3 Part of switch circuit in SE

the network. From the facts described above, the cell transfer delay time can be reduced by using the photonic switch.

5. Conclusions

A photonic lattice type self-routing network with small transfer delay was proposed. Further study is however needed to devise a method for preventing cell collision in the network, a method for detecting failure, and alternative routing to avoid the failure position.

References

[1] J.S.Turner, "New directions in communications," International Zurich seminar, A3, p.25 (1986).
[2] N.F.Maxemchuk, "The Manhattan street network," GLOBECOM'85, 9.1, p.255 (1985).
[3] P.S.Cross, R.V.Schmidt, R.L.Thorton, and P.W.Smith, "Optically controlled two channel integrated-optical switch," IEEE Trans., Vol.QE-14, 8, p.577 (Aug., 1978).

Part IX

Space-Division Switching

Lithium Niobate Devices and Their Role in Photonic Switching

M. Izutsu[1,*], *N.F. Whitehead*[2,*], *K. Murakami*[3,**], *R.A. Thompson*[4,**], *L. Thylén*[5,**], *M. Kondo*[6,**], *and G.L. Tangonan*[7,**]

[1]Osaka University, Toyonaka, Osaka 560, Japan
[2]British Telecom Research Laboratories, Martlesham Heath, Ipswich IP5 7RE, UK
[3]Fujitsu Laboratories Ltd., 1015 Kamikodanaka Nakahara-ku, Kawasaki 211, Japan
[4]University of Pittsburgh, Pittsburgh, PA 15260, USA
[5]Ericsson Telecom AB, Fibre Optic Research Centre, S-12625 Stockholm, Sweden
[6]NEC Corporation, 4-1-1 Miyazaki, Miyamae-ku, Kawasaki 213, Japan
[7]Hughes Research Laboratories, Malibu, CA 90265, USA
*Coordinators
**Contributing Authors

Abstract. This is a report summarizing the points cited and found through the panel session held in PS'90 to discuss the prospects for the use of lithium niobate devices in photonic switching. The current research status is reviewed by contributing experts who also give their insights into possible roles of the devices.

1. Introduction

N.F. Whitehead and M. Izutsu

Lithium niobate switching devices have been under development in laboratories worldwide for some years. A wide range of systems experiments have also been reported, and the field may now be viewed as possibly the most mature of the photonic switching technologies. The contributing authors review these developments and possible alternatives to lithium niobate, and assess what progress may be made towards real systems and network applications.

Integrated optic components in lithium niobate are considered which exploit a range of techniques for realising electrically controlled crosspoints and passive waveguide crossovers and branches. Devices may be characterized in terms of insertion loss, crosstalk performance, polarization and wavelength sensitivity, and electronic drive voltage and speed. In addition progress on fabrication and process control of lithium niobate for manufacture, and on component reliability needs to be considered.

The strengths of the technology are indicated by the reported switching matrices of various designs and sizes connecting up to 16×16 waveguides on a single substrate, and by demonstrations of matrices using multiple interconnected substrates. Systems architectures have been developed which exploit features favourable to implementation in lithium niobate, and systems experiments conducted which use spatial switching, high speed time multiplexed switching, switching of dense WDM and wavelength selective switching, all using lithium niobate components.

Springer Series in Electronics and Photonics, Vol. 29 **Photonic Switching II**
Editors: K. Tada · H.S. Hinton © Springer-Verlag Berlin, Heidelberg 1990

Clearly lithium niobate continues to provide a useful medium for photonic switching systems experiments, but it may be questioned whether the technology will play a major role in future telecommunications networks. Prospects for systems products exploiting lithium niobate need therefore to be considered in the light of developments in optoelectronic integration and switching, leading to alternative approaches in a number of the major applications. It may be that the major product markets in photonic switching will be made by alternative technologies unless lithium niobate can move into systems products in an effective and timely way.

This is a summary of the panel session held to discuss the prospects for the use of lithium niobate devices in photonic switching. K.Murakami and M.Izutsu were asked by the program committee of PS'90 to organize this. Feasible and critical issues involved in the subject were addressed by invited panelists, K.Murakami, R.A.Thompson, L.Thylén, M.Kondo and G.L.Tangonan at the session coordinated by N.F.Whitehead and M.Izutsu.

2. Photonic Switching System Requirements and Lithium Niobate Switches

K.Murakami

A combined time-division and wavelength-division system can efficiently realize a large switching capacity. High-speed switching capability is the most important requirement for Lithium Niobate (LN) devices for use in such switching systems.

Photonic switching system requirements. The high speed and wide bandwidth of the optical switching system make it ideal for broadband ISDN (B–ISDN). A particularly important target for such a system will be to accommodate large line capacity video information. The bandwidth required by high-definition video is about 10,000 times wider than the 64 Kb/s voice bandwidth. If we consider 100,000 channels of 600 Mb/s, the required throughput of the broadband photonic switching system will have to be 60 Tb/s. A combined time-division and wavelength-division system can efficiently realize such a large switching capacity.

The optical switch is a key component in the photonic switching system. A switching speed of more than 6 Gb/s is required when the number of input lines is 128 and the wavelength-division multiplexity is 10. Switch size is sufficient for a 16×16 matrix switch. Before such a system will be practical, optical memory and a wavelength conversion laser will have to be developed.

LN Applications. To construct a high-speed photonic switching system, basic LN characteristics must be considered. These include drive voltage vs chip size, insertion loss, crosstalk, DC drift polarization-dependence, and refractive effects.

Subject 1: High-speed switching. High-speed switching capability is the most important requirement for LN devices for use in time-division switching systems. A new electrode structure, such as a traveling electrode, has to be designed to enable high-speed switching of the electrically controlled optical switch. However, the speed of this switch is limited by the electrical drive circuit. Therefore, it is also important to develop an optically-controlled optical switch.

Subject 2: Low insertion loss and low crosstalk. A low insertion loss optical switch and optical amplifier must be developed to realize a large switching capacity.

The high gain of the optical amplifier effectively compensates for optical loss. Experiments and system configurations using a traveling-wave amplifier (TWA) have been reported [1–3]. However, the amplifier noise and insertion position of the amplifier will be problems in a photonic switching system.

Therefore, a low insertion loss optical switch is required. In the conventional matrix switch (crossbar type), since the number of switch elements is path-dependent, both insertion loss and crosstalk increase. In the $m \times m$ matrix switch, the worst path is through $2m-1$ switch elements, and the best path is through only one switch element. As a solution, an optical space switch structure with path-independent insertion loss is obtained by changing the physical topology of the conventional matrix switch [4]. The number of switch elements through which all signals pass would be the same and insertion loss would be reduced by about half. This structure also has low crosstalk.

Subject 3: Stable devices. The polarization-independent optical switch makes the use of single-mode optical fiber possible without resorting to polarization control. A photonic space-division switching system using a polarization-independent matrix switch has been reported [5]. Temperature changes shift device drive points and ultraviolet light affects their characteristics. Therefore, considerations for DC drift and refractive effects are also very important to make these devices stable. For example, semi-insulating Si has been used to resolve these problems [6].

3. Architectures for Photonic Space and Time-Switching Systems

R.A.Thompson

Lithium niobate technology allows serial digital data in photonic form to be directly switched under electronic control without requiring that the device be switched at the bit-rate. This property is a natural match to classical time-multiplexed circuit-switching and direct time switching, as long as data is not bit-interleaved. In this section, four general architectural concepts, applicable to the design of photonic switching systems, are abstracted from two research systems. The object research systems are: a photonic time-space-time system called DiSCO (Distributed Switching with Centralized Optics) with a time-multiplexed photonic switching network at its center stage [7] and a hierarchical system of photonic timeslot interchangers built from photonic switches and varying lengths of optical fiber as recirculating delay lines [8]. We hope to show that lithium niobate technology may be more ready than is generally thought for certain applications and with appropriate architectures.

Optical Path. The high insertion loss from coupling an optical fiber to a chip of lithium niobate motivates the integration of successive stages of switching onto a single chip and suggests butt coupling chips instead of using inter-chip fiber. The insertion loss and the additive crosstalk noise from individual switches suggest that systems should be architected such that photonic signals traverse a minimal number of switches as they propagate through the system. Insertion loss from bends and crossovers in diffused waveguides should influence the choice of switching network architecture. Another consideration is the problem faced by receivers if signal strength varies from one packet or timeslot to the next. Variation in path loss may be a worse problem than worst-case path loss. This consideration motivates architectures that ensure that photonic signals traverse the same number of switches and waveguide crossovers in all paths and even motivates adding dummy switches and dummy crossovers to make all paths equivalent.

Among nonblocking switching networks, planar architectures, active splitter/combiner architectures, and crossbar architectures each have certain advantages. Benes architectures have a constant number of switch crossings with a low number of stages and total switch count. Timeslot interchangers using many variable-length stubs have fewer stages but more switches than those using variable-delay modules made from fiber loops [8]. The feed-forward binary design of the variable-delay module improves the original unary design with a reduced and constant number of switch crossings.

Dilation. An insidious noise source in photonic systems is crosstalk from imperfect switching of signals that arrive simultaneously at the two inputs of a switch. The classical solution is to require near-perfect switching by tightly controlling polarization, wavelength, and switching voltage. A better solution is to design the network structure or system operation to prevent simultaneous arrival of photonic signals at the two inputs of any switch. Avoiding this principal cause of crosstalk relaxes requirements on crosstalk per switch, polarization, wavelength, and switching voltage such that existing lithium niobate technology is practical.

Dilation provides a tool for transforming a general switching network into one in which coexisting paths never share common switches. While the directional coupler is a general 2×2 switch, it is operated such that only one input-to-output connection is allowed at any time. End-to-end crosstalk in dilated networks exceeds the crosstalk performance of any individual switch. Crosstalk is avoided in timeslot interchangers by using a multiple-switch variable-delay module so that no switch is simultaneously input and output.

Signal Distribution. Photonic signal distribution between one fiber and N fibers can be passive or active. Passive distribution with $1 \times N$ star couplers accomplishes 1-to-N broadcast or N-to-1 capture, but with considerable optical loss. Devices incur a 3db loss for every 1-to-2 split of the photonic signal with a small amount of excess loss depending on device quality. Active distribution uses integrated 1-to-N or N-to-1 tree structures of 1×2 switches (switches with one unused port). They connect the common fiber to only one of the N fibers on the other side. The devices incur a small loss per switching stage but a large fiber-to-chip coupling loss. For $N \leq 4$ passive distribution is less lossy and for $N > 8$ active distribution is less lossy. However, system operation and crosstalk may be overriding considerations. In timeslot interchangers, active distribution is necessary on either side, but crosstalk is much better if both sides are active.

Architectural debate on DiSCO concerned the provision of (1) one laser per module, (2) one laser per board, or (3) one laser per transmission rate. In Case (1), the laser's rate is the peak rate required of any service and no photonic distribution is needed between a module and its links to the photonic center stage. But electronic distribution of signals must be changed if the rate is ever changed. In Case (2), there is no electronic distribution of signals but a large active photonic distributor is needed at the uplink and downlink. Case (3) appears to be the best compromise and is the case used in our research prototype, with one laser for voice-grade signals and another (at 1.6 Gbps) for video.

Hierarchical Structure. Hierarchical design is a powerful architectural tool that could help circumvent physical size limitations even in prototypes that initially appear to have little growth potential or generalization. Examples are provided for both the space and time domains.

The DiSCO center-stage is reported as an 8×8 switching network with a dilated Benes architecture. Time-multiplexing the entire system (as a Time-Space-Time system) is a form of hierarchical design that provides a multiplier of 256 (number of timeslots in a frame). The effective switching network has 2048 channels on each side, where a channel consists of a segment of time equal to 488 ns out of every 125 μs on each of eight fibers. At 1.6 Gbps, each timeslot can hold over 600 bits and each channel carries an average of about 5 Mbps.

The construction algorithm for the Benes architecture is recursive and the algorithm for the dilated Benes network follows suit. Recursion is another form of hierarchical design. The reported 8×8 dilated Benes network has six stages of switches all constructed on two butt-coupled chips of lithium niobate. While this technique might be extended to a 16×16 dilated Benes network with eight stages, larger networks will probably require more than two chips. While some faint hope exists for implementing a practical dilated Benes network as large as 64×64, physical design is very difficult and a path selection algorithm is computationally complex. Simply extending the Benes recursion offers little hope for ever implementing a 256×256 switching network.

The Clos architecture is a generalized three-stage structure that has r $n \times m$ switches in the left and right stages and m $r \times r$ switches in the middle stage with perfect shuffle interconnection between adjoining stages. Clos architectures are rearrangeably nonblocking if $m \geq n$. The $N \times N$ Benes architecture is a special case with $n = m = 2$ and $r = N/2$. A Clos architecture with $n = m = r = 8$ is 64×64 overall and could be implemented with 24 8×8 switching subnetworks. Each 8×8 switch has its own controller. Interstage fibering would be complex but manageable. Continuing up the hierarchy, eight 64×64 networks could be used in the left and right stages of a 512×512 network. The middle stage requires 64 8×8 networks. Interstage fibering is admittedly a major problem, but the promise of hierarchical design is demonstrated.

A similar technique was suggested for the timeslot interchanger with variable-delay modules. Reference [8] describes a practical timeslot interchanger that handles 100 timeslots and suggests using three similar stages, after the Clos architecture, to handle 10,000 timeslots. More layers of nesting allow 10,000 timeslots to be handled by more stages but where each stage is simpler.

4. $LiNbO_3$ Switch Arrays: Status Review and Prospects

L.Thylén and P.Granestrand

Optical space switches will play an important role in the broadband fiber optics transport network, since they constitute a switching fabric which is compatible with the optical fibre and preserve the frequency and code transparency of the fibre. The transparency feature is generally lost if a temporal switching is introduced. In this paper we deal with the low rearrangement speed applications with potential near term applications. The size requirements for these applications are shown in Table 1. (Slow means a rearrangement speed of hours or longer, fast means call dura-

Table 1. Requirements on switches in possible applications of optical space switching. MDF = Main Distribution Frame, DCC = Digital Cross Connect, DDCsub = DCC at subscriber. "Bitrate" refers to routed data.

Switch	Size	Bitrate	Slow/Fast
Protective	(2-20)x(1-20)	620 Mbit/s- 2.4 Gbit/s	Slow
MDF/DCCsub	1000x500	620 Mbit/s	Slow
Access Switch	1000x500	140 Mbit/s	Fast
DCC	(20-50)x(20x50)	620Mbit/s- 2.4 Gbit/s	Slow

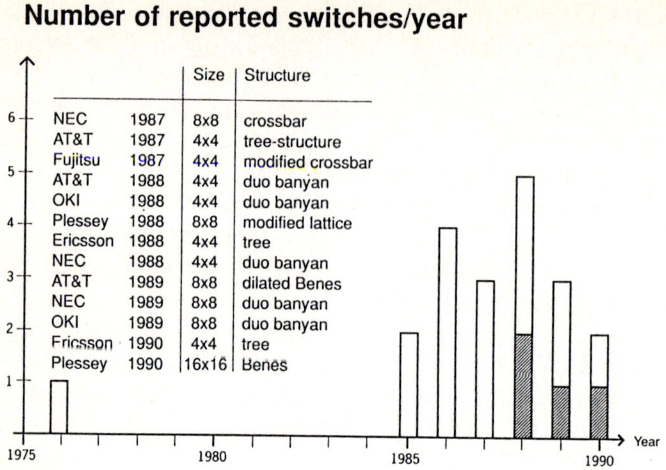

Fig 1. Number of published switch arrays ($\geq 4\times4$) vs year of publication for polarization dependent as well as polarization independent (shaded) $LiNbO_3$ switch arrays. Reported $LiNbO_3$ switch matrices ($\geq 4\times4$) since 1987.

tion). As a comparison, Fig. 1 shows the reported $LiNbO_3$ switch arrays over the past few years and the number of reported switches $\geq 4\times4$, vs the year of publication. Sizes have increased to 16×16 [9] (rearrangeably nonblocking, polarization dependent); polarization independent switch matrices up to 8×8 have been reported [5]. A packaged 4×4 matrix based on the digital switch [10] has been reported [11] (Fig.2), where the "digital" response of the switch element permits simple realization of polarization and wavelength independence, and stable switch arrays. This is an example of the combination of technology and architecture in order to achieve certain performance. It can be estimated that the maximum switch size that can be brought down on one $LiNbO_3$ chip is on the order of 20×20 [12]. The possibilities for creating significantly larger arrays will critically depend on the development of *optical interconnection schemes* such that several chips can practically be cascaded together. Given the possibility to cascade chips and to use optical amplifiers (possibly integrated into the $LiNbO_3$), current or improved (loss, crosstalk) $LiNbO_3$ technology meets the systems requirements of the applications of Table 1 regarding *number of ports, blocking properties, insertion loss, crosstalk, speed of rearrangement, instantaneous bandwidth, total bandwidth, ease of control and polarization independence*. The digital switch is especially advantageous to use with regard to the last three properties. However, with regard to *stability* current $LiNbO_3$ technology is not likely to be adequate in the above applications with their extreme reliability requirements. The stability issue has been the subject of intense research, e.g. in the RACE OSCAR project (R1033) and significant progress has been made towards

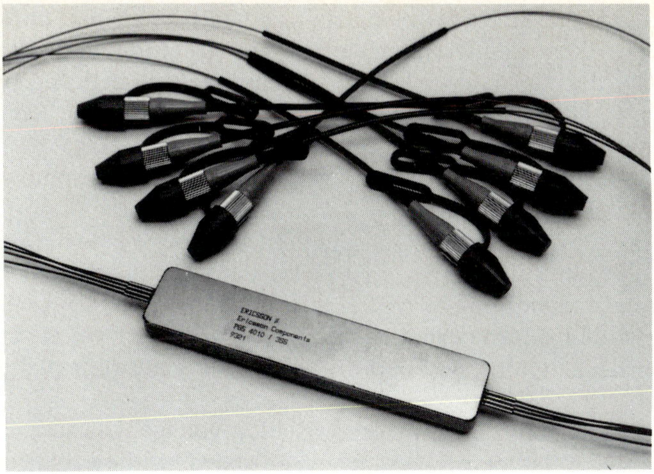

Fig 2. 4×4 polarization independent switch array with tree structure, based on 1×2 digital switch elements.

making the switches comply with telecom requirements. Stability in $LiNbO_3$ is associated with fundamental charge carrier transport phenomena, where more research is required. With an improved $LiNbO_3$ technology in this respect, it is concluded that $LiNbO_3$ will be a viable candidate for space switching in the transport network.

The work reported here was partially supported by the European RACE programme (R1033 OSCAR project).

5. $LiNbO_3$ Optical Matrix Switches for Photonic Space-Division Switching Networks

M.Kondo

Photonic space-division(SD) switching system, constructed with optical waveguide matrix switches, is attractive for exchanging or distributing broadband signals such as HDTV signals, because of its transparent characteristics [5]. The $LiNbO_3$ optical waveguide switch is suitable for such a purpose, since it has low loss and is relatively insensitive to differences in optical signal wavelengths. Performances required for $LiNbO_3$ optical matrix switches to realize the SD switching systems and their feasibility are discussed.

The SD switching network can be constructed by using Clos network architecture [13]. The maximum line capacity depends on matrix size, structure, insertion loss, and optical crosstalk for the matrix switches. The matrix size is limited by wafer dimensions and a switch element length required for obtaining a pertinent switching voltage. In Simplified Tree Structure (STS) [14], the number of switch elements connected in tandem along the path is small and a large matrix size can be possible, compared with other reported non-blocking matrix switch structures. The line capacity limitation by the matrix switch insertion loss will be relieved by utilizing optical amplifiers. However, a 32-line network, which is considered to be a practical size, can be constructed, using 3-stage 8×8 matrix switches with about 10-12 dB insertion loss, without optical amplifiers.

Allowable optical crosstalk in a switch element depends on switching network size and its structure [15,16]. In tree structures including STS, the increase in crosstalk, according to matrix size increase, is small, compared with crossbar structure. For example, a-15 dB crosstalk level is sufficient for a 32-line SD switching system with 3-stage Clos network, using STS 8×8 matrix switches.

Switching speed practically depends on switching voltage, because of driving circuit limitations. However, μsec or msec speed is allowable for the previously mentioned SD systems applications.

Polarization dependency is an important problem which must be overcome for practical use. It is desired to remove such polarization dependency, to enable construction of a switching network without polarization maintaining fibers and polarization controllers, as well as to connect the network directly to subscriber optical fiber cables. Several kinds of polarization independent switching devices have been reported. They all need higher applied voltages, because small electro-optic coefficient must be used. A uniform directional coupler switch controlled by uniform Δβ electrodes, where complete coupling lengths for TE and TM modes are coincident, has a relatively small voltage-length product [17].

A polarization independent strictly non-blocking 8×8 optical matrix switch was developed using the directional coupler switches[5]. The 8×8 switch is constructed with four STS 4×4 switches, connected by four distributing 1×2 switches and four selecting 2×1 switches. In this switch, less than 12 dB insertion loss and less than -18 dB optical crosstalk were obtained. These values satisfy the required level for a 32-line SD network. Furthermore, using LD amplifiers, the possibility of achieving a 128-line network was estimated [3].

For practical applications, reduction is required in insertion loss deviation between optical paths. Also, improvements have to be made in packaging and long term test.

6. Optoelectronic Switching – An Alternative to $LiNbO_3$ Arrays?

L.Tangonan

The advent of integrated optical switches, notably the $LiNbO_3$ variety of switch arrays, has proven to be a major advance for the designers of telecommunications systems. Certainly the impressive demonstrations of advanced 8×8 and 16×16 arrays are powerful indicators of the strength of this technology.

One may question however whether the $LiNbO_3$ will play a major role in the development of several new distribution architectures. These new architectures are based on passive distribution networks and multiplexing techniques like dense wavelength division multiplexing DWDM or sub-carrier multiplexing SCM (or a combination of DWDM and SCM). The multiplexing techniques can be thought of as our best attempts to exploit the full spectrum (microwave, optical or both) attainable with the high bandwidth of the fiber transmission medium. Thus it is not surprising that there will be a growing need for optoelectronic switches and circuits expressly designed for these new architectures.

For broadband networks based on DWDM, wavelength switchable receivers are necessary. The wavelength tunable receivers may have a tuned filter (EO, AO, or mechanical) in front of a standard receiver circuit, or an array of switchable detectors or receivers used with a fixed grating demux. For advanced SCM receivers, advanced OEIC's with built in tuning circuitry or circuits exploiting detector mixing will certainly have major impact.

Array based OE componentry – lasers, amplifiers, detectors in array form – will be utilized in many new applications. Byte wide interconnects, which are becoming popular for interprocessor and computer applications, require these array based componentry today. OEIC transmitters with multiple lasers combined onto a single output line will be a crucial component for many DWDM systems. Arrays of transmitters and receivers will become necessary for advanced distribution networks with multiple rings or high redundancy.

References

1. T. Tahara, Y. Sakita, Y. Tamura and T. Ishida, "Loss-compensated all-optical multi-stage switching system using TWA & LN-Switch," OEC'88 3A2-2, 1988.
2. M. Gustavsson, L. Thylén and A. Djupsjobacka, "Travelling wave semiconductor laser amplifiers for simultaneous amplification and detection : system experiments," Topical Meeting on Photonic Switching FD4, 1989.
3. M. Fujiwara, T. Kajitani, M. Itoh, H. Nishimoto and S. Suzuki, "Studies on semiconductor laser optical amplifiers for line capacity expansion in photonic space-division switching system," IOOC'89 20D2-1, 1989.
4. T. Shimoe, K. Hajikano and K. Murakami, "A Path-independent-insertion-loss optical space switching network," ISS'87 C12.2, 1987.
5. S. Suzuki, H. Nishimoto, M. Iwasaki, S. Kajitani, M. Kondo, M. Shikada, M. Ashibe and F. Akashi, "HDTV Photonic space-division

switching system using 8×8 polarization independent $LiNbO_3$ matrix switches," Topical Meeting on Photonic Switching FE1, 1989.
6. H. Nakajima, "Temperature-stable, low-loss Ti:$LiNbO_3$ devices," IOOC'89 19D3–3, 1989.
7. Thompson, Anderson, Camlet and Giordano, "Experimental modular switching system with a time-multiplexed photonic center stage," OSA Topical Meeting on Photonic Switching, March 1989.
8. Thompson, "Architectures with improved signal-to-noise ratio in photonic systems with fiber-loop delay lines", IEEE JSAC, August 1988.
9. P.J. Duthie, M.J. Wale, I. Bennion, "Size, transparency and control in optical space switch fabrics: a 16×16 single chip array in Lithium Niobate and its applications", Submitted to PS'90, Kobe, Japan, April 1990.
10. Y. Silberberg, P. Perlmutter and J.E. Baran, "Digital optical switch", Appl. Phys. Lett., vol.51, pp.1230, 1987.
11. P. Granestrand, B. Lagerstrom, P. Svensson, L. Thylen, B. Stoltz, K. Bergvall, J.E Falk and H. Olofsson, "Integrated optics 4×4 switch matrix with digital optical switches", Electron. Lett. vol.26, pp.4–5, 1990.
12. L. Thylén, "Integrated optics in $LiNbO_3$: Recent developments in devices for telecommunications", J Lightwave Techn., vol.6, pp.847, 1988.
13. C. Clos, Bell System Tech. J. 32, pp.406,1953.
14. H. Nishimoto, S. Suzuki and M. Kondo, Electron. Lett. 23, pp.1167,1988.
15. S. Suzuki, M. Kondo, K. Nagashima, M. Mitsuhashi, H. Nishimoto, T. Miyakawa, M. Iwasaki and Y. Ohta, NEC Research & Development 87, pp.44, 1987.
16. R. A. Spanke, IEEE Communications Magazine 25, pp.25, 1987.
17. M. Kondo, Y. Ohta, Y. Tanizawa, T. Aoyama and R. Ishikawa, Electron. Lett. 23, pp.1167, 1987.

Size, Transparency and Control in Optical Space Switch Fabrics: A 16 × 16 Single Chip Array in Lithium Niobate and Its Applications

P.J. Duthie, M.J. Wale, and I. Bennion

Plessey Research Caswell Ltd., Allen Clark Research Centre,
Towcester, Northants NN128EQ, UK

Abstract. Parallel planes of large single-polarisation single-wavelength optical switches are compared to interconnected smaller transparent switches for large fabrics. The first 16x16 $TiLiNbO_3$ switch is reported.

1. Large Optical Space Switches

Optical space switches have attracted attention for their exceptional data path transparency. Applications now being studied include cross connect, time-multiplexed space switching and customer access connections [1]. Practical implementation must consider the switch fabric size available. This is limited by loss and crosstalk, optical carrier transparency and control. System application need not, however, be concerned with the properties of the component optical switch modules, as parallel subfabrics may more efficiently handle multiple wavelengths, polarisations and routing requirements. We should therefore consider the tradeoffs available between transparency, blocking and size in an optical switching module intended for interconnection in a larger switching fabric. The range of functionality available can be illustrated by considering what is possible using 3" lithium niobate wafers. For polarisation-independent operation at both 1.3 and 1.55 μm 4x4 is possible [2]. For single-wavelength signals of arbitrary polarisation, 8x8 has been demonstrated [3]. And in this work we report experimental work on a 16x16 single-wavelength single-polarisation switch. The signals can be divided between parallel subfabrics by low loss fibre-based wavelength and polarisation multiplex devices to restore fabric transparency, together with high-speed 2x2 commutators to improve blocking properties [4]. Given a requirement for large space switch fabrics, what is the most efficient combination of these components?

2. Parallel Fabrics

Switching networks may be constructed either from a single plane of multiple interconnected modules containing small but transparent switches (figure 1) or from larger switching modules arranged in parallel planes using one or all of wavelength and polarisation multiplexing and high speed commutators to provide the required network function (figure 2). If the small switching units connected in a multistage network are each wide sense non-blocking, then the blocking properties of the resulting network depend on the number of stages. For wide

 © Springer-Verlag Berlin, Heidelberg 1990

Figure 1

Figure 2

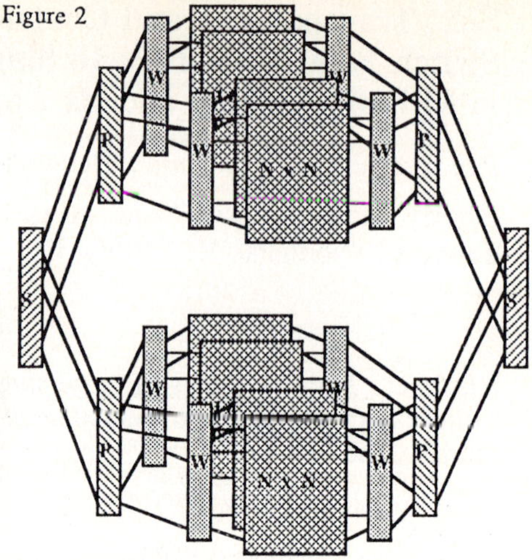

Figure 1 : Interconnected transparent switch modules : (a) rearrangeably non-blocking
(b) wide sense non-blocking

Figure 2 : Parallel switch planes providing wavelength and polarisation transparency together with wide sense non-blocking routing.

sense non-blocking operation of the network, five stages must be used. For example for a 16x16 fabric, twenty 4x4 units are required. For rearrangeably non-blocking operation three stages are required, corresponding to twelve units in the example. Figure 1 illustrates this. If, instead, the parallel plane design is used for a single-polarisation single-wavelength rearrangeably non-blocking switch module, then four units together with attendant multiplexers are required for a rearrangeably non-blocking network, and eight for wide-sense non-blocking operation where commutators are employed. This is shown in figure 2. As the wavelength and polarisation multiplex functions can be realised in fibre, and since the total crosstalk in decibels will be half that of the individual multiplex components, it is reasonable to assume that the multiplexing will make minimal contribution to network loss and crosstalk budgets. Optimisation of commutator loss, where only one optical gate is required, may be expected to yield losses in the region of 25% of those for large switch arrays. The crosstalk due to the commutators is likely to be comparable to that of the switch ranks within the space switch modules. With parallel planes, parallel control logic can be used to avoid increasing control complexity. Using these assumptions gives the results shown in table 1.

These figures show that for large space switch fabrics, a parallel arrangement of large space switch modules can offer significant advantage over a multistage network of smaller more transparent modules.

Alternative architectures for integrated optical switch arrays have been usefully reviewed by Reason [5], who has classified them by physical size, crosspoint count, interconnection complexity, loss and crosstalk. Where the largest possible

Table 1

Basic switch module and fabric composition	4x4 3-stage (rearrangeable)	16x16 (rearrangeable)	4x4 5-stage (wide sense)	16x16 (wide sense)
Number of modules	12	4	20	8
Number of control gates	144	56	240	112
Total loss	3xModule loss	1xModule	5xModule	1.5xModule
Total crosstalk	9 ranks	7 ranks	15 ranks	7+2 ranks

Figure 3 : 16x16 Benes network

switch matrix on a single chip is required, the preferred architecture is that due to Benes [6], shown schematically in figure 3. In order to study the feasibility of using parallel planes to realise large switch matrices we have built a 16x16 Benes optical space switch.

3. Design and Fabrication of an Experimental Switch

Realising a large Benes switch on a single 3" wafer does not permit much length per element and so relatively high control voltages are inevitable. The interconnection topology is not well suited to a reflective architecture [7], while a two-chip arrangement [8] is not preferred at this time because of interfacing difficulties. We therefore use a single chip double-ended arrangement.

The interconnection length is determined by the minimum intersection angle acceptable for crosstalk and loss (7°), the minimum bend radius for low loss without extra process steps (35mm) and the fibre pitch suitable for 80μm fibre (84μm). With a chip length of 70mm, 7mm remains for switching couplers. Each pair of electrodes has a length of 2mm since the switching action is obtained in the region where the waveguides are beginning to diverge at each end of the device. The resulting circuit layout is shown in figure 4.

Z-cut lithium niobate wafers were titanium metallised to a thickness of 900Å by thermal evaporation. Waveguides with nominal widths of 7μm were defined by photolithography and wet etching. After diffusion at 1050 degrees for 10 hours in wet flowing oxygen, a silicon dioxide buffer layer was deposited by CVD. The waveguides were edge polished before the definition of aluminium electrodes. A completed chip is shown in figure 5.

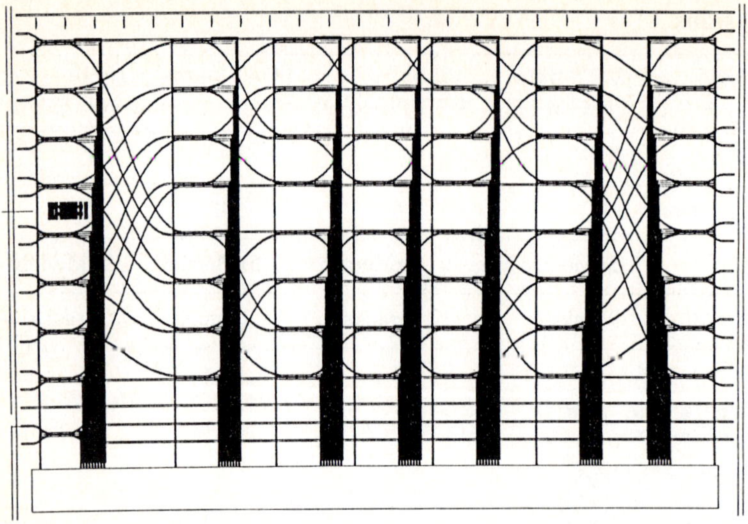

Figure 4 : 16x16 Benes circuit layout

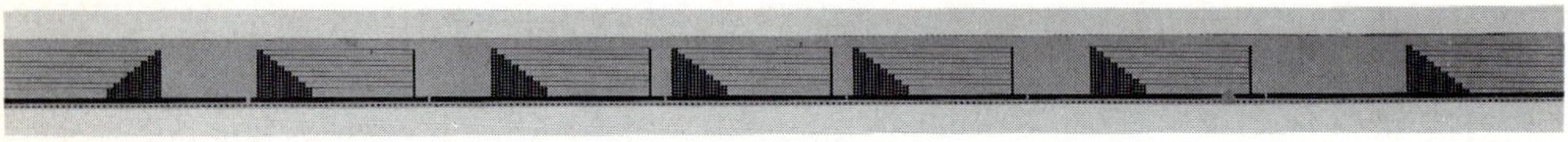

Figure 5 : 16x16 Benes optical switch array

4. Assessment and Control

Chips were selected which possessed a coupler gap most suitable for reverse delta beta operation at 1.3μm wavelength with TM polarisation. At this stage the single waveguide loss was measured as 4dB, or 0.6dB/cm. Single crosspoints on each chip were then wirebonded to permit electro-optic testing. Crosstalk was better than 20dB, the measurement being limited by scattered light. These crosspoint switches had an excess loss of 2dB, owing to the combination of curvature loss and metal overlay loss. Chips with switching voltages in the 40-50V range measured on single couplers were then mounted on sub-carriers, allowing connection of all 56 array gates plus ground planes. Characterisation of signal routing within the array showed gate switching voltages varying by 5 to 10V according to the manner in which the waveguides diverge from the ends of the directional couplers, routing loss variation corresponding to 0.3dB per interconnection bend plus 0.3dB per crossover and an extinction loss of 0.8dB per gate with applied voltage. Initial measurements of crossovers show crosstalk below -25dB.

Switch control potentially limits performance, size and cost. Components reported hitherto have required separate DACs for each gate. We have implemented an alternative arrangement exploiting the gate capacitance as an analogue voltage memory, whereby a commutative electrical controller sequentially addresses each gate's single control line, either refreshing the existing state, or changing the gate state between cross and bar. This scheme requires only one DAC to provide independent control of all 56 gates.

5. Conclusions

For large optical space switch fabrics, a parallel arrangement of large space switch modules offers advantage over a multistage network of smaller more transparent switches. We have demonstrated a 16x16 optical space switch suitable for the construction of large networks.

Acknowledgments. Thanks are due to D. Brambley for photomask fabrication; to R. Gibbs, J. Hendy and E. Hindson for device fabrication; and to F. Randle, M. Owen and D. Eddolls for assistance with component packaging. This work was partly funded by GEC-Plessey Telecommunications Ltd and the EEC under RACE programme 1033 'OSCAR'.

References

1. N. Whitehead et al, 'Photonic switching systems experiments in the European RACE programme',paper WA2, Photonic Switching, Technical Digest, Optical Society of America, Washington DC 1989, pp3-5.
2. P. Granestrand and L. Thylen, 'Analysis of switching employing a 4x4 switch matrix : crosstalk requirements and system proposal', European Conference on Optical Communication, 1989, paper WeA15-4, pp276-279.
3. S. Suzuki et al, 'HDTV photonic space-division switching system using 8x8 polarisation independent LiNbO3 matrix switches', Photonic Switching, Technical Digest, Optical Society of America, Washington, DC 1989, pp3-5.
4. J.R. Erickson et al, 'High speed hitless switching using Ti:LiNbO3', ibid, pp89-91
5. R.J. Reason, 'Optical space switch architectures based on lithium niobate crosspoints, Br.Telecom. Technol.J., Vol.7, No.1, Jan 89, pp83-91
6. V.E. Benes, 'Optimal rearrangeable multistage connecting networks', Bel. Syst. Tech. J., Vol.43, pp1641-1656, July 1964.
7. P.J. Duthie, M.J. Wale and I. Bennion, 'A new architecture for large integrated optical switch arrays', Photonic Switching, Technical Digest, Optical Society of America,Washington DC, 1987, pp146-150.
8. J. Veselka et al, 'Low-voltage low-crosstalk 8x8 Ti:LiNbO3 switch for a time-multiplexed switching system', Optical Fibre Communications, Technical Digest, Optical Society of America, Washington DC, 1989, paper THB2.

Photonic SD-Switching for Subscriber Networks in B-ISDN

T. Matsunaga[1], *M. Yamaguchi*[1], *and M. Okuno*[2]

[1]NTT Communication Switching Laboratories
[2]NTT Opto-electronics Laboratories, 3-9-11 Midori-cho,
Musashino-shi, Tokyo 180, Japan

Abstract. Application of photonic SD-switching to fiber optic subscriber network is proposed. Experimental system having a newly developed silica-based 8x8 switch matrix is reported.

1. Traffic Concentration in Subscriber Network

Fiber optic transmission systems are being introduced into subscriber loops carrying POTS (Plain Old Telephone Service), leased circuits and ISDN (Integrated Services Digital Network) basic services. This is in response to customer demands for higher quality, more reliable services and superannuation of metallic cables, and to overcome congestion in underground facilities and take advantage of the significant cost reductions of single mode fiber cable and opto-electronic devices. The high bandwidths of the fiber optic subscriber loops will stimulate demands for broadband telecommunication services, and then the ATM (Asynchronous Transfer Mode)-based B-ISDN (Broadband-ISDN) will be constructed. After the installation of fiber optic transmission lines in this scenario, there apparently exists a transition period during which the network must support a variety of user-network interfaces ranging from POTS to ATM. Furthermore, continuing technological progress will yield a new interface and another transition period will begin even after the ATM-based interface becomes wide-spread.

SD (Space division) switching using low loss and low crosstalk switches is one of the candidates for concentrating subscriber loop traffic that has a variety of interfaces during the transition period. As optical signals do not suffer from distortion in optical switches, no pulse reshaping or regeneration is needed. Therefore, photonic SD systems support multiple bit-rates in their switching networks. Even the co-existence of analog and digital signal in the system is possible.

We propose a photonic subscriber loop office having an SD switching network as shown in figure 1. Services like analog CATV other than the B-ISDN

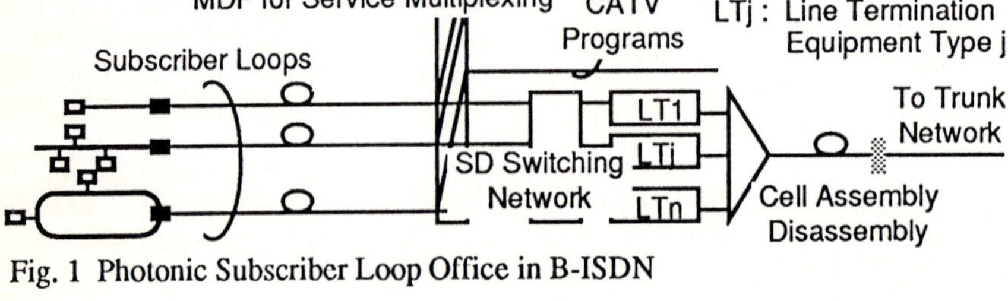

Fig. 1 Photonic Subscriber Loop Office in B-ISDN

Springer Series in Electronics and Photonics, Vol. 29 **Photonic Switching II**
Editors: K. Tada · H.S. Hinton © Springer-Verlag Berlin, Heidelberg 1990

services are assumed to be WD-multiplexed at the front-end of the photonic SD switching network. The bit-rate transparency enables us to terminate various kinds of fiber optic subscriber lines at the back-end of the switching network. Thus, customers can share many kinds of termination equipment, thus reducing customers' and carriers' expenses. Behind this line termination equipment, transmitted data from customers is converted into the form defined for the trunk network.

2. Experimental System

We have reformed our original SD switching system[1] to demonstrate multiple bit-rate switching using a crossbar-type 8x8 switch matrix.

2.1 System Configurations

Figure 2 shows the configuration of our experimental system. On the SM fiber lines, bidirectional signals are WD-multiplexed with 1.29 μm and 1.31 μm wavelengths. It supports four types of customers, G4 facsimiles with X. 21 interface (64 kbps), IEEE 802.3 CSMA/CD LANs (10 Mbps), NTSC TVs (32 Mbps) and HDTVs (400 Mbps), by having four types of trunks, BWT-64K, BWT-10M, BWT-32M and OWT-400M, and a call detector. The call detector scans all customers, detects call originations, identifies their call types and sends the information to the controller. Customers will have proper interface units with O/E and E/O converters, wavelength multiplexers and electrical circuits for network control functions like dialing, when connecting terminals to the network. In this system, customers can change their terminals freely because this does not entail any change of hardware or software in the switching office.

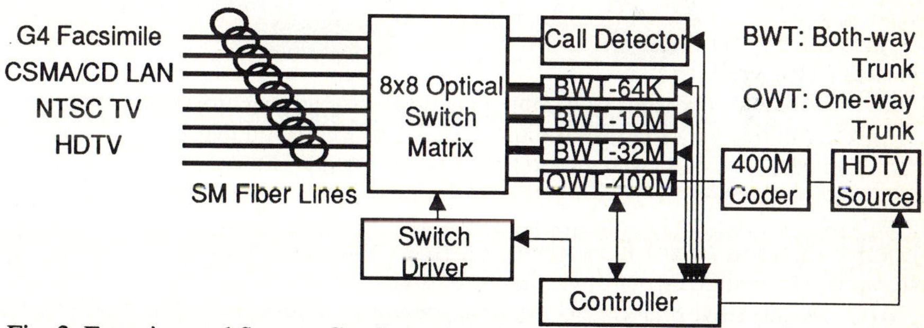

Fig. 2 Experimental System Configurations

2.2 Switching Devices

A crossbar-type polarization-independent switch matrix with 8 inputs and 8 outputs is fabricated using silica-based waveguide techniques[2]. Figure 3(a) shows its arrangement made of many 2x2 switch elements. A symmetrical Mach-Zehnder interferometer with thin film heaters is used as this element as shown in the figure 3(b). Switching is based on the thermo-optic effect in the glass waveguide. Of the 112 elements, 64 elements forming a diamond area in the figure 3(a) are active and controlled by the heating power to be in either a cross state or bar state, while the other 48 elements at the four corners are inactive and fixed at cross state. These

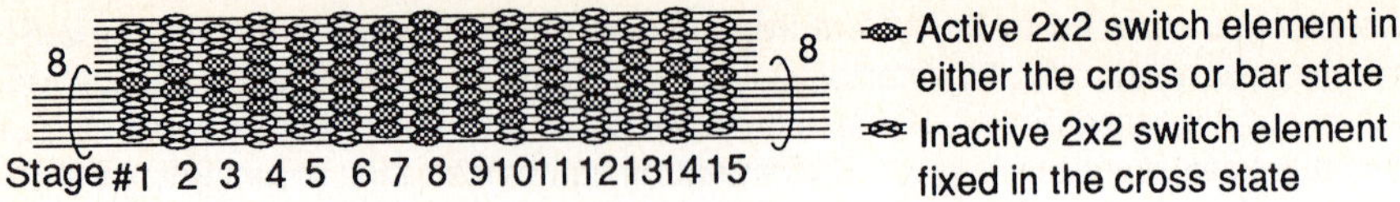

(a) Switch Matrix Arrangement

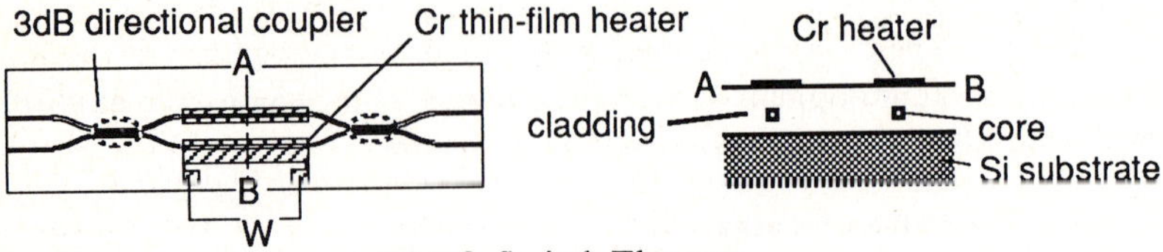

(b) TO-Switch Element

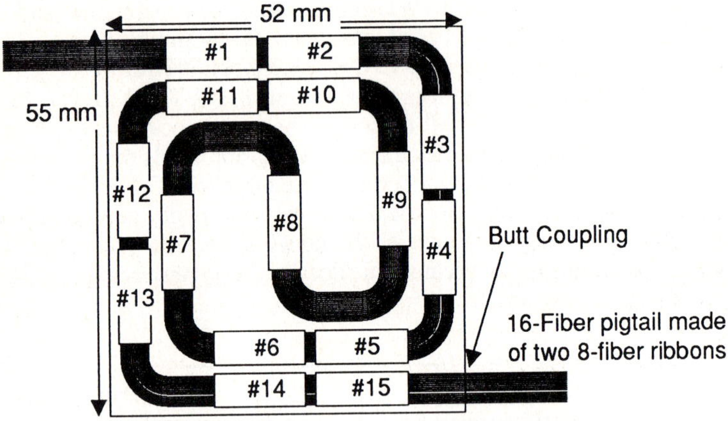

(c) Waveguide Layout on Silicon Substrate

Fig. 3 Silica-Based 8x8 Polarization Independent Switch Matrix

inactive elements reduce the variance of losses along different passes. Figure 3(c) shows its waveguide layout on a silicon substrate.

A pigtailed 8x8-optical crossbar switch is mounted on a printed circuit-board having electrical circuits on its backside. SC-type single-mode connectors are adopted to interface with other optical devices in the system for the purpose of high packing density. Adaptors for multifiber-connector to SC-connectors are also mounted on the switch package board.

3. Experimental Results

3.1 Switch Package Board

Insertion loss of the switch package board was measured for every combination of input and output pair. In the measurement, optical power, whose wavelength was 1.31 μm, was launched from ports on the trunk side of the switch, T(0) through

T(7). The loss includes insertion loss of multifiber-connectors, SC-connectors and adaptors. The maximum, minimum and average insertion losses were 24.3 dB, 13.1 dB and 17.4 dB, respectively. The loss was around 7 dB more than for a bare switch (unpigtailed switch) test, in which the loss consisted of waveguide loss and coupling loss with a fiber at each facet. The loss for TE-polarized light wave was on average 0.9 dB smaller than for TM-polarized light wave by the measurement of a bare switch.

Accumulated crosstalk to each output port on the line side of the switch, L(0) through L(7), when every input port on the trunk side has equal optical power and no heating power is applied to the switch, was also measured. The maximum, minimum and average accumulated crosstalks were - 20.8 dB, - 25.7 dB and - 23.3 dB, respectively, where the values are normalized by the optical power at an input port. In the measurement of a bare switch, the crosstalk for TE-polarized light was on average 1.1 dB smaller than that for TM-polarized light. The average ratio of signal to accumulated crosstalk was 5.9 dB.

The rise time and the fall time of switching response were each 1.3 ms.

It was also confirmed in some elements that crosstalk on zero voltage can be reduced by several dB by applying a small voltage. This is due to the difference in lengths between the two arms of a Mach-Zehnder interferometer.

3.2 System Performance

The connection of HD-video calls is the most critical in the system because they have the highest bit rate and hence, the smallest loss budget. The minimum detectable power for the bit error rate for a 400-Mbps 8B1C NRZ signal of HD-video is measured to be - 40.2 dBm. Loss measurements on various points of the system indicate that the experimental system has a loss margin of 19 dB.

The call detector successfully detects call origination and identifies the terminal type for every medium. The experimental system successfully demonstrated G4 facsimile communication, file transfer between two terminals on different coaxial cables in a CSMA/CD LAN, video-phone communication and HD-video reception and program selection.

References

[1] A. Himeno, M. Yamaguchi, Y. Yamada, T. Matsunaga; Experimental Optical Switching System Using Space-Division Matrix Switches Gated by Laser Diodes, Conf. Rec. GLOBECOM'88, pp. 928-932, Nov. 1988.

[2] N. Takato, K. Jinguji, M. Yasu, H. Toba, M. Kawachi; Silica-Based Single-Mode Waveguides on Silicon and their Application to Guided-Wave Optical Interferometers, J. Lightwave Tech., Vol. 6, No. 6, pp. 1003-1009, June 1988.

High-Isolation Optoelectronic Switch Matrix

M.C. Veilleux [1], *R.I. MacDonald* [1], *F. Gouin* [2], *and J. Noad* [2]

[1] Alberta Telecommunications Research Centre, 4245 97th Street, Edmonton, Alberta, Canada T6E5YE

[2] Communications Research Centre, Department of Communications, Ottawa, Ontario, Canada K2H8S2

A 10 x 10 optoelectronic switch matrix with 1.8 GHz bandwidth, 65 dB crosspoint isolation, 0 dB insertion loss and variable coupling is reported.

1. Introduction

Optoelectronic switching matrices initially used photodiodes as the switching elements [1]. Recent work however, centres on optoelectronic crosspoints that do not require bias in the off-state , simplifying the design of a complete matrix. Photoconductors and metal-semiconductor- metal diodes meet this requirement, and are relatively easy to integrate with other circuit elements [2].

A small (3x3) switching fabric using GaAs photoconductors has demonstrated optoelectronic switching of frequencies up to about 1 GHz [3]. More recently [4], a full 8x8 matrix has been reported based on metal-semiconductor-metal (msm) Schottky photodiodes for digital signals at rates to about 200 MB/s. The msm and photoconductor devices have very similar properties to optoelectronic switches. One distinction is that the msm device contains opposed junctions, while ideally the photoconductor contains no junctions. Since photovoltages are generated in unbiassed junctions, the isolation of the msm depends on exact device symmetry. In consequence, msm optoelectronic switches have typically 50dB isolation between on and off-state response, whereas up to 70 dB is observed in photoconductors.

2. Crosspoint Photoconductors

We report here a 10 x 10 matrix designed to explore applications of wideband, high isolation optoelectronic

Springer Series in Electronics and Photonics, Vol. 29 Photonic Switching II
Editors: K. Tada · H.S. Hinton © Springer-Verlag Berlin, Heidelberg 1990

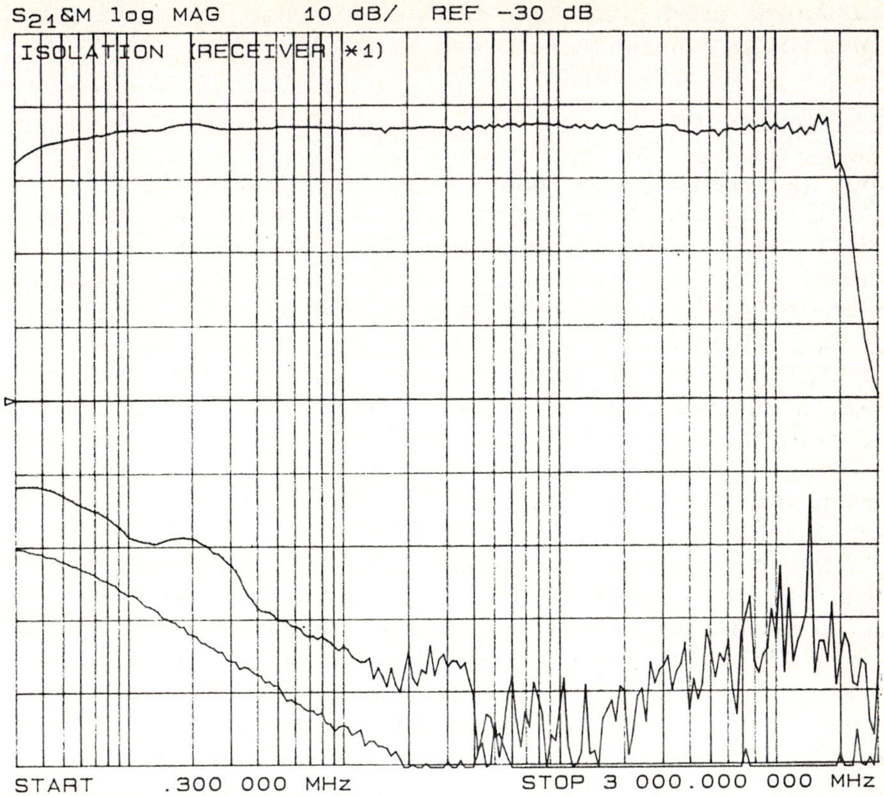

Fig.1 Upper trace: frequency response of one through channel of the matrix. Insertion loss is 0dB +/- 1dB from 1MHz to 1.8GHz.

Middle trace: response of the same channel with the crosspoint turned off. Isolation exceeds 65 dB from 10MHz to 1GHz.

Bottom trace: receiver noise level.

switching. The crosspoints are photoconductors with thin photosensitive channels of undoped GaAs grown on semi-insulating substrates. At low illumination levels the channels are depleted of charge by surface effects [5]. When equalized, the photoconductors exhibit responsivity of 0.2 A/W from 10MHz to 1.8 GHz, noise equivalent power of about 2×10^{-11} W/Hz $^{0.5}$, and switching isolation over 65 dB, as illustrated in Fig 1. The frequency response shown in this figure was measured with a signal of 450 microwatt average power and a modulation factor of 0.7. The photoresponse can be modulated over the 65dB isolation range by varying the bias voltage between zero and about 10 volts.

3. Construction

The ten transmitters use inexpensive GaAlAs lasers with 780nm operating wavelength, pigtailed with 50/125 micron multimode graded index fibre. Input signals drive the lasers directly through SMA connectors. Laser cooling is by means of air forced through the rf shielding baffles.

The signals are distributed by optical fibre power splitters. The ten transmitters and the power splitters are housed in a single 8"x17" rack and the 100 fibres from the power splitters are led to a similar rack holding the ten receivers. The entire 10x10 matrix with its power supply and control electronics is housed in one 19"x48" cabinet.

The receivers are mounted in pairs on circuit cards. Each receiver is shielded in a solid aluminium case. Ten fibres, one from each transmitter enter through a small aperture, and are attached to monolithic arrays of photoconductors spaced at 200 micron intervals, with one common terminal. Fibre alignment is by means of silicon V-grooves. Each receiver incorporates a four stage wideband amplifier with ~50dB gain, which raises the signal to the same level as the input drive to the laser.

The matrix is designed with the unusual feature of providing not only crosspoint switching, but weighted coupling between all inputs and outputs. The computer driven controller permits application of variable, bipolar bias voltages to each of the photoconductors. Because it has full fan-in capability on the output lines, the switch array is capable of performing the product of an input vector of signals with a matrix of weights. This design permits studies of the application of optoelectronic matrices to wideband signal processing [6] and high-speed neural network implementation.

The entire matrix throughput path with optoelectronic conversion in the transmitter, optical power division, detection and amplification has a dynamic range of 120 dB normalised to a 1 Hz bandwidth. It is thus expected to be capable of switching digital signals at rates over 3Gb/s with bit error rates below 10^{-9}. In analogue switching applications, it would be suitable for relatively narrowband signals on high frequency carriers - for example a 30 MHz bandwidth analogue high definition video channel would have a 45 dB SNR. Many could be switched simultaneously in frequency division multiplex.

4. Conclusions

The crosspoint isolation is very high in comparison to the SNR, implying that very large dimension matrices are possible, with improved SNR. We believe that optoelectronic switches of considerably greater size and performance could be made with present technology. Both direct modulation of lasers, and interdigital detector responses with bandwidths in excess of 10GHz have been demonstrated. With integrated optical signal distribution and improved detectors, arrays of hundreds, rather than tens of lines are foreseeable. Optoelectronic matrix switching will continue to challenge both electronic and optical switching techniques for communication and signal processing in the future.

We are now investigating optoelectronic integration techniques that will be essential to constructing larger matrices of this type.

5. References

1. R.I.MacDonald, E.H.Hara "Optoelectronic Broadband Switching Array", Elect.Lett. 14, pp502- 503, 1978

2. R.I.MacDonald, E.H.Hara, R.H.Hum,"Fast Photoconductive Optoelectronic Broadband Switch with Low Control Voltage", Electron. Lett. 17, pp611- 612, 1981

3. R.I.MacDonald, R.H.Hum, R.Kuley, D.K.W.Lam, J.Noad, "Experimental Demonstration of an optoelectronic switch matrix for satellite- switched TDMA signals in the 0.3-4.0 GHz band", CRC Technical Note No. 717, Communications Research Centre, Department of Communications, Ottawa, Canada 1984

4. S.R.Forrest, G.L.Tangonan, V.Jones "A Simple 8x8 Optoelectronic Crossbar Switch", IEEE J. Lightwave Technol. LT-7, pp. 607- 614 1989

5. D.K.W.Lam, R.I.MacDonald, J.P.Noad, B.A.Syrett, "Surface-depleted Photoconductors" IEEE Trans. Electron Devices ED-34, pp1057-1059,1987

6. R.I.MacDonald "Switched Optical Delay- Line Signal Processors",IEEE J.Lightwave Technol., LT-5 pp. 856-861, 1987

Optical Switches Using Liquid Crystals

T. Kurokawa[1] *and T. Yoshizawa*[2]

[1]NTT Opto-electronics Laboratories, 3-1 Mironosato Wakamiya, Atsugi-shi, Kanagawa-ken 243-01, Japan
[2]NTT Telecommunication Networks Laboratories, 1-2356 Take, Yokosuka-shi, Kanagawa-ken 238-03, Japan

Abstract. We show the design, fabrication and characteristics of optical switches using twist-nematic and ferroelectric liquid crystals.

1. Introduction

An optical switch for reconfiguring optical fiber networks is very attractive because of its broadband transmitting features[1]. A matrix switch with millisecond switching speed is required to constitute an optical space division exchange unit. Specifically, a 2x2 optical switch with low loss and low crosstalk is quite important not only as the basic unit but also as the exchange-bypass switch. The mechanical and waveguide switches investigated so far are not in practical use yet because the requirements for the switches are very severe, that is; low loss, low crosstalk, low driving power, independent operation on wavelength and polarization, self-maintaining of switched state and extension for a number of terminals.

Twist-nematic liquid crystal (TNLC) is a promising material for optical switches because of low driving power and high reliability[2]. Moreover, recently developed ferroelectric liquid crystal(FLC) has the attractive features of high speed and bistability[3]. In this paper, we discuss the feasibility of optical switches using TNLC and FLC.

2. TNLC optical switch

Although the LC switch was proposed long ago[4], a practical switch for long-wavelength single-mode fibers has not yet been developed. Polarization rotation in a twist-nematic LC is used for switching the light beam. When the bias voltage is not supplied to the TNLC cell, the polarization of the transmitted light through the cell is rotated by 90 degrees, and vice versa. Then, the propagating direction of the polarized beam is switched with the polarized beam splitter(PBS) that is positioned after the LC cell, as shown in Fig. 1(a).

A high extinction ratio is obtained when the thickness of TNLC layer is adjusted corresponding to the following condition,

$$d = \frac{m\lambda}{2\Delta n} \qquad (m = \sqrt{3}, \sqrt{15}, \cdots),$$

where d is the LC layer thickness, λ the wavelength, Δn and the birefringence of LC.

Springer Series in Electronics and Photonics, Vol. 29 **Photonic Switching II**
Editors: K. Tada · H.S. Hinton © Springer-Verlag Berlin, Heidelberg 1990

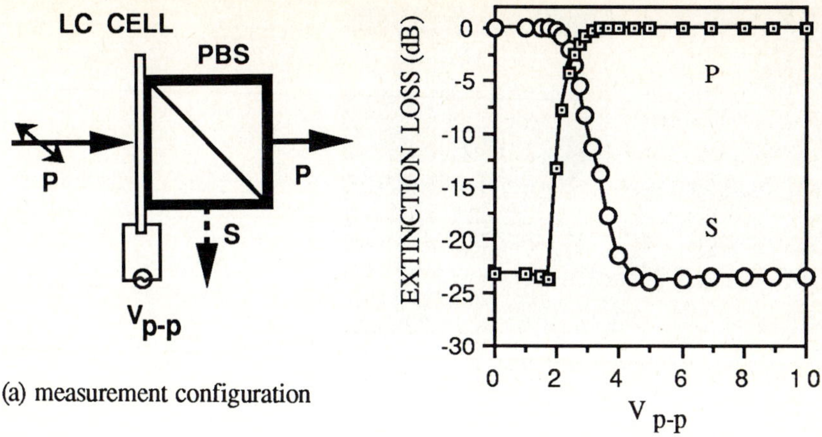

(a) measurement configuration

(b) measurement results

Fig.1 Measurement of the extinction ratio with TNLC cell

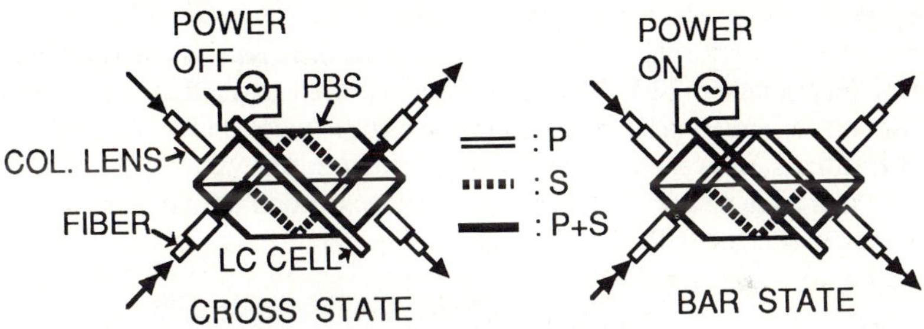

Fig.2 Schematic structure of LC switch

The thickness tolerance and wavelength tolerance is within ±7% for achieving a 20 dB extinction ratio. The extinction ratio for the TNLC cell with an 8.5 μm LC thickness was measured using a 1.3μm LD, as shown in Fig. 1(b). The threshold voltage was quite low, 3V, and the extinction ratio more than 20 dB was obtained in the range of 1.18-1.41μm.

The 2x2 switch based on the polarization division configuration shown in Fig. 2 was fabricated for 1.3μm single-mode fiber networks. The collimated beam from a single-mode fiber is divided by a polarizaion beam splitter(PBS) into two beams with orthogonal linear polarizations. These beams pass through a TNLC cell of a common polarization rotator and are recombined by a second PBS. A picture of the switch is shown in Fig. 3. The insertion loss and crosstalk were 1.4 dB and 26 dB, respectively. The polarization dependence was negligibly small, less than 0.1 dB and the back-reflection loss was more than 23 dB. It was driven at 10 ms switching speed with 5 V dc since it had an IC oscillator inside.

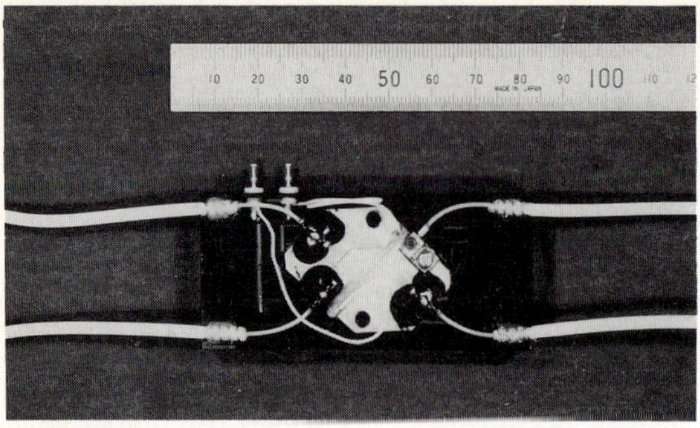

Fig.3 Fabricated 2x2 switch with TNLC

3. FLC switch

The FLC has some promising features, such as fast switching and bistability that conventional nematic LCs lack. The surface-stabilized FLC switches from one stable state to the other in response to the polarity of the applied electric pulse. For an application of the FLC with the same configuration as the TNLC switch, the tilt angle of the FLC must be 22.5 degrees to obtain a high extinction ratio. The switching was driven with a positive and negative electric pulse with 5 V_{p0} height and 500μs width, as shown in Fig. 4. The switching speed was 80μs and the switched state was maintained for several hours. The extinction ratio gradually fell from 20 dB initial value to 10 dB after 10 hours.

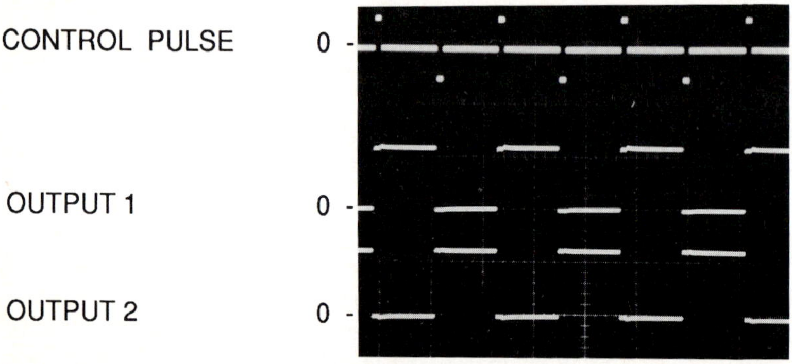

Fig.4 Switching operation of FLC switch

4. Extension to matrix switch configuration

The matrix switch with multi-terminals is quite important to reconfigure the fiber network. The matrix switch is obtained by the cascaded integration of LC cells and PBSs. An example of a 4x4 switch is shown in Fig. 5. Individual drive of the LC layer in each light path provides the nonblocking switching. Fiber collimators can

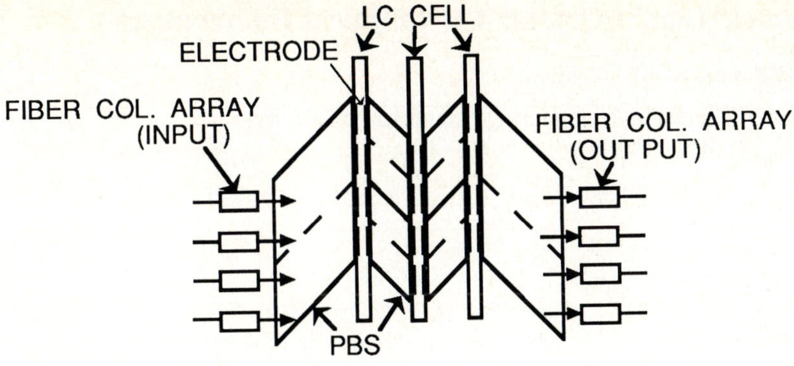

Fig.5 Schematic of 4x4 matrix switch

be positioned in two dimensional array since the LC cell has a planar structure. This means a matrix switch with a number of switching elements could be integrated in the block.

5. Conclusion

We showed the design, fabrication and characteristics of optical switches using liquid crystals. The TNLC switch for single-mode fibers had a low loss and high extinction ratio. The FLC switch showed promising features such as faster switching speed and self-maintaining of switching state. Furthermore, an extension to the matrix switch was discussed.

The authors would like to thank S. Fukushima for his help.

References

1. J. E. Midwinter: IEEE J. Lightwave Tech. **6**, 1512 (1988)
2. H. Seki, T. Uchida, Y. Shibata: Proceeding of the SID **25**, 275 (1984)
3. N. A. Clark, S. G. Lagerwall: Appl. Phys. Lett. **36**, 899 (1980)
4. E. Wagner, J. Cheng: Appl. Opt. **19**, 2921 (1980)

Ferroelectric Liquid Crystal 4 × 4 Optical Interconnect

L.R. McAdams and J.W. Goodman

Department of Electrical Engineering, Stanford University, Stanford, CA 94305, USA

We report the design and experimental results of a 4x4 ferroelectric liquid crystal double crossbar switch. We argue that for asynchronous data this switch can be scaled to accept 50 to 100 inputs.

Ferroelectric liquid crystal (FLC) polarization rotators have been investigated for use in free space photonic switching networks. Two approaches have been reported[1, 2]; however, in both cases the size of the networks has been limited by crosstalk accumulation. This crosstalk has its roots both at the device level, where the nonideal nature of the FLC polarization rotators causes crosstalk, and at the architecture level, because the network is built of a cascade of 2x2 switches. We propose to overcome both these limitations by constructing a NxN network from 1xN switches[3]. Our approach relaxes the constraints on the polarization rotators and results in a fault tolerant network exhibiting double crosstalk rejection.

The basic 1xN switch consists of an alternating series of FLC polarization rotators and polarizing beamsplitters. Each pair forms one output port in the switch. Linearly polarized light incident on the switch is transmitted through all the unselected output ports unchanged. At the desired output port, the FLC polarization rotator is set to rotate the incident light 90°, and the light is reflected out of the switch by the polarizing beamsplitter. The FLC polarization rotators are aligned so that in one state the director is parallel to the incident light. This alignment causes the extinction ratio at the output ports to depend only on the scattering from the zig-zag walls within the FLC cell, not on the FLC cell thickness, tilt angle, or operating wavelength.

A NxN double crossbar network can be formed from N 1xN active fanout switches and N Nx1 active fanin switches. Our experimental layout of a 4x4 network using one polarizing beamsplitter and two 4x4 arrays of FLC polarization rotators is shown in Figure 1. Its operation can be understood with the aid of Figure 2. The inputs enter the bottom of the array of row switches. Each input spirals upward in the row switch until arriving at the desired row where the FLC polarization rotator rotates the polarization of the input and passes it through the polarizing beamsplitter to the column switch. The column switch rotates the polarization of the input back to its original state and walks the input across the row to the output.

The loss and the crosstalk were measured at each output for each state of the switch. The loss ranged from -1.3 dB for the shortest path through the

Springer Series in Electronics and Photonics, Vol. 29 Photonic Switching II
Editors: K. Tada · H.S. Hinton © Springer-Verlag Berlin, Heidelberg 1990

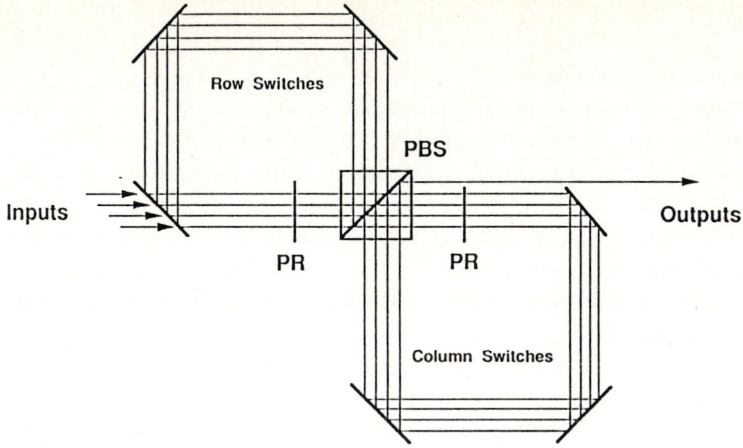

Figure 1: Experimental Layout of 4 x4 Network

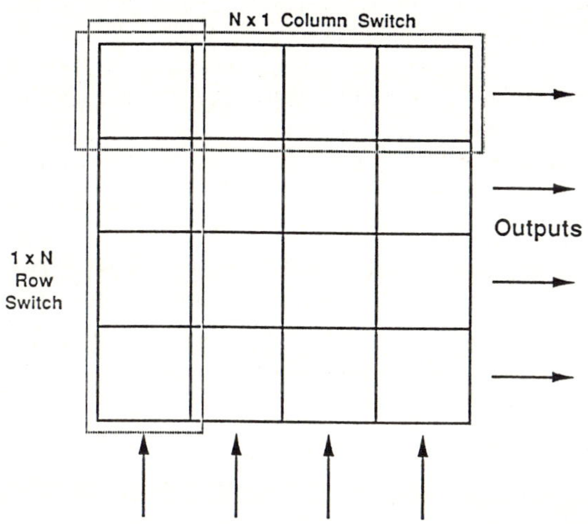

Figure 2: Topology of 4 x 4 Network

switch to -6.9 dB for the longest path, yielding an average throughput per loop of -.76 dB. An interesting property of the architecture is that if a FLC polarization rotator fails, causing half the light passing through it to be rotated and half not, a connection is not lost, the throughput is just reduced 3 dB. In this sense, the switch is fault tolerant. The crosstalk ranged from 20 to 28 dB with an average value of 23.4 dB. Half of this spread is due to variation in the FLC polarization rotators' performance and half is due to the crosstalk's dependence on input port. The input port that is closest to the output ports in Figure 2 produces a signal that is stronger relative to the noise contributions than the input port which is furthest from the output ports. The average value of the crosstalk has the expected form,

$$\frac{s}{n} = \frac{1}{(N-1)\epsilon^2},$$

where N is the size of the network and ϵ is the normalized extinction of a FLC polarization rotator between crossed polarizers. The switching time for the polarization rotators was 40 μsec at a drive voltage of ± 10 volts.

With the results of the 4x4 switch as a basis, it is interesting to estimate how large a network is possible. For this we assume that the input channels carry asynchronous, statistically independent digital data, and the receivers are AC coupled. We calculate the crosstalk for a signal that enters the switch through the input that is furthest away from the outputs. In this case, the signal, s, is given by

$$s = \alpha^{N+1}.$$

The noise is the square root of the weighted sum of the variances of the light leaked from the undesired channels. The intensity of the leaked light is a binary random variable with values 0 and ϵ^2 and variance $\epsilon^4/4$. Thus the noise, n, can be expressed as

$$n = \left[\frac{\epsilon^4}{4}\sum_{i=2}^{N}\alpha^i\right]^{1/2} = \frac{\epsilon^2\alpha}{2}\sqrt{\frac{1-\alpha^{N-1}}{1-\alpha}},$$

where α is the throughput per loop. The worst case crosstalk as a function of network size, N, is

$$\frac{s}{n} = \frac{2\alpha^N}{\epsilon^2}\sqrt{\frac{1-\alpha}{1-\alpha^{N-1}}}.$$

This expression is plotted in Figure 3. Curve A shows the size of the network that is possible with our experimental values of α and ϵ. Curves B and C show the increase in network size for improved values of α and ϵ. With modest improvements, a network with 50 to 100 inputs is possible.

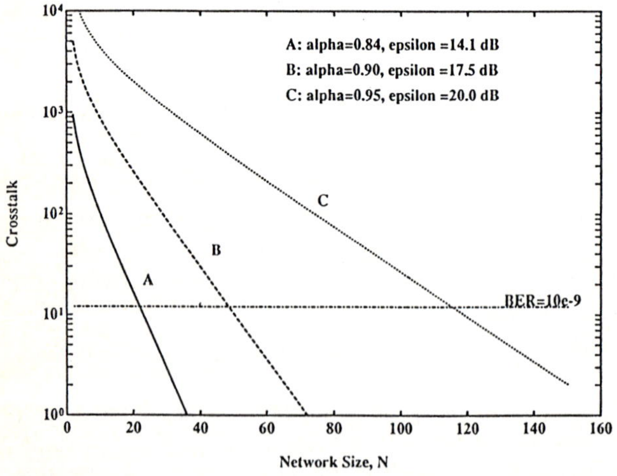

Figure 3: Plot of Network Size vs. Crosstalk

References

[1] L. R. McAdams, R. N. McRuer and J. W. Goodman, Appl. Opt., **29**, 1304 (1990).

[2] K. M. Johnson, M. R. Surette and J. Shamir, Appl. Opt., **27**, 1727 (1988).

[3] R. A. Spanke, IEEE J. of Quan. Elec., **QE-22**, 964 (1986).

AlGaAs Grating Surface-Emitting Beam Deflector for a 1 × N Space-Division Optical Switch

K. Yamashita, H. Nagata, Y. Kuroda, and K. Tone

Optoelectronics Technology Research Corporation
c/o Tsukuba Research Laboratory, Nippon Sheet Glass Co. Ltd.,
5-4 Tokodai, Tsukuba City, Ibaraki 300-26, Japan

Abstract. A new type of beam deflector for a 1×N space-division optical switch has been fabricated. The deflection is controlled by a carrier induced refractive index change of the semiconductor waveguide. The deflected beam is emitted normally with a grating coupler. The maximum number of resolvable spots is 10 for an injection current of about 630 mA.

1. Introduction

A beam deflector can be used in a 1×N space-division optical switch if the deflector can resolve at least N spots. Thus, using only N devices, an N×N optical crossbar can be realized. In fact, optical crossbars have been demonstrated using acousto-optic deflectors.[1],[2] We have proposed a new type of AlGaAs grating surface-emitting beam deflector [3] which is expected to be suitable for integration with light sources. Such a surface-emitting type is also suitable for optical interconnection between stacked boards. In this paper, a device with a new structure is reported to have improved thermal properties.

2. Device Principle

The device is basically a grating output coupler with a waveguide whose effective refractive index can be changed. The angle θ of the output beam from the normal is given by

$$k_0 \sin \theta = \beta - K, \tag{1}$$

where k_0 (= $2\pi/\lambda$, λ is the wavelength) is the wavenumber of the output beam, β (= $n_{eff} k_0$) is the propagation constant of the guided wave with the effective refractive index n_{eff}, and $K(=2\pi/\Lambda)$ is the wavenumber of the grating with period Λ. If

Springer Series in Electronics and Photonics, Vol. 29 Photonic Switching II
Editors: K. Tada · H.S. Hinton © Springer-Verlag Berlin, Heidelberg 1990

n_{eff} is changed, deflection should occur with constant wavelength.[4] It is derived from eq.(1) that the change of θ is approximately equal to the change of n_{eff} if θ is small.

We employ the band-filling effect due to carrier injection to change the refractive index because a large index change is expected in the wavelength region where the absorption is relatively small.

3. Device structure and fabrication process

The device structure is shown in Fig.1. The AlGaAs waveguide has the same structure as the previous one.[3] The waveguide was grown by two-step atmospheric pressure metal organic vapor phase epitaxy. A second-order grating (period: 0.25 μm) was fabricated by holographic interference lithography after the first-step growth. A ridge structure is formed to confine the injected current and guided wave. The output beam is taken out through the substrate by a groove-like window which was opened by selective chemical etching which was stopped by the AlGaAs layer (aluminum composition, x = 0.42). The device was finally cleaved to the lengths L of 0.5 and 2 mm and mounted junction-side-down.

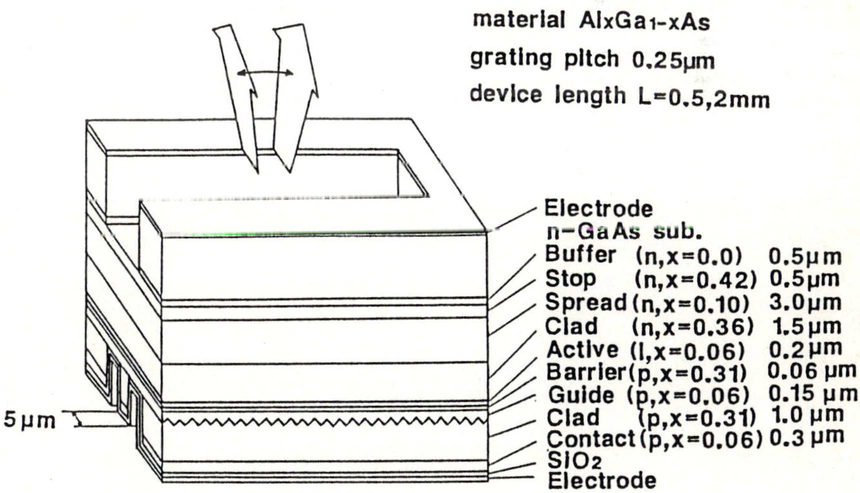

Fig.1 Structure of the fabricated device.

4. Device properties

The evaluation system for deflection properties has been reported elsewhere.[3] Figure 2 shows typical far-field patterns (FFPs) from the device at an input wavelength λ of 857 nm. The left curve corresponds to the FFP at zero injection current (I = 0) and the right one corresponds to that at 350 mA. A deflection angle of 0.38 degree is obtained in this condition, and a divergence angle of about 0.06 degree. Several side peaks observed in each FFP are due to the multi-modo oscillation of the laser diode under pulsed operation. The deflection angle versus injection current characteristic is shown in Fig.3 for various input wavelengths. The deflection angle increases sublinearly with increasing injection current and is larger at shorter wavelengths because

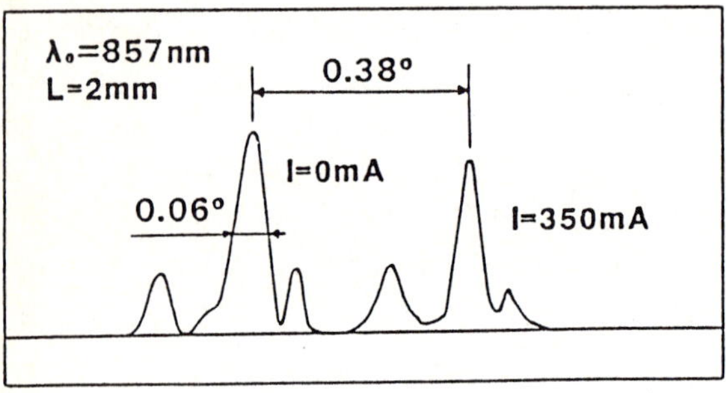

Fig.2 Typical far-field patterns of output beams from the device.

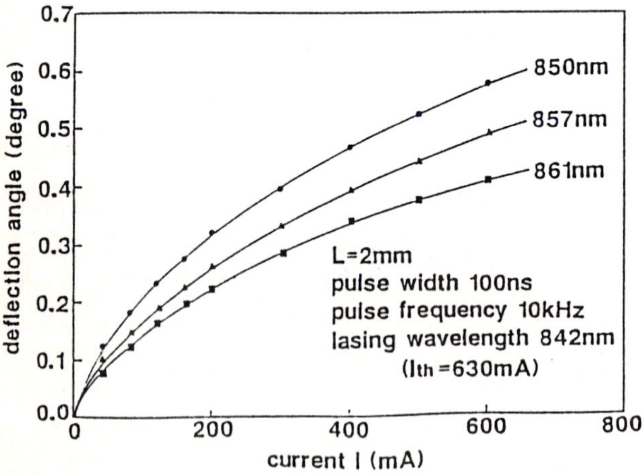

Fig.3 Deflection angle versus injection current characteristics for wavelengths of 850, 857 and 861 nm.

the refractive index change is largest when the energy of the input light approaches the bandgap energy. The device starts lasing at a wavelength of 842 nm for an injection current larger than 630 mA. The maximum deflection angle of about 0.6 degree is obtained at 850 nm and this corresponds to 10 resolvable spots. The injection current is reduced to 20 % of that of the previous device [3] for the same deflection angle because of current confinement with the ridge structure.

5. Summary

A new type of surface-emitting beam deflector has been fabricated. The deflection is controlled by changing the effective refractive index of the AlGaAs waveguide by current injection. The maximum deflection angle of about 0.6 degree and ten resolvable spots are obtained for an injection current of 630 mA.

References

1. P.C.Huang, W.E.Stephens, T.C.Banwell and L.A.Beith, Electron. Lett. **25**, 252 (1989).
2. D.O.Harris and A.VanderLugt, Optics Lett. **14**, 1177 (1989).
3. K.Tone, H.Nagata and K.Yamashita, Appl.Phys.Lett. **56**, in press.
4. B.C.Svensson, C.T.Seaton, U.J.Gibson and G.I.Stegeman, Appl. Phys.Lett. **53**, 941 (1988).

Application of Single-Stage Electronic Networks Concepts to Photonic Switching

A. Jajszczyk

The Teletraffic Research Centre, The University of Adelaide,
P.O. Box 498, Adelaide, S.A. 5001, Australia

Abstract. The application of single-stage electronic network concepts to photonic switching is studied. Structures of point-to-point and multicast directional-coupler-based switch elements are proposed. Some representative examples of optical switch matrices employing such switch elements are given. These matrices are nonblocking and contain only two active stages independently of their capacity.

1. Introduction

Single-stage electronic switching networks combining a multiplicity of smaller switches have very attractive features, e.g. small attenuation (or low delay in the case of time-division switches), simple control procedures, and nonblocking properties also for multicast connections. Their main disadvantage is a rapid growth of the number of required switches with an increase of the capacity. However, in optics it is much more important to minimize signal attenuation and maximize signal-to-noise ratio than to attempt to minimize the number of switch elements [1].

In this paper we study the applications of single-stage electronic networks concepts to photonic switching. Several architectures of single-stage electronic networks have been proposed. The simplest structure consists of $\lceil N/n \rceil$ modules, each containing $\lceil N/n \rceil$ switches, where N is the total network capacity, n is the capacity of a single element, and the symbol $\lceil x \rceil$ denotes the smallest integer greater than or equal to x [2]. For some values of N and n, networks containing fewer elements have been reported [3], [4]. In [5] single-stage networks characterized by a regular interconnection pattern between elements have been described.

2. Switch Matrix Structures

The allowed connection patterns in a single directional coupler are presented in Fig. 1a, and such patterns for an electronic 2×2 switch (excluding multicast connections), in Fig. 1b. We can construct a simple optical switch, shown in Fig. 2, which can realize all states of Fig. 1b. An alternative solution can be obtained by using a single directional coupler and two optical gates [6]. The multicast capability preserving nonblocking properties can be achieved by using switches of Fig. 3.

Now, we present some representative examples. Fig. 4 shows a 3×3 switch matrix employing the idea of [5]. Instead of hard wired connections in the

Springer Series in Electronics and Photonics, Vol. 29 Photonic Switching II
Editors: K. Tada · H.S. Hinton © Springer-Verlag Berlin, Heidelberg 1990

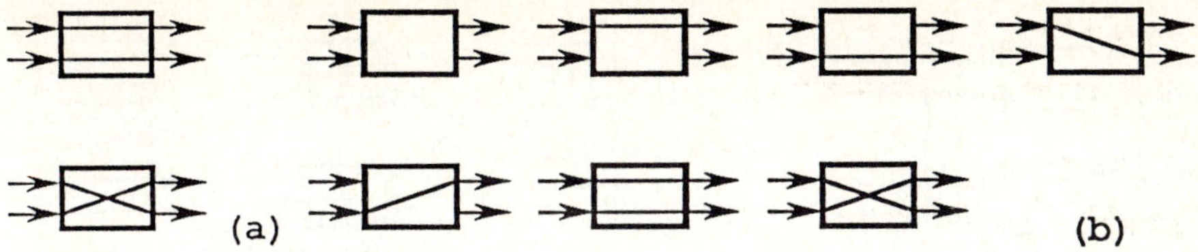

Figure 1: Allowed connection patterns; (a) in a directional coupler, (b) in an electronic switch

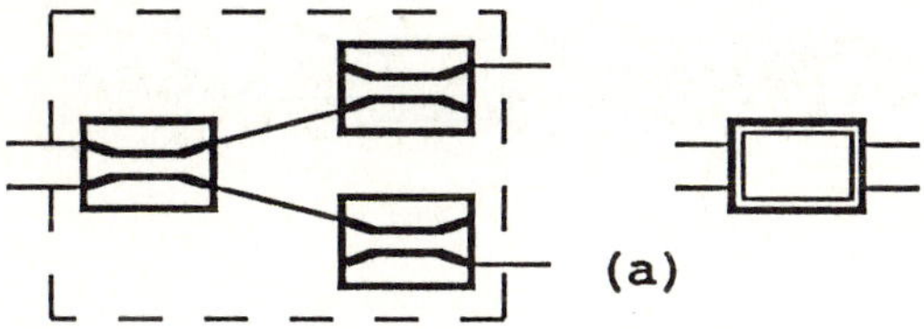

Figure 2: A 2×2 electronic-like optical switch; (a) structure, (b) symbol

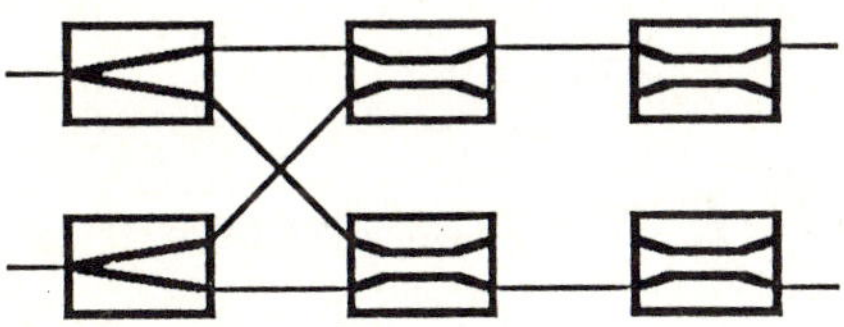

Figure 3: A multicast 2×2 switch

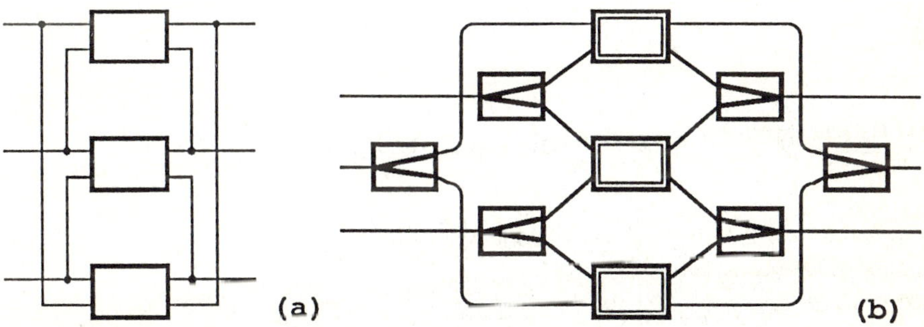

Figure 4: A 3×3 switch matrix; (a) electronic, (b) optical

electronic matrix we use passive splitters and combiners. In Fig. 5 an $N \times N$ switch is presented [2]. An alternative, *nested*, structure shown in Fig.6 can be used. In this case basic elements of Fig. 2 are replaced by complete networks of Fig. 5 or other nested structures.

We can obtain networks of any capacity by straightforward replacements of electronic switches and hard wired connections in the single-stage electronic networks by electronic-like optical switches and passive splitters or combiners. The obtained switch matrices are nonblocking in the strict sense. The total number of switch elements is $3 \cdot C$, where C is the number of switch elements

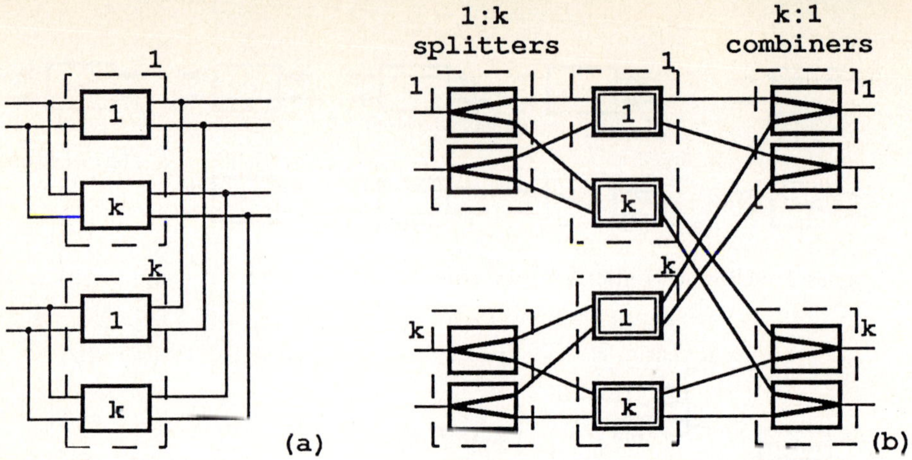

Figure 5: An $N \times N$ switch matrix; (a) electronic, (b) optical

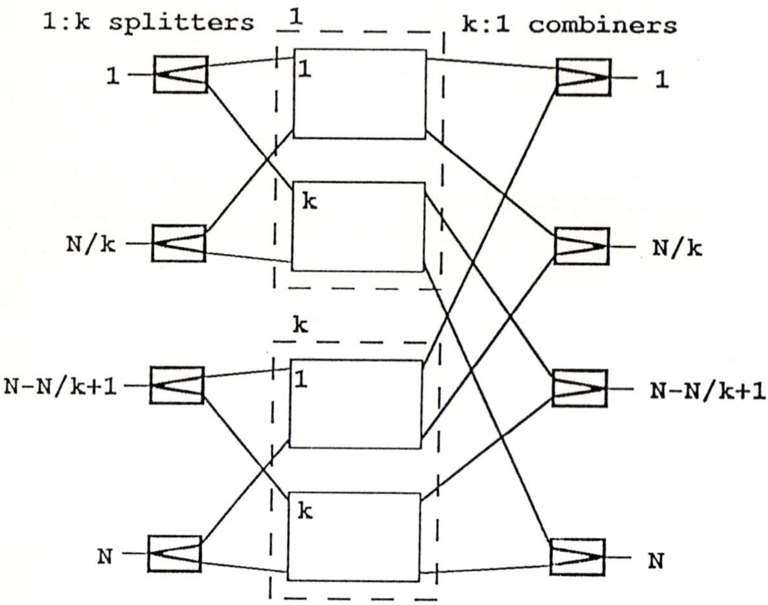

Figure 6: An $N \times N$ nested switch matrix

in the corresponding electronic switching network. For example, for simplest single-stage matrices of Fig. 5 this number is $3(N/2)^2$, that is fewer elements are required than in the case of a crossbar switch. The number of stages of active switch elements is equal to 2 independently of the switch matrix capacity. However, it should be noted that additional attenuation is caused by passive splitters and combiners. The proposed switch matrices have regular structures and can be also used as submodules in multistage optical networks as in the case of electronic solutions [7].

References

1. A. Selvarajan, J. E. Midwinter: Opt. Quantum Electron. **21**, 1 (1989).
2. A. Jajszczyk: In *Proc. 11th Int. Teletraffic Congress*, ed. by M. Akiyama (North-Holland, Amsterdam 1985), p. 1061.
3. A. Jajszczyk: Electron. Lett. **21**, 909 (1985).
4. A. Jajszczyk: Electron. Lett. **25**, 460 (1989).
5. A. Jajszczyk: Electron. Lett. **25**, 209 (1989).
6. A. Himeno, H. Terui, M. Kobayashi: J. Lightwave Technol. **LT-6**, 30 (1988).
7. A. Jajszczyk: Electron. Lett. **20**, 683 (1984).

Two-Dimensional Photonic Switching Network Using Waveguide Optics

S. Suzuki[1] *and J.W. Goodman*[2]

[1]C & C Systems Research Laboratories, NEC Corporation, 4-1-1 Miyazaki, Miyamae-ku, Kawasaki 213, Japan
[2]Department of Electrical Engineering, Stanford University, McCullough 150, Stanford, CA 94305, USA

Abstract. The proposed practical two-dimensional photonic switching network using waveguide optics can achieve switching capability for several hundreds of signals in small size.

1. Introduction

Parallel operation is very effective for increasing the capability of computer and communications systems. In pursuing this parallelism, much attention has been placed on free-space optics and several two-dimensional (2D) free-space photonic switching networks, which are indispensable for optical parallel systems, are proposed[1].

However, 2D free-space photonic switching networks are in the early stage of development and it is difficult to fabricate practical switching networks. On the other hand, waveguide optics have been relatively well developed and practical photonic switching networks using waveguide optics, such as $LiNbO_3$ matrix switches[2], have recently been reported. This paper proposes a 2D photonic switching network, which consists of waveguide optics and is suitable for near future implementation.

2. Switching network structure

Figure 1 shows a new 2D photonic switching network. It consists of M sheets of printed optical waveguide boards and an optical back board. Each printed optical waveguide board has N pairs of input/output ports and can make optical connection between any input and output ports. Optical connection between different printed optical waveguide boards can be established via the optical back board. As a result, this 2D photonic switching network can make optical connection between N·M pairs of input/output ports.

The structure of each printed optical waveguide board with three pairs of input/output ports is shown in Fig. 2. Among fifteen VSTEP's (Vertical-to-Surface Transmission Electron-Photonic device)[3], optical input signals are injected into the VSTEP-1, 6 and 11, which are used as an optical non-linear amplifier. Optical signals, amplified by the VSTEP-1, 6 and 11, are

Springer Series in Electronics and Photonics, Vol. 29 Photonic Switching II
Editors: K. Tada · H.S. Hinton © Springer-Verlag Berlin, Heidelberg 1990

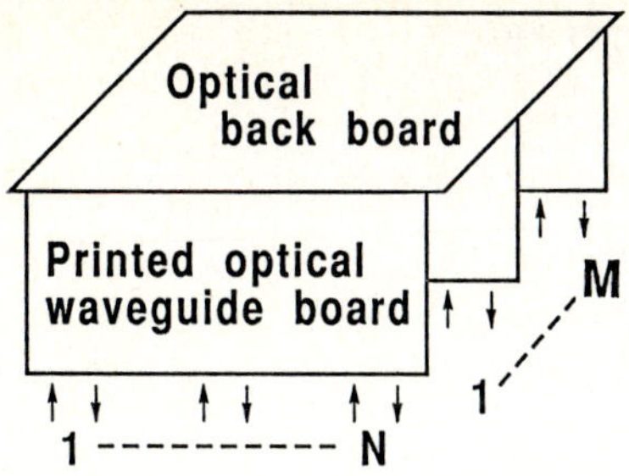

Fig. 1: 2D photonic switching network

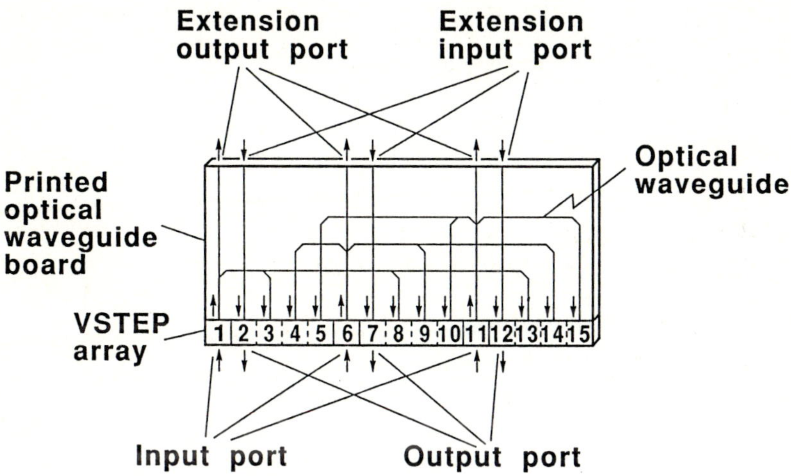

Fig. 2: Printed optical waveguide board

distributed to other VSTEP's and extension output ports through optical waveguides. The VSTEP-2, 3, 4 and 5 are electrically connected and can be used as a 4×1 optical switch controlled by an electric bias signal, like an opto-electronic switch[4]. Thus, one optical signal, sent from the VSTEP-1, 6, 11 and an extension input port, can be selected and emitted from the VSTEP-2. Two more 4×1 optical switches are also fabricated using the VSTEP-7 -10 and VSTEP-12 -15. As a result, this printed optical waveguide board can be used as a 3×3 optical switch with extension input/output ports and broadcast capability.

The optical back board can also be realized using VSTEP's and optical waveguides, as shown in Fig. 3. It shows an example for connecting three sheets of printed optical waveguide boards. Optical signals from the extension output ports of the 1st printed optical waveguide board are injected into VSTEP-b, d and f in a VSTEP array-1, respectively. The VSTEP-b, d and f amplify the optical signals and send them to a VSTEP array-2 through optical waveguides. The VSTEP array-2 also amplifies the optical signals and sends them to a VSTEP array-3. In the same way,

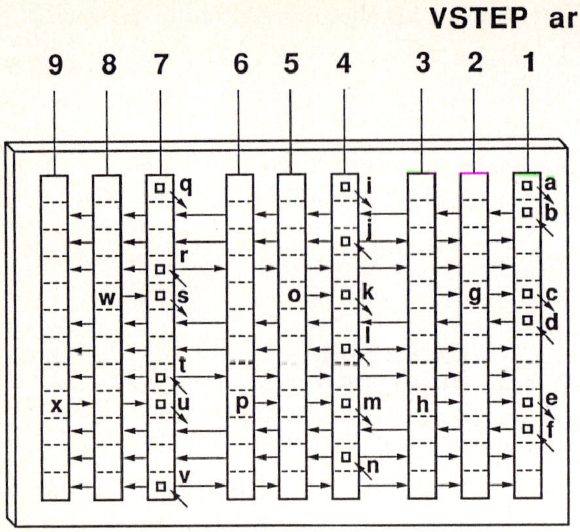

Fig. 3: Optical back board

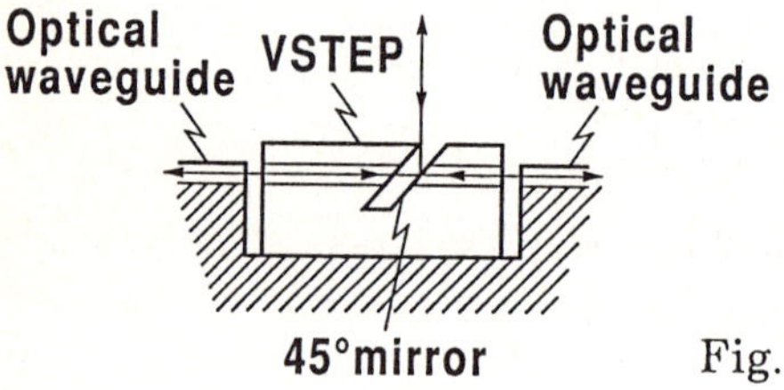

Fig. 4: VSTEP in optical back board

the optical signals are sent to VSTEP-4, ···, -9. Optical signals from the 2nd and 3rd printed optical waveguide boards are injected into VSTEP-j, l, n in the VSTEP array-4 and VSTEP-r, t, v in the VSTEP array-7, respectively. These signals are also amplified and sent to other VSTEP arrays. As a result, every VSTEP array is injected with optical signals from all printed optical waveguide boards.

In the VSTEP array-1, one optical signal can be selected and emitted from a VSTEP-a to one extension input port of the 1st printed optical waveguide board by changing electric bias signals. The VSTEP array-2 also selects one optical signal and emits it from a VSTEP-g. The output optical signal from the VSTEP-g can be sent to the printed optical waveguide board through a VSTEP-c. The selected signal in the VSTEP array-3 is also sent to the printed optical waveguide board via VSTEP-e. Independently selected optical signals are similarly sent to the 2nd and 3rd printed optical waveguide boards from VSTEP-i, k, m and q, s, u, respectively.

Figure 4 shows a schematic view of the VSTEP in the optical back board. Vertical to surface light transmission capability is provided by a 45° mirror monolithically integrated into the device[3].

3. Possible port number estimation

The maximum input/output port number of the proposed switching network will be limited by device size of the VSTEP, crosstalk between optical waveguides and optical transmission loss in the switching network.

The size of the VSTEP will be able to be reduced to as small as $30\mu m \times 30\mu m$. As for the printed optical waveguide board in Fig. 2, a $45mm \times 45mm$ board can be cut down from a 3-inch $LiNbO_3$ wafer. The total length of $N \cdot (N+1)$ pieces of VSTEP's must be shorter than 45mm. Thus, N must be less than 38.

Optical crosstalk between optical waveguides with 30-μm separation and 135-mm length is negligible, less than -60 dB, assuming 7-μm width $LiNbO_3$ single mode waveguides fabricated by titanium diffusion.

Allowable optical transmission loss in the switching network depends on optical signal speed. Switching energy of the VSTEP is around 10 fJ. Then, the required input power is 1 μW and 10 μW for 100-Mbps and 1-Gbps signals, respectively. The allowable optical transmission loss between VSTEP's in the printed optical waveguide, shown in Fig. 2, is therefore 23 dB and 13 dB, respectively, assuming 2-mW VSTEP output power and a 5-dB VSTEP-to-waveguide coupling loss. The optical transmission loss is the sum of an optical power splitting loss, waveguide propagation loss and excess loss at each branch of the waveguide. However, the waveguide propagation loss is negligible for the $LiNbO_3$ waveguide and the excess loss can be reduced to 0.3 dB. As a result, N can reach 32 and 10 for 100-Mbps and 1-Gbps signals, respectively.

As for M, it will be mainly limited by the thickness of the printed optical waveguide board, because the optical transmission loss in the optical back board can be compensated by every VSTEP array. Assuming a 45 mm $\times$45 mm optical back board and 2mm pitch spacing between printed optical waveguide boards, 22 is possible for M. Consequently, the total port number $N \cdot M$ can reach 704 and 220 for 100-Mbps and 1-Gbps signals, respectively.

4. Conclusion

A 2D practical photonic switching network using waveguide optics, which is suitable for near future implementation, was proposed. It can achieve switching capability for several hundreds of signals in small size and is suitable for optical parallel computer and communications systems.

The authors thank Mr. M. Kondo for calculation of optical waveguide crosstalk characteristics.

References

[1] H. S. Hinton IEEE J-SAC Vol. 6, No. 7, P. 1209 (1988)
[2] S. Suzuki et al. Photonic Switching '89 FE1
[3] Y. Tashiro et al. Appl. Phys. Lett. 54(4), P. 329 (1989)
[4] R.I. Macdonal et al. Electron. Lett. No. 16, P. 502 (1978)

Part X

Systems Considerations

Multidimensional Photonic Switches

N.J. Parsons[1], *N.F. Whitehead*[2], *and M. Erman*[3]

[1]GEC-Marconi Research Centre, Chelmsford, CM2 8HN, UK
[2]British Telecom Research Laboratory, Martlesham Heath, Ipswich 1PS 7RE, UK
[3]Laboratories d'Electronique Philips, 22 Avenue Descartes, F-94453 Limeil Brévannes, France

The use of time- and wavelength- multiplexed switching in optical networks is discussed in the context of the European RACE programme. Some examples of recent developments in optical switching technology are highlighted.

1. Introduction

Deployment of optical fibre in telecommunications networks has drastically reduced the cost of data transmission, which in turn has stimulated demand for new broadband services. In Europe, the RACE programme is developing the technology and infrastructure for Integrated Broadband Communications Networks in which some scenarios predict that each customer will have access, at a price, to services with an aggregate bit rate of up to 600 Mbit/s. A major challenge is to realise switching technologies capable of handling the consequent increase in data throughput, of several orders of magnitude, at the local switching centre. This paper examines the potential impact of optical technologies in this area and outlines some of the RACE activities on high-capacity optical switches.

In an all-optical network (Fig. 1) the switching functions required fall broadly into two classes, those associated with the transmission medium for routing and network management, and those embedded within the switching machine for processing signals. Different requirements are placed on the switching technology by each class: routing switches are primarily slow devices of moderate complexity which need to be network-compatible, environmentally-stable and non-blocking. Some European work on optical switches for routing applications is described elsewhere in these proceedings[1,2]. By contrast, processing switches support a large number of channels, normally on multiplexed physical paths, and are reconfigured rapidly. Control functions are closely integrated with the switch, but less demand is placed on environmental stability.

When considering the use of photonics to perform these operations, some practical difficulties can arise since efficient means of data synchronisation, storage and control are also required. Optics can provide a high-capacity flexible interconnect medium, but it has difficulty in

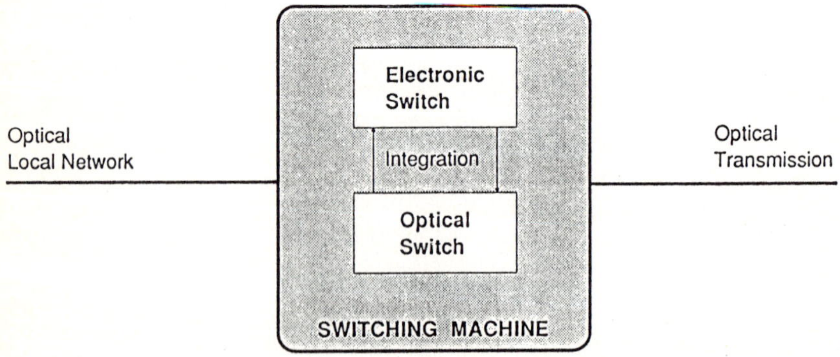

Figure 1: Switching in an All-Optical Network

Springer Series in Electronics and Photonics, Vol. 29 **Photonic Switching II**
Editors: K. Tada · H.S. Hinton © Springer-Verlag Berlin, Heidelberg 1990

performing the necessary logic and memory functions. The embedded signalling information must therefore be either read at the switching point using some form of logical processing, transmitted in parallel on a separate channel, or be translated into an optically-readable label such as wavelength or spread-spectrum code.

It should be emphasised that advances in broadband electronic switching continue to provide benchmarks against which the performance of photonic technologies will be judged. One recent example is a 16x16 GaAs MESFET chip handling 1.7 $Gbits^{-1}$ per channel in a 3x4 mm chip consuming 800mW of power[3] In addition to primary performance characteristics such as throughput, granularity and delay, new switch technologies will also need to satisfy requirements on ease of interfacing and control, and on reliability and cost.

2. Photonic ATM Switching

In recent years the interest of many telecommunications systems designers has turned to Asynchronous Transfer Mode (ATM) transmission and switching. In ATM systems the information streams are first packetized into fixed cells, each comprising an information field and a header. A standard has recently been agreed at CCITT for ATM cells with a 48 octet information field and a 5 octet header. ATM allows multiple rate transport across the network by maintaining a virtual circuit between the ends of each service connection. Transmission rates within an ATM network can be totally independent of the source rates; and multiple services such as voice, video and data can be carried simultaneously on the same fabric.

Switching of ATM cells implies the need to access address information within the headers and make routing decisions accordingly. Other functions such as cell buffering and header translation must also be carried out. ATM switching architectures can be centralised with a single switching fabric; or distributed in a ring or bus structure. Optical technologies bring some significant advantages to ATM systems including high capacity interconnection media, new methods for encoding routing information, and options for self-routing switching fabrics.

An example of a network in which optical processing is used in an ATM environment is the Very High Speed Optical Loop (VHSOL) proposal[4]. Figure 2 shows the basic principle, where each node has access to an optical ring via an electro-optic switch. Data is transmitted in cells preceded by headers containing address and synchronisation bits. The switch performs three functions: a low-power optical tap to **listen** for correct address, a high-power tap to **receive** data and to modulate empty cells to **transmit** data. Header processing can be performed by optical correlation in the listen mode. A group access node (REG) generates empty packets, performs policing functions and interfaces with a backbone network.

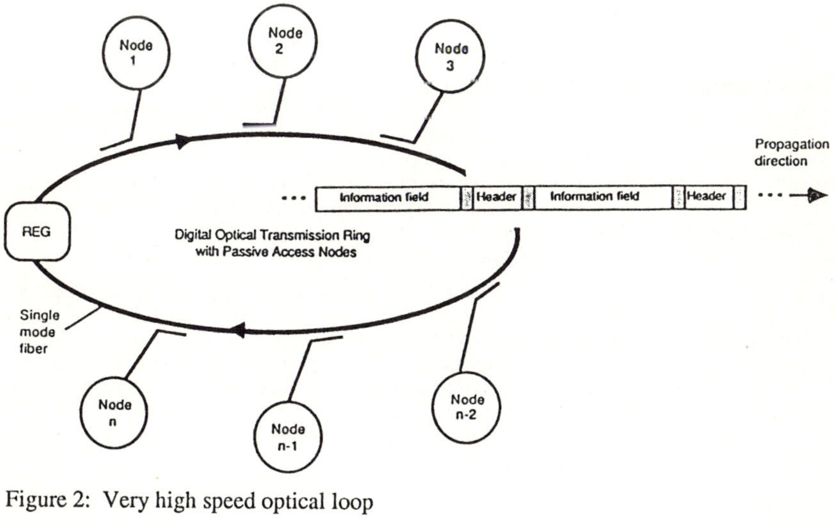

Figure 2: Very high speed optical loop

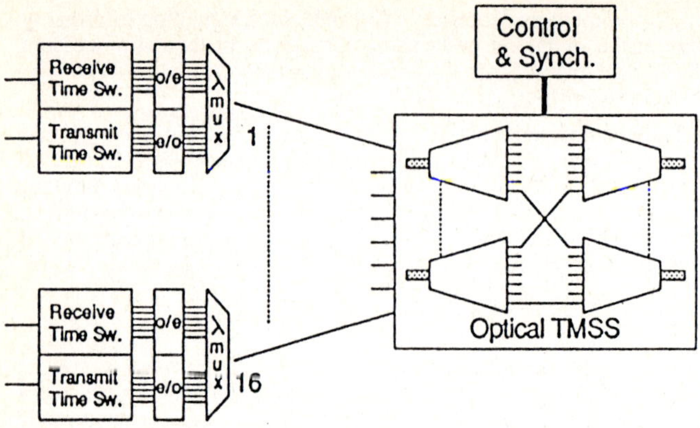

Figure 3: Optical time-multiplexed space switch

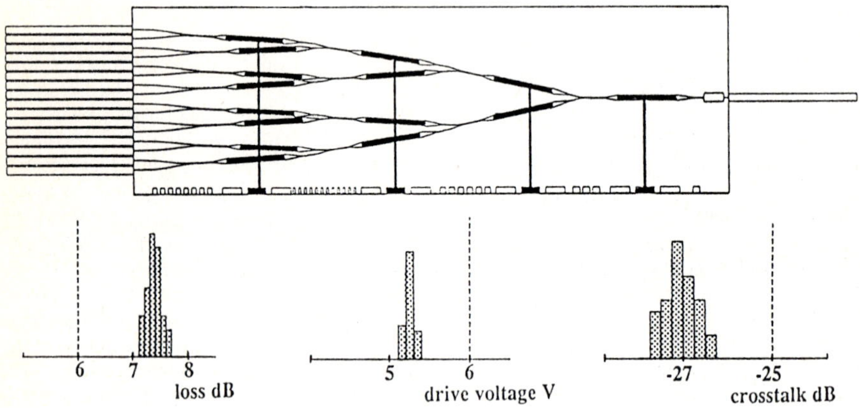

Figure 4: Structure & performance of a 1x16 optical switch in lithium niobate

Fast optical space switches can also be used in a dual-layer' ATM switch[5], or as the centre stage in a synchronous time-space-time network[6]. In each case cells are buffered and multiplexed into synchronous timeslots and the optical switch is reconfigured between each timeslot (Figure 3). Though the reconfiguration rate is limited by the speed of drive electronics and timing jitter, the data throughput can be several orders of magnitude higher. For example, the wavelength transparency of the optical switch (typically 50 nm) can be exploited to send data in cell-parallel format. Wavelength-multiplexing can also be used to carry synchronisation and signalling information between the centre stage and the peripheral nodes. The latter may be remotely located if careful attention is paid to timing and control.

Figure 4 shows the structure and performance of a lithium niobate optical switch designed to form part of a fast 16x16 passive splitter-active combiner network[7]. The 16x1 optical combiner is made up of interferometric units cells which combine a low drive voltage of 5.2V with separable dc bias. This enables each rank of the binary tree to be driven in parallel from standard CMOS buffers, considerably simplifying the control electronics. The switching speed varied from 2ns to 4.5ns as the number of switches driven in each rank varied from 1 to 8. Good reproducibility of insertion loss and extinction ratio was obtained; the dc bias voltages were supplied from a laser-trimmed resistor array.

The packaged dual 1x16 switch module with integral drive electronics is shown in Figure 5.

Figure 5: Photograph of packaged dual 1x16 optical switch with integral drive electronics

3. Wavelength Switching

The emergence of frequency-stable optical sources has generated many proposals for wavelength-multiplexed switching[8], mainly based upon passive star (shared medium) topologies. The limited number of wavelength channels available from current technologies requires efficient strategies for channel allocation and contention resolution if maximum use is to be made of the network capacity.

A useful building block in such schemes is a wavelength converter, which may, for example, act as a gateway between adjacent passive networks or as a wavelength switching stage in a multidimensional wavelength-space-wavelength network. Figure 6 shows one method of implementing this function[9], in which signals are demultiplexed onto input channels of an array or orthogonal waveguides and selectively detected, remodulated and multiplexed onto output wavelength channels. Some electronic processing may be performed at each intersection, from simple thresholding functions to logical exchange-bypass operations. The degree of complexity possible depends upon the performance and maturity of integrated optoelectronic technologies; recent work has made progress in the integration of GaAs electronics and InP waveguide modulators[10].

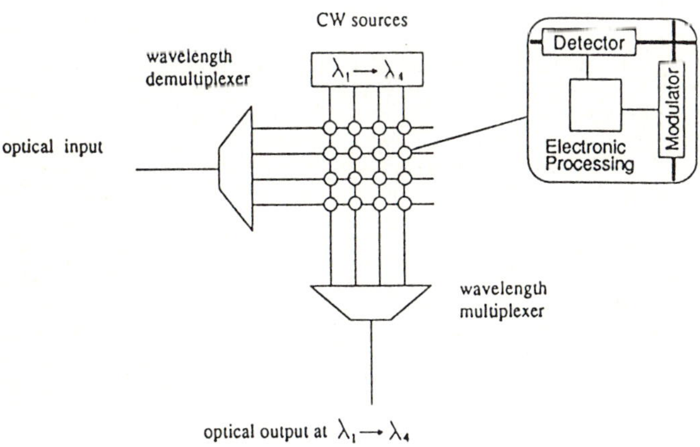

Figure 6: Optoelectronic wavelength switch

Figure 7: SEM picture of GaAs MESFET heteroepitaxially grown on an InP buried-waveguide modulator. The InP regrowth layer has been etched away in the lower part of the picture to reveal the strip-loaded waveguide.

Figure 7 shows a GaAs MESFET structure heteroepitaxially grown on an InP waveguide modulator. No change in waveguide propagation loss was observed from the initial value (before heteroepitaxy) of 0.5dB/cm. MESFET characteristics were close to those grown on GaAs and a phase modulation efficiency for the combined structure of 26°/V.mm at 1.3μm wavelength was obtained. Similar results have been achieved when MESFET integration was performed using the epitaxial lift-off technique[11].

4. Summary

Very high throughputs can be obtained using photonic switching techniques, though the electro-optic interface is a major bottleneck. The success of photonics for switching in the core network will depend upon development of technologies for the close integration of electronic and optical functions to form high performance intelligent building blocks. Full exploitation of the transparency of optical media will require appropriate combinations of space-,time-, and wavelength-division techniques to form multidimensional switched optical networks.

Acknowledgements

Thanks are due to our colleagues in the RACE programme, particularly Adrian O'Donnell and Piet Demeester. This work is partially supported by the European Commission under RACE Project 1033, OSCAR.

References

1. L. Thylen & P. Granestrand "$LiNbO_3$ switch arrays: status review and prospects" this conference, paper 14 A-3.

2. P. Duthie et al."Size transparency and control in optical space switch fabrics; a 16x16 single chip array in lithium niobate and its applications", ibid., paper 13 A-3.

3. G.J. Scott & C.S. Anderson, "A GaAs 16x16 MESFET crosspoint switch at 1700 Mbit/s", Proc. ECOC'88 p.210.

4. T. Martinson, "Passive access to a very high speed optical loop - the VHSOL concept", to be presented at EFOC/LAN 90 Munich, June 1990, paper 3.3.3.

5. A. de Bosio et al. "Deterministic and statistic circuit assignment architectures for optical switching system", Topical Meeting on Photonic Switching, Lake Tahoe, March 1987 p.35.

6. R.A. Thompson et al."An experimental modular switching system with a time-multiplexed photonic centre stage", Topical Meeting on Photonic Switching, Salt Lake City, March 1989, paper WA2.

7. I.M. Burnett and A.C. O'Donnell "Synchronous optical switching in a time multiplexed network", to be presented at EFOC/LAN 90, Munich, June 90, paper 3.3.1.

8. M.S. Goodman "Multiwavelength networks and new approaches to packet switching", IEEE Comms., 27 (10), 27, October 1989.

9. N.J. Parsons & N.F. Whitehead, "Optical switching in telecommunications networks", Proc. ECIO Paris, April 1989, paper 1141-25.

10. P. Demeester et al. "Integration of GaAs MESFETs and InP waveguides for optical switching network", ECOC'89 paper ThA19-4.

11. A Yi-Yan et al."Semiconductor-grafted integrated optics", Proc. IPR '90, paper MI-1.

Photonic Switching Network Utilizing Wavelength Multiplexing

M. Miyagi and M. Akiyama*

Faculty of Engineering, University of Tokyo, 7-3-1 Hongo,
Bunkyo-ku, Tokyo 113, Japan
*Present address: Central Research Laboratory, Hitachi Ltd.,
1-280 Kokubunji, Tokyo 185, Japan

Abstract. A photonic switching network which capitalizes on space division and wavelength division techniques is proposed. Non-blocking function of this network is proved and its advantages are shown.

1. Introduction

A wavelength multiplexing technique is useful to transmit multiple optical signals by a single physical medium like an optical fiber or optical waveguide. Although it is expected that large capacity photonic switching networks such as a 3-stage switching network can be made utilizing a wavelength division technique [1], this technique cannot be applied effectively because wavelength multiplexed signals must be demultiplexed to connect the switches of different neighbouring stages by links. We propose a photonic switching network which capitalizes on the wavelength multiplexing technique to extend the line capacity. Moreover, the proposed switching network is strictly non-blocking both for point-to-point and multicast switching, which will be useful for broadband communication.

2. Concept and features of proposed switching network

When applying the wavelength multiplexing technique to a photonic switching network, we believe that the following two factors are important, namely the wavelength multiplexed signals should be demultiplexed the smallest possible number of times and the wavelength multiplexing technique should be used for multiple link connections. To achieve these concepts, the proposed switching network is based on a combination of space division and wavelength division techniques as shown in Fig.1. It consists of a two-stage space division photonic switching network with wavelength converters at the inlets and variable wavelength filters at the outlets. In this proposed network the multiple inlets of a 1st stage switch can be logically connected to the multiple outlets of a 2nd stage switch simultaneously (see Fig.1). The signals of the multiple inlets are converted into different wavelengths, multiplexed in the 1st stage switch as shown, connected through the common link, separated in the 2nd switch, and selected by the wavelength filter to meet the desired connection.

The wavelength converters convert the input wavelength into another one [2] and the variable wavelength filter selects the desired wavelength randomly from among the

Springer Series in Electronics and Photonics, Vol. 29 **Photonic Switching II**
Editors: K. Tada · H.S. Hinton © Springer-Verlag Berlin, Heidelberg 1990

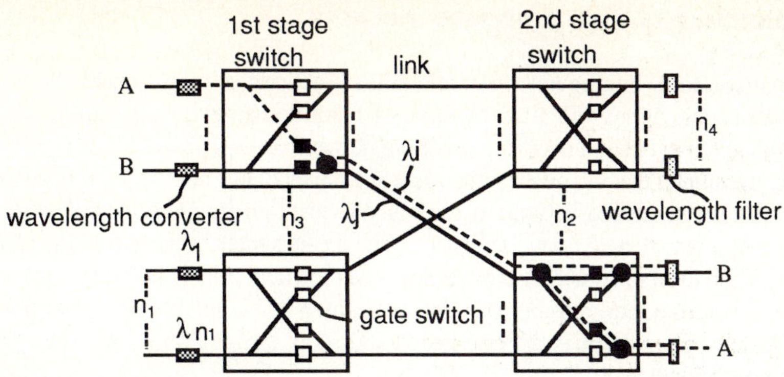

Fig.1 Proposed photonic switching network utlizing wavelength multiplexing technique. Signals A and B are wavelength-multiplexed as λi and λj on a link.

multiplexed input [3]. An optical space switch is a split and combine structure and consists of a passive splitter, a gate switch, and a passive combiner [4]. It has the same switching function as the ordinary crossbar matrix switch. (The number of required gates is equal.) The wavelength converter, gate switch, and variable wavelength filter can all be made as a kind of laser diode [2][3][4].

The amount of hardware is decreased by switching the multiplexed signal and using the link as a wavelength multiplexed path, because the laser diode gate switch has such a broad bandwidth that the wavelength multiplexed signal can be handled all at once and the devices for wavelength division, such as converters or filters, handle the multiple wavelength signals with one device. Furthermore, a non-blocking architecture capable of point-to-point and broadcast connection is possible (described in the next section) through this configuration.

3. Discussion

3.1 Non-blocking function

The space division optical switch sizes are defined as n_1 by n_2 for a 1st-stage switch, n_3 by n_4 for a 2nd-stage switch, and the number of these switches is defined as n_3 and n_2 for each stage (see Fig.1). Output signals of the wavelength converters at a 1st-stage switch are the fixed wavelengths $\lambda 1$ through λ n1. Since the maximum number of connections that should occupy a certain link at every logical connection pattern, which is equal to $n_0 = \min(n_1, n_4)$, is smaller than the maximum number of signals that a link can handle at a time, which is equal to n_1 owing to wavelength multiplexing, no blocking occurs on the link. Therefore, if all the gate switches of the 1st-stage switch are switched into the on-state, it is clear that all the inlets will be connected to the n_3 gate switches of every outlet, which means the switching network is strictly non-blocking for multicast and point-to-point connections.

3.2 Evaluation of the proposed network

The comparison to switching networks of similar functions, namely non-blocking for point-to-point connections which are crossbar and clos networks for space division [5], and the 1st stage wavelength division switching network as described in [6], is done as follows. It is assumed that the given switch size and the available number of wavelengths are n, and that the switching network is evaluated by the total number of switching elements, namely a wavelength converter, a filter, and a gate switch. It is also assumed that the space switch used to construct the crossbar and clos networks is the same switch as used here. When the total number of inlets is defined as N, and the total number of switching elements, which is an evaluation index, is defined as F, the following equations are obtained:

$F = 2N (N^{1/2} + 1)$	(proposed network)
$F = N^2$	(crossbar network)
$F = 3N (2N^{1/2} - 1)$	(clos network)
$F = 2N$	(1st stage wavelength division network)

Note that in the proposed network $n_1 = n_2 = n_3 = n_4 = n$ is assumed because it is convenient to use switches of the same size. Fig.2 shows the results of this evaluation. The proposed network is advantageous in that it has a greater number of inlets and, at the same time, a fewer number of switching elements compared to the other networks for a given n.

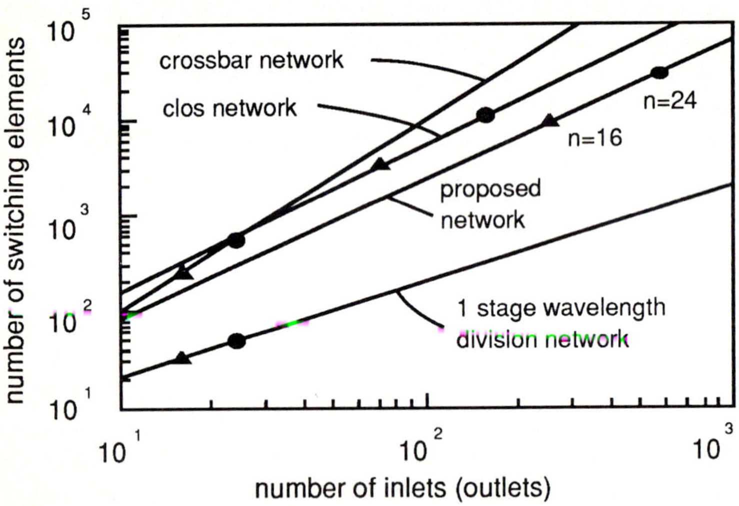

Fig.2 Evaluation results. n denotes space switch size and number of available wavelengths. Proposed network has the greatest line capacity and fewest number of elements for a given n.

4. Conclusion

We have proposed a photonic switching network utilizing the wavelength multiplexing technique. The concept that this technique can be applied effectively to a switching function is described and the non-blocking feature of the proposed network is proved. Hardware comparison shows this photonic switching network is advantageous compared to the other switching networks. The authors expect practical devices which realize this switching network to be available in the near future.

References

1. S.Suzuki, K.Nagashima, and B.Hirosaki,"Evolutional optical switching systems for broadband communications networks", International Conference on Communications, 45.3, p.1565 (1987).
2. H.Kawaguchi, K.Oe, H.Yasaka, K.Magari, M.Fukuda, and Y.Itaya,"Tunable wavelength conversion using a multi-electrode DFB LD with a saturable absorber", Extended Abstracts of the 19th Conference on Solid State Devices and Materials, B-10-2(LN), p.521 (Tokyo, 1987).
3. T.Numai, S.Murata, and I.Mito,"1.5mm tunable wavelength filter using a phase-shift-controlled distributed feedback laser diode with a wide tuning range and a high constant gain", Appl. Phys. Lett., p.1859 (1989).
4. A.Himeno, H.Terui, and M.Kobayashi,"Guided-wave optical gate switch", J.Lightwave Technol., p.30 (1988).
5. C.Clos,"A study of non-blocking switching networks", The Bell System Technical Journal, p.406 (1953).
6. H.S.Hinton,"Architectural considerations for photonic switching networks", IEEE

High-Frequency Operation of an All-Optical Synchronization Circuit

K. Takayama, K. Habara, and A. Himeno

NTT Communication Switching Laboratories,
3-9-11, Midori-cho, Musashino-shi, Tokyo 180, Japan

Abstract. 3.2GHz operation of an all-optical synchronization circuit using a semiconductor laser amplifier is demonstrated. This circuit utilizes mode competition between two independent linearly polarized lights.

1. Introduction

All-optical network systems are required for high-speed broadband communications. To realize such systems, all-optical devices such as optical logic function devices are necessary. In addition, all-optical systems require all-optical synchronization and clock regeneration circuits [1].

Recently, we proposed an all-optical synchronization circuit using a semiconductor laser amplifier (SLA) [2]. Synchronized intensity-modulated clock output was obtained by injecting intensity-modulated input light into an all-optical self-excited oscillator. This oscillator utilizes mode competition between the transverse electric (TE) wave and the transverse magnetic (TM) wave. 300MHz all-optical synchronization was demonstrated.

In this paper, 3.2GHz all-optical synchronization and self-excited oscillation at over 10GHz are reported. These were achieved by shortening the cavity length of the self-excited oscillator.

2. Configuration

The all-optical synchronization circuit is shown schematically in Fig. 1. The all-optical self-excited oscillator consists of an SLA comprising a Fabry-Perot semiconductor laser with two antireflection coated cleaved facets, an external cavity and a quarter-wave plate. A TE wave is converted to a TM wave by the quarter-wave plate, and vice versa. Thus, the oscillator operates alternately in the TE and the TM waves, as shown in Fig. 1, if enough current is injected into the SLA. The oscillator output signal is obtained by choosing the TE or TM component of the emitted light. The self-excited oscillation frequency f_0 can be tuned by adjusting the cavity length. It is expressed approximately as

$$f_0 = c / 4L,$$

where c is the velocity of light and L is the optical path length of the cavity. This self-excited oscillator requires no

Springer Series in Electronics and Photonics, Vol. 29 **Photonic Switching II**
Editors: K. Tada · H.S. Hinton © Springer-Verlag Berlin, Heidelberg 1990

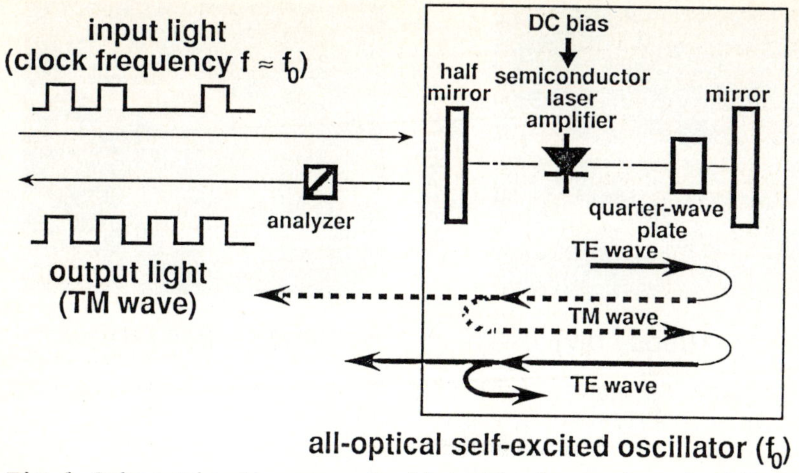

Fig.1 Schematic diagram of all-optical synchronization circuit

electrical circuits except a direct current source for the SLA. Thus, the circuit is all-optical and a high operation frequency is expected. The tunable frequency is wide because self-excited oscillation frequency can be widely tuned by adjusting the cavity length.

The input light is intensity-modulated, as shown in Fig. 1. Synchronization is realized when the clock frequency of the input light signal is near the self-excited oscillation frequency.

3. Experiment

The experimental set-up for the all-optical synchronization circuit is shown in Fig. 2. To shorten the cavity length, a Fabry-Perot semiconductor laser with one antireflection coated facet was used as an SLA. The other facet was used as a half mirror. We chose the TM component of emitted light as output light. The threshold current for laser oscillation was 33mA.

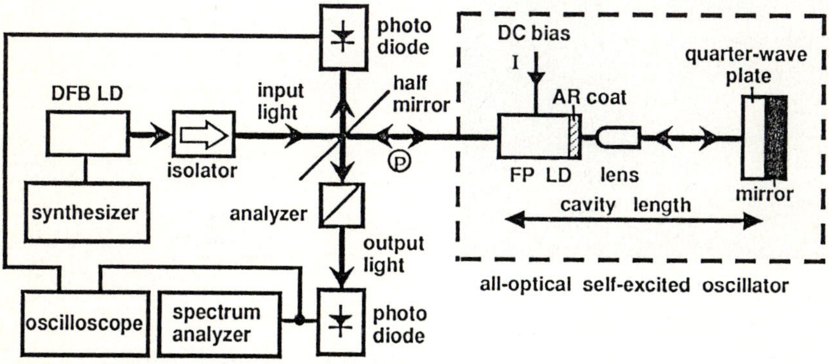

Fig.2 Experimental setup

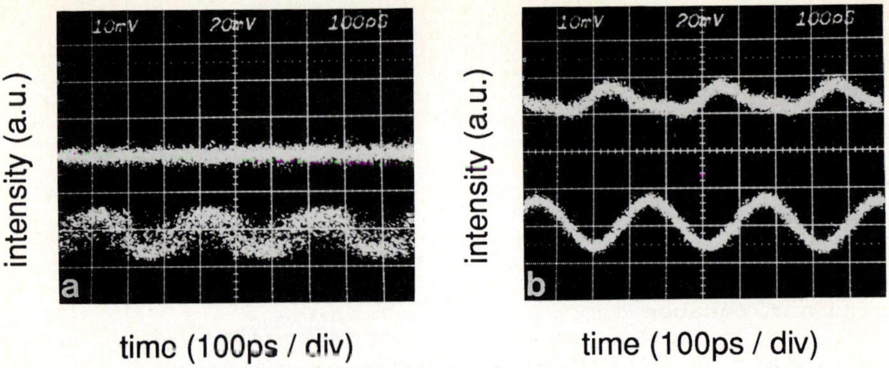

Fig.3 Waveforms of input light (upper) and output light (lower)
(a) in the self-excited oscillation state
(b) in the synchronized state

Self-excited oscillation at about 3.2GHz was observed at an injection current of $I \geq 53$mA. It was determined by about 2cm cavity length. The temperature of the SLA was stabilized at 20±0.01°C. The input light was a TM wave emitted by a DFB LD. The intensity of the input light and the output light is the intensity at point P, shown in Fig. 2.

Figure 3 shows the waveforms of the input light (upper) and output light (lower) (a) in the self-excited oscillation state (without input light) and (b) in the synchronized state. The current injected into the SLA was 54mA. The waveform in the self-excited oscillation looks noisy because the self-excited oscillation frequency fluctuates. The center frequency of the self-excited oscillation was 3.198GHz. The intensity of the output light was 519μW.

When the input light was injected, the waveforms of the output light and the input light were synchronized. The clock

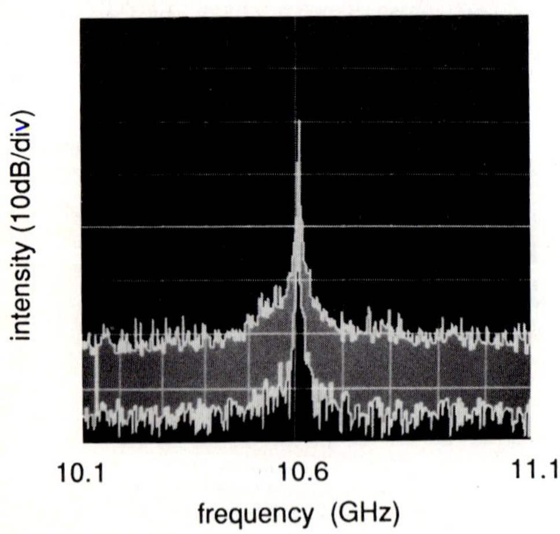

Fig. 4 Power spectrum of output light in the self-excited oscillation at over 10 GHz

frequency of the input light was 3.201GHz. The intensities of the input light and the synchronized output light were 230μW and 533μW, respectively. The lock-in range was within about 30MHz of the self-excited oscillation frequency when the intensity ratio of the input light to the output light in the self-excited oscillation state was -2.8dB. By decreasing the frequency difference between the input light and output light in the self-excited oscillation state, the input light intensity required for synchronization is decreased. The minimum required intensity ratio in the experiment was -16dB.

The self-excited oscillation frequency was mainly determined by the cavity length. High-frequency operation is expected because the carrier density change in the SLA is small. Self-excited oscillation at over 10GHz was observed experimentally by further shortening the cavity length to an optical path length of 7mm. Figure 4 shows the power spectrum in self-excited oscillation at higher than 10GHz.

4. Conclusion

An all-optical synchronization circuit was demonstrated at 3.2GHz. The lock-in range was within about 30MHz of the self-excited oscillation frequency when the power ratio of the input light to the output light in the self-excited oscillation state was -2.8dB. The minimum intensity ratio of the input light to the output light required for synchronization was -16dB.

Self-excited oscillation at over 10GHz was also demonstrated by further shortening the cavity length.

Acknowledgments The authors wish to thank Dr. Ikutaro Kobayashi for helpful discussions.

References

1. K. HABARA and K. TAKAYAMA : 'An optically triggered synchronization circuit', ECOC'88, PP167-170, 1988.
2. K. TAKAYAMA and K. HABARA : 'An all-optical synchronization circuit using a polarization self-excited oscillator', IOOC'89, 21D4-4, 1989.

Consideration of Optical Channel Number and Data/Frame Phase Tolerance for the Photonic Ring Highway

N. Fujimoto, H. Rokugawa, K. Yamaguchi, T. Nakagami, and S. Yamakoshi

Fujitsu Laboratories Ltd., 1015, Kamikodanaka,
Nakahara-ku, Kawasaki 211, Japan

This paper discusses the maximum channel number and allowable data/frame phase tolerance in a photonic ring highway and gives experimental results using a new high-speed bistable laser diode.

1. INTRODUCTION

B-ISDN requires more than several dozen gigabit-per-second throughputs in the feeder loop, which connects a central office and remote terminals having drop/insert function. Optical signal processing (OSP) can give us the network capacity and flexibility B-ISDN requires [1]. We have proposed an OSP-based broadband subscriber loop called the "photonic highway" [2] using photonic access nodes (PANs) that process optical signals directly. Each PAN has an optical drop/insert function and can synchronize optical frame signals using optical sampling memory with bistable laser diodes (BSLDs) [2].

This paper discusses the maximum number of channels on the ring and gives experimental results on the tolerable phase difference between optical data and frame for PANs using BSLDs.

2. OPTICAL CHANNEL ASSIGNMENT

We focus on the photonic ring network as shown in Fig. 1 [3]. Here, we describe briefly our proposed optical channel assignment architecture [3]. The master PAN at the CO supervises the ring operation and controls the optical drop/insert function of the slave PAN over a common wavelength λ_C. Drop/insert channel assignment is based on optical sampling memory, which retains only desired data by superimposing the optical channel selection signal (optical frame signal) on the data sequence. For example, in Fig. 1, channel 3 is dropped at node #2 and channel 5 is dropped at node #3. In the #1 PAN, the optical frame signal is superimposed on data (channel 3) to be dropped at the next #2 PAN. And the optical frame signal is extracted from the data sequence with the superimposed frame signal by the optical frame extraction circuit [3] at the #2 PAN. Based on control information by λ_C, the optical variable delay circuit (OVD) (Fig. 2) gives adequate delay to the extracted optical frame signal. The delayed optical frame pulse is superimposed on the data sequence so that desired data (channel 5) will be dropped at the next #3 PAN.

Springer Series in Electronics and Photonics, Vol. 29 **Photonic Switching II**
Editors: K. Tada · H.S. Hinton © Springer-Verlag Berlin, Heidelberg 1990

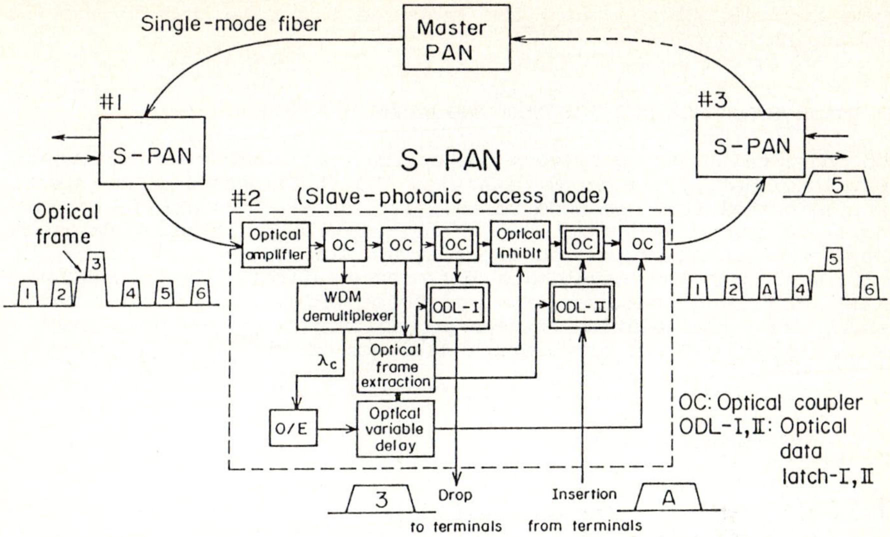

Fig. 1 Configuration of the photonic ring highway and PAN

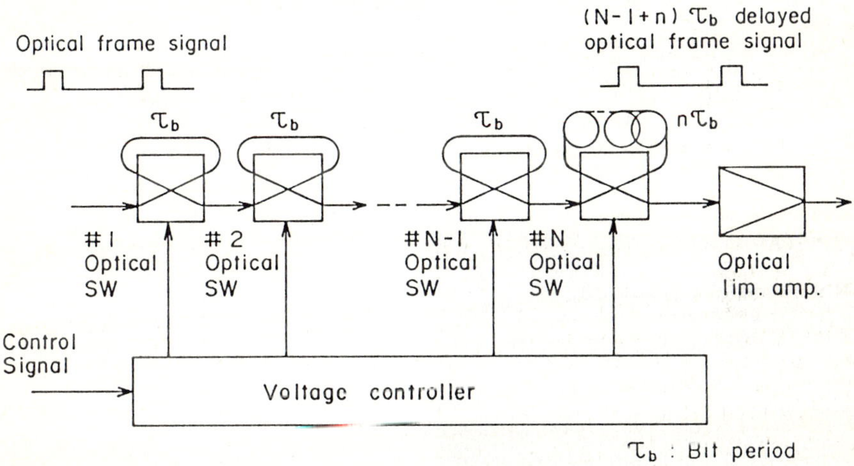

Fig. 2 Optical variable delay circuit

3. NUMBER OF CHANNELS ON THE RING

The above architecture and node configurations enable any optical signal to be dropped and inserted optically at any node. The maximum number of multiplexed channels is determined by the loss budget of OVD. The optical limiter amplifier uses a BSLD and requires -10 dBm as a minimum input level. The input level of the OVD is estimated to be at -1.8 dBm from the optimum level share. The total loss of N optical switches should be less than 8 dB. Therefore, the allowable number of switches is 8 when the loss of each is 1 dB. In Fig. 2, if n is also 8, the total number of delay combinations will be 16. Then the maximum multiplexed channel number is 16. The bit rate in the ring is 2.488 Gbit/s when the required bit rate per channel is 155.520 Mbit/s (STM-1).

4. PHASE TOLERANCE BETWEEN DATA AND FRAME

The phase relationship between the optical data and frame for ODL is very important. The optical data with the superimposed frame signal is transmitted from one node to the next. If the wavelengths of data and frame are slightly different, the relative position of data and frame will change owing to the propagation delay difference. We calculated how the wavelength difference affects the propagation delay difference between data and frame (Fig. 3). The calculated results show the relation between propagation delay and node spacing is 10.5 psec/km for a wavelength difference of 15 nm.

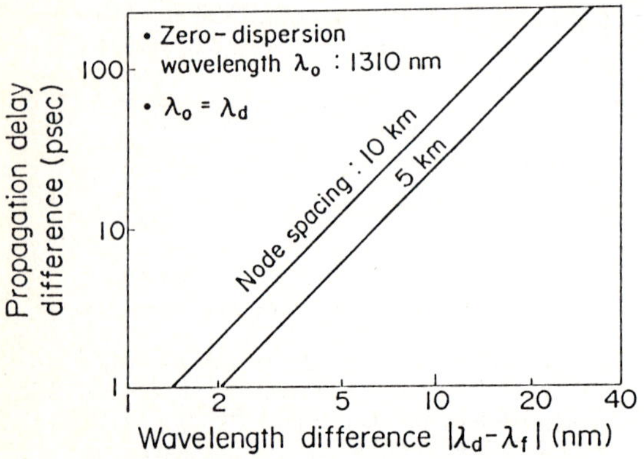

Fig. 3 Calculated difference in propagation delay

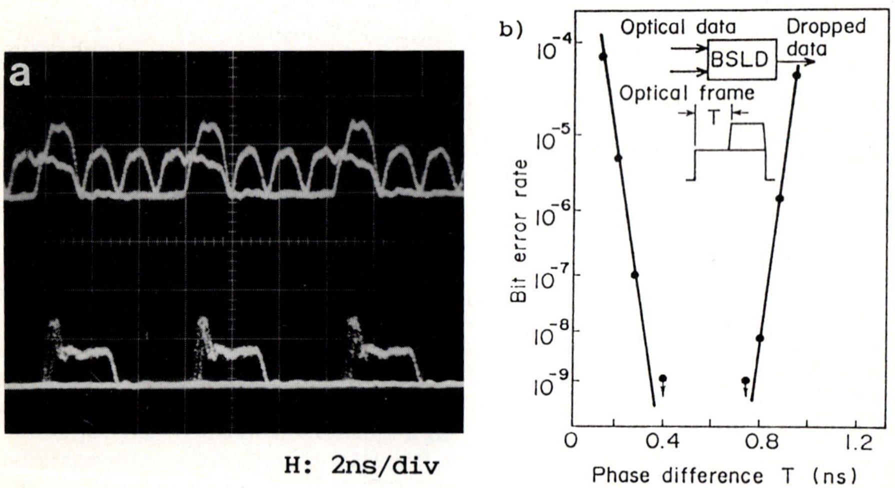

Fig. 4 Experimental results a) Upper: Bistable LD input (phase difference between data and frame: 0.5 ns), Lower: Dropped data, b) Bit error rate characteristics of the dropped data.

We measured the phase tolerance of ODL using a new reset method [4] of BSLD by shifting the phase of the frame. A wide tolerance of 410 psec in order to keep the error rate below 10^{-9} was obtained (Fig. 4). This means 0.26 bits for 622 Mbit/s. Therefore, the maximum allowable wavelength difference between data and frame is less than 15 nm for a node spacing of 10 km, according to the calculated results, to realize a propagation delay difference less than 105 ps (=0.26/2.488Gbit/s), when we assume the maximum allowable difference is the same at 2.488 Gbit/s. This wavelength difference is obtainable.

5. CONCLUSION

We have discussed the maximum number of channels and allowable phase tolerance between optical data and frame in the photonic ring highway and gave experimental results using the BSLD. These results show our proposed ring system can be constructed using BSLDs.

Acknowledgments: We wish to thank Drs. H. Takanashi, T. Sakurai, and K. Murano, Mr. Touge, and others for their valuable guidance.

References

1. M. S. Goodman, H. Kobrinsky, and K. W. Loh, "Application of wavelength division multiplexing to communication network architectures", Proc. ICC'86, pp. 931-933, 1986.
2. N. Fujimoto, H. Rokugawa, K. Yamaguchi, S. Masuda, and S. Yamakoshi, "Next generation optical transmission network based on optical signal processing technology: photonic highway", CLEO'89, ThO3, pp. 338-339, 1989.
3. N. Fujimoto, H. Rokugawa, K. Yamaguchi, S. Masuda, and S. Yamakoshi, "Photonic Highway: Broad-band Ring Subscriber Loops Using Optical Signal Processing", IEEE J. Lightwave Tech., Vol. 7, No. 11, p.1798, 1989.
4. T. Sanada et al., "New reset method for optical bistable laser diodes", Extended abstracts of the 21st. conference on solid-state devices and materials, Tokyo, pp. 321-324, 1989.

Line Capacity Consideration for a Photonic Space-Division Switching System with Switch Matrices and Optical Amplifiers

M. Fujiwara [1], *S. Suzuki* [2], *and H. Nishimoto* [3]

[1] Opto-electronics Research Laboratories, NEC Corporation, 4-1-1 Miyazaki, Miyamae-ku, Kawasaki 213, Japan

[2] C&C Systems Research Laboratories, NEC Corporation 4-1-1, Miyazaki, Miyamae-ku, Kawasaki 213, Japan

[3] Fiber Optic Communications Development Division, NEC Corporation, 1753 Shimonumabe, Nakahara-ku, Kawasaki 211, Japan

Abstract. Line capacity value for a photonic space-division switching system with optical amplifiers was theoretically examined, taking crosstalk and amplifier noise into account. The estimated line capacity value has been shown to exceed 10^3.

1. Introduction

Photonic space-division(SD) switching systems are promising and practical for providing future broadband services. A 32-line photonic SD switching system, using polarization independent $LiNbO_3$ 8x8 switch matrices, has already been demonstrated[1]. For applying such a switching system to the MAN (Metropolitan Area Network) area, the line capacity value is required to be in excess of 10^3. Introducing travelling-wave optical amplifiers (TWAs) to photonic SD switching systems appears to be very attractive for constructing a transparent switching system with large line capacity[2]-[4]. However, the possible line capacity of a photonic SD switching system with TWAs has not been clarified yet. This paper describes theoretical considerations for the line capacity of a photonic SD switching system employing TWAs.

2. Line Capacity Consideration

Figure 1 shows the structure of a photonic SD switching system with TWAs. The line capacity values for the photonic switching systems would be limited by both TWA noise and optical crosstalk characteristic of the switching network. To analyze the line capacity value for the switching network shown in Fig. 1, these two characteristics were examined.

2.1 Optical Crosstalk

The optical crosstalk characteristic in a photonic SD switching network is determined by the crosstalk value of each photonic crosspoint and also by the network architecture. Figure 2 shows the calculated results of the SNR degradation due to optical crosstalk, for strictly non-blocking switching network architectures[5][6]. Required optical SNR is assumed to be greater than 11.5 dB to achieve a bit error rate (BER) of 10^{-12}. The crosstalk value of each crosspoint was assumed to be -25 dB. The Simplified-Tree-Structure (STS), which is used in the polarization independent $LiNbO_3$ 8x8 switch matrix[1], has excellent crosstalk avoidance capability[6]. Therefore, STS-

Springer Series in Electronics and Photonics, Vol. 29 **Photonic Switching II**
Editors: K. Tada · H.S. Hinton © Springer-Verlag Berlin, Heidelberg 1990

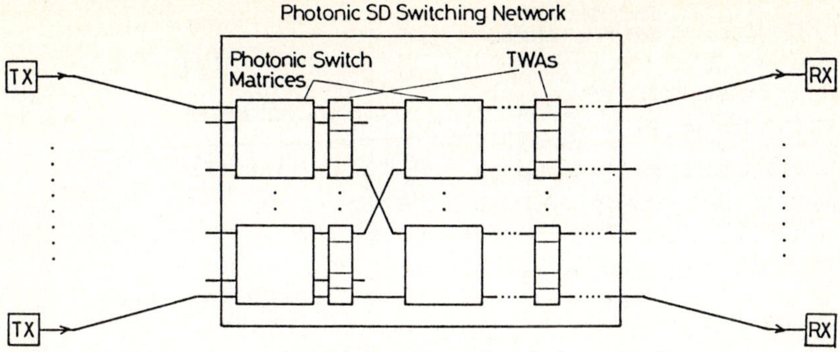

Fig. 1. Photonic switching network with switch matrices and TWAs

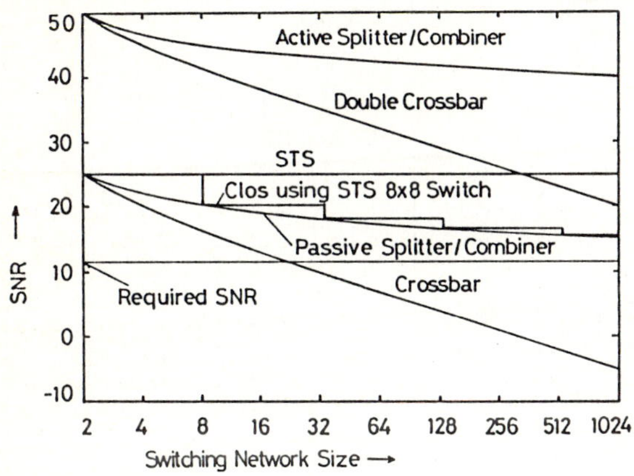

Fig. 2. Variation of signal to noise ratio due to optical crosstalk with switch network size

architecture switch matrices can be used to construct a larger line capacity switching network by using a multi-stage network, such as the Clos network. As shown, it is possible to achieve line capacity exceeding 10^3 by using the Clos network with STS-architecture 8x8 switch matrices, without being subject to significant degradation in the SNR characteristic.

2.2 TWA Noise

If introducing TWAs to the switching network, their spontaneous emission has to be taken into account. Spontaneous emission effects in the network shown in Fig. 1 were calculated using the network model shown in Fig. 3. Here, the attenuated intensity-modulated signal due to each cascaded switch matrix loss L (including coupling loss with TWA) is recovered by individual TWA gain G_i (G_i=L=20 dB). The SNR calculation was made after Ref. [7], assuming a 5.7 dB noise figure ($n_{sp}\chi$) for TWAs.

The SNR just after the k-th TWA for a 1.2 Gb/s signal is shown in Fig. 4, as a function of the number of cascaded TWAs, k. The parameters used are optical input power to TWA (P_i) and bandwidth of optical filter (Δf) for spontaneous emission noise reduction. A 30 nm filter bandwidth is analogous to a non-filtered condition. The required electrical signal SNR was assumed to be greater than 23 dB, to obtain a BER of 10^{-12}. As

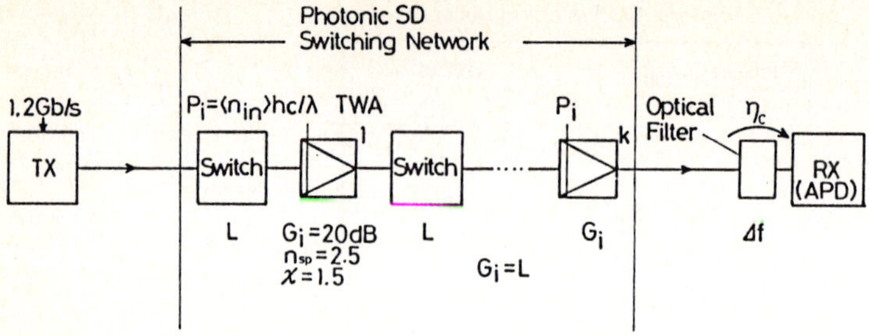

Fig. 3. Switching network model used in the calculation

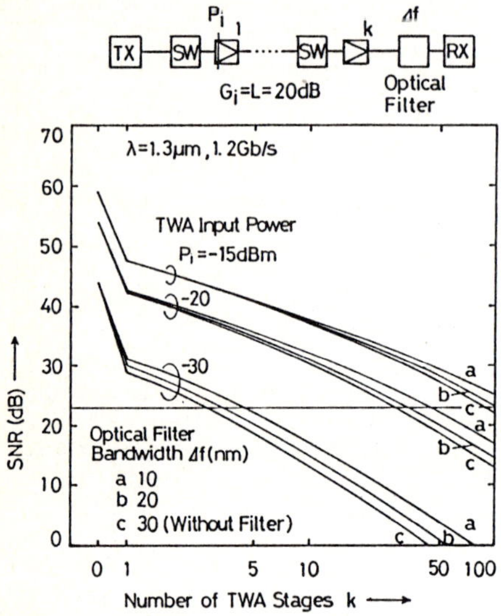

Fig. 4. Signal to noise ratio just after the k-th TWA

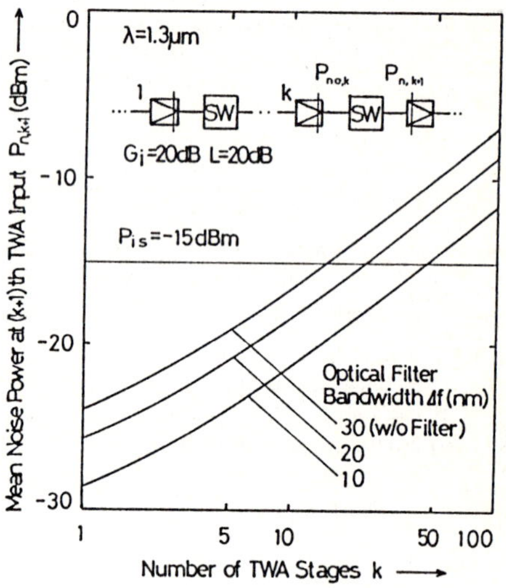

Fig. 5. Mean noise power for the (k+1)-th TWA input

shown, the k value is restricted mainly by the size of P_i. More than 20-stages of cascaded connections in switch matrices and TWAs are possible, by keeping P_i above about -20 dBm, even without an optical filter. When applying a photonic switching system to the LAN or MAN area (within several tens sq.-km area), there is no problem with regard to keeping P_i above -20 dBm, because optical fiber loss would be low.

Another issue to be considered is the gain saturation due to spontaneous emission. Figure 5 shows the calculated results for mean noise power for the (k+1)-th TWA input. Assuming saturation input power for the TWA to be -15 dBm, the numbers of cascaded connections in TWAs are restricted to less than 15, without an optical filter. However, the number can be improved to more than 20 by using a 20 nm bandwidth optical filter.

Assuming utilization of the Clos network, 20-stage connections in 8x8 switch matrices results in a line capacity of approximately 10^6. Therefore, there is no problem regarding the TWA, in achieving a photonic SD switching network with more than 10^3 line capacity.

3. Conclusion

The possible line capacity value for the photonic SD switching system with photonic switch matrices and TWAs was theoretically examined, taking crosstalk and TWA noise into account. It has been shown that the expected line capacity value for such a switching system exceeds 10^3. By develop optical interconnection technologies to construct multi-stage connections, a practical photonic switching system with large line capacity would be possible.

References

1. S. Suzuki, H. Nishimoto, M. Iwasaki, T. Kajitani, M. Kondo, M. Shikada, M. Ashibe, and F. Akashi, Topical Meeting on Photonic Switching '89, Salt Lake City, U.S.A., March 1-3, 1989, FE1
2. L. Thylén, P. Granestrand, and A. Djupsjöbacka, Topical Meeting on Photonic Switching '87, Incline Village, U.S.A., March 18-20, 1987, PDP8
3. T. Tahara, Y. Sakita, Y. Tamura, and T. Ishida, Second Optoelectronics Conference (OEC'88), Tokyo, Japan, Oct. 2-4, 1988, pp.36-37, 3A2-2
4. M. Fujiwara, T. Kajitani, M. Itoh, H. Nishimoto and S. Suzuki, IOOC'89, Kobe, Japan, July 18-21, 1989, **3**, pp. 182-183, 20D2-1
5. R. A. Spanke, IEEE Commun. Magazine **25**, pp.42-48, 1987
6. H. Okayama, A. Matoba, R. Shibuya, and T. Ishida, IEEE J. Lightwave Technol. **7**, pp.1023-1028, 1989
7. T. Mukai, Y. Yamamoto, T. Kimura, IEEE J. Quantum Electron. **QE-18**, pp.1560-1568, 1982

Optical Switched-Star Networks for Computer-System Interconnection

A. Varma

IBM Research Division, T.J. Watson Research Center, Yorktown Heights, NY 10598, USA

Abstract. This paper describes three designs for switched-star networks for computer-system interconnection based on an optical crossbar switch and an electronic controller. These switch designs demonstrate tradeoffs between hardware cost, bandwidth, and latency.

1. Introduction

The use of optical techniques has been widely recognized as a solution for overcoming some of the fundamental problems in data communication and switching. A promising application of optical technology is in the coupling of high-performance computers to form powerful multiprocessors. A central switching network, connected to the computers through high-speed fiber-optic links, is a suitable candidate for interconnection in such systems. These networks can provide high bandwidth and low latency, together with modular expansion. The number of processors in the system is relatively small, not more than 100, allowing the use of a non-blocking switched-star topology.

Current application of optical techniques in multiprocessors is primarily limited to the use of optical fibers for data communication between the processors, or between processors and a central switching system. The switching system is completely electronic and this requires expensive optoelectronic components to perform the conversion of signals. Further, the available system bandwidth is limited by the speed of the electronic components in the switch. An all-optical implementation of the switch would eliminate the problem of optoelectronic conversion. Several techniques for implementing such optical non-blocking crossbar networks are described in the literature [1, 2, 3]. Optical switches possess several desirable features such as high data bandwidth, large amount of inherent parallelism, small size and power requirements, and relative freedom from mutual interference of signals. However, the control function of the switching system is well beyond the capability of current optical technology. In addition to setting up the path for the desired connection, the controller is also required to resolve conflicts among requests for connection and administer the protocols to establish, maintain, and terminate connections through the switch. These functions require the use of an electronic controller.

In this paper our interest is on the design of high-speed switching systems based on an optical switch and an electronic controller. we study three distinct designs for switched-star networks based on an optical switch with electronic control. These architectures represent tradeoffs between hardware cost, bandwidth, and latency. All the systems use fiber-optic data links between the processors and the switching system, as well as an electronically-controlled optical crossbar similar to the switches described in [1, 2]. The control functions are implemented in electronics. One is then faced with the problem of transferring the control information necessary to set up a connection from a network node to the controller. This can be accomplished by means of a set of links

Springer Series in Electronics and Photonics, Vol. 29 **Photonic Switching II**
Editors: K. Tada · H.S. Hinton © Springer-Verlag Berlin, Heidelberg 1990

dedicated for that purpose (outband signalling), or on the same links used for data-transfer (inband signalling). The following designs demonstrate some of the tradeoffs involved.

2. Design I: A Circuit-Switch

In this system, shown in Figure 1, each processor is connected to the switch by means of a pair of high-speed data links and a pair of additional control links which are used solely for communicating the setup information to the switch. The data links are connected directly to the switch with no interface components. The control links are connected to the switch-controller via control ports. Because the control messages are usually very short, the control links can be operated at a much lower speed than the data links without any appreciable impact on performance. For example, the data links may be operating at 1 Gb/s and the control links at 10 Mb/s. Each control port contains hardware for optoelectronic conversion, serialization/deserialization, and buffering. These functions are comparatively simple and inexpensive because of the low speed of the control links and the short messages used for signalling.

Operation of the system is limited to circuit-switched protocols. A node wishing to communicate with a destination sends a connect request as a control message to the switch through its control link. The switch controller decodes this message and sets up the optical path through the crossbar, thereby creating a physical path between the high-speed links associated with the source and destination nodes. The controller simultaneously notifies the destination node through its control link about the connect request, so that the interface adapter at the destination port can set itself up for the transfer of data. On receipt of a favorable response from the destination, and after the setup operation of the optical crossbar is complete, the controller sends a response message to the source node as acknowledgement. The source node, on receiving the acknowledgement, sends the data over its data link to the destination. An overhead is paid in terms of the time required for the handshake and the setup time of the optical crossbar. However, the all-optical data path allows very high bandwidths to be realized. For large blocks of data, the setup time is offset by the higher bandwidth available in the data path. This design is therefore suited to systems where communication is in terms of large blocks of data.

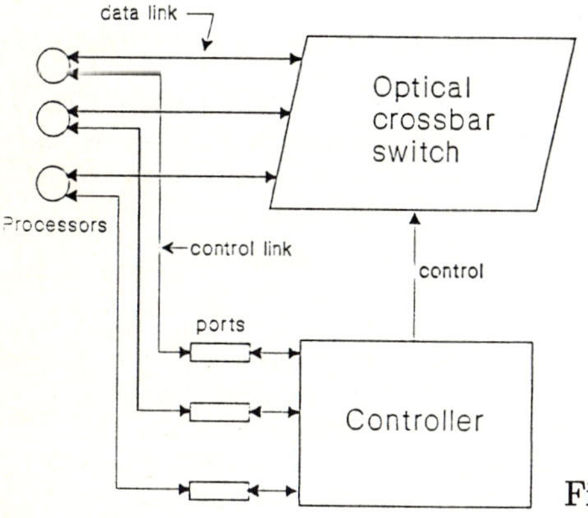

Figure 1: A circuit-switch.

3. Design II: A Message Switch

In this system, shown in Figure 2, a single pair of links is used to connect each processor to the switch. At the switch, each link is connected to a port adapter. The port adapter in this case is substantially more complex than that in the previous design. Since the setup information is now carried by the same set of links, hardware must be provided in the port to separate this information and process it. The port adapter deserializes the incoming data, extracts the setup information, reformats the data and re-serializes it before sending to the switching fabric. The re-serialized data is converted to an optical signal and passed through the optical switching fabric to the port adapter corresponding to the destination. The port adapter at the destination deserializes the data, reformats with any control messages to be inserted, and sends to the destination through the outgoing fiber. The data thus go through two conversions of optical-to-electrical signal and vice-versa, thereby increasing the cost of opto-electronics. Additionally, the bandwidth of the system is limited by the cost of buffering and other port functions that need to be implemented in electronics. However, this design eliminates the need for the extra pair of links in Design I, as well as the handshaking overhead.

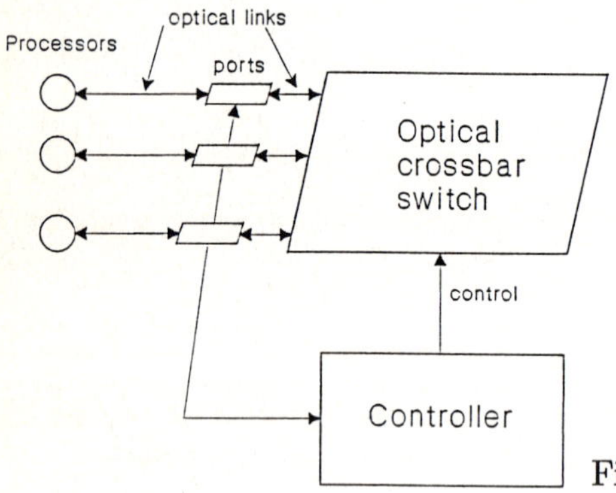

Figure 2: A message-switch.

The system is meant for use as a message-switch. Because of the elimination of the handshaking overhead, much higher utilization of the link-bandwidth can be obtained as compared to Design I.

The only advantage of this design over an all-electronic switch is in the simplicity of routing the data signals from the port-adapters to the switching fabric. Low-cost plastic fibers can be used for this purpose in place of high-speed electronic interconnects in an all-electronic implementation.

4. Design III: A Hybrid Switch

The hybrid system, shown in Figure 3, augments the control paths in Design I with an electronic crossbar switch. The control path can now be used for data-transfer as well, thereby reducing the large latency experienced by short messages in Design I. The system consists of two switching planes, an electronic plane and an optical plane. The optical plane is an optical crossbar network controlled electronically. The electronic plane is a conventional electronic crossbar network. The high-speed data links are connected to the optical plane and the low-speed control links are connected to the electronic plane through the

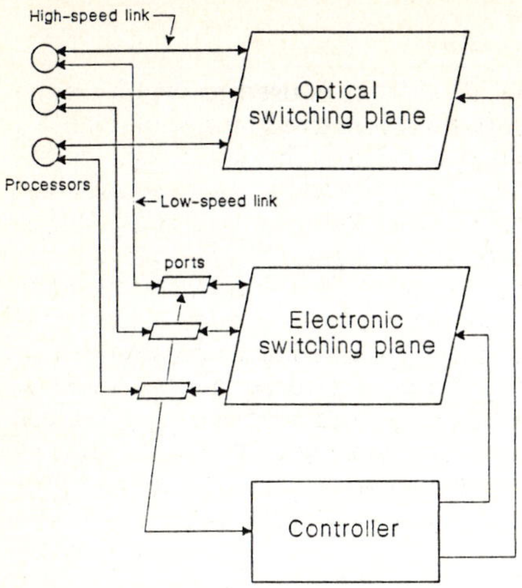

Figure 3: A hybrid design.

ports. Optoelectronic conversion is performed only for the low-speed links. A single electronic controller controls both switching planes. Setup information is sent through the low-speed links whereas data can be sent through either of the links. Each low-speed port has hardware for serialization/deserialization and buffering.

The system can be operated in one of two modes, circuit-switched or message-switched. The mode can be selected independently for each communication session and is specified by a special bit in the header of the message. The circuit-switched mode is identical to the operation of Design I, and is suitable for large blocks of data because of the large bandwidth of the data path. Message switching is done through the low-speed links. In this mode, the address of the destination node is encoded in the header of the message and is sent by the source node to the low-speed port of the switch. All incoming data is buffered at the low-speed ports. The port generates a request for the designated destination to the controller. If the desired destination is available, the controller then sets up the electronic switching plane and allows the data to be transmitted to the destination. The optical switching plane and the high-speed links do not participate in the data transfer and hence the available bandwidth is that of the low-speed link. Therefore, the message-switching mode is suited to small blocks of data.

Ideally, the choice of the two operating modes should be made to minimize the overall communication time. The total communication time depends on several parameters, such as the bandwidths of the fast and slow links, message-length, length of each link, and the setup times of the switching planes.

5. Comparison

A comparison of the three designs is given in Table 1 with respect to the number of links, communication time for long and short messages, and hardware cost of the switching system (excluding links), for a system with N nodes. The bandwidths of the high-speed links in all the three designs are assumed to be identical for the comparison. Design II has the highest cost because of the complexity of the port adapter, while Design I has the lowest cost. The hybrid system falls in between. For short messages, the speed of the link is not critical and Designs II and III provide good performance; the setup time

Table 1: Comparison of the three designs.

Design	number of duplex links	hardware cost	transfer time (short msgs)	transfer time (long msgs)
I	$2N$	low	high	low
II	N	high	medium	low
III	$2N$	medium	low	low

of the electronic crossbar in Design III is lower, resulting in slightly lower communication time. However, Design I performs poorly with short messages because of the control overhead. With long messages, all the three designs perform almost equally well because the transfer-time is primarily determined by the link-bandwidth. However, the control overhead is absent in Design II, which could result in a lower transfer time as compared to I and III.

References

1. A. A. Sawchuk, B. K. Jenkins, C. S. Raghavendra, A. Varma, "Optical Crossbar Networks," *IEEE Computer*, Vol. 20, No. 6, June 1987, pp. 50–60.

2. A. R. Dias, R. F. Kalman, J. W. Goodman, A. A. Sawchuk, "Fiber-Optic Crossbar Switch with Broadcast Capability," *SPIE Vol. 825, Spatial Light Modulators and Applications II*, August 1987.

3. A. Himeno, M. Kobayashi, "4 × 4 Optical-Gate Matrix Switch," *Journal of Lightwave Technology*, Vol. LT-3, No. 2, 1985, pp. 230–235.

Optoelectronic Switching for Radar Steering

G.L. Tangonan, W. Ng, W. Walston, and J.J. Lee

Hughes Research Laboratories, 3011 Malibu Canyon Road,
Malibu, CA 90265, USA

Abstract. We describe the use of fiber delay lines for radar steering and the implementation of laser switching to achieve true time delay steering of a dual band phased array transmitter.

1. Optoelectronic Steering of Radars

A growing tendency in active aperture radar designs is the use of transmit/receive modules which have wide operating frequencies and wide instantaneous bandwidth. In addition, the trend for many radar aperture designs is to build large and even conformal arrays. The combination of these two requirements [wide bandwidth and large apertures (>10 wavelengths)] motivates the development of wideband beam steering concepts. Photonics, which provides a natural means of sending high speed analog signals, is considered a prime candidate for new beam steering approaches.

The steering of a phased array antenna is normally accomplished by using phase shifters. The phase of the individual radiating elements is adjusted to form a strong coherent sum of the outputs of the elements in a particular direction. This commonly used technique is, however, incapable of coping with the extremely wide band operation of future radars both in the range of frequencies of operation and the instantaneous bandwidth. As the radar frequency is varied the beam also steers. The frequency excursion (instantaneous bandwidth) is therefore limited to keep the beam steering angle a small fraction of the beam width (<1/10). The alternative steering technique, which eliminates beam steering effects, is true time delay steering. In this technique a selectable delay line is placed before each radiating element (or subarray). By choosing the delay versus element position properly, a strong coherent sum of wideband signals can be steered toward any direction. This true time delay technique is frequency independent, so beam steering is eliminated. In addition, a common feed for the radar can be used with an extreme wideband capability.

Two approaches are being pursued to the end of developing wideband true time delay beam steering. The two approaches utilize optoelectronic switching of either lasers or detectors. A schematic of the two approaches configured to work as a 3 bit time delay module is given in Figure 1. The laser switched approach, shown in Figure 1(a), uses an array of lasers as the switch elements. The RF signal feeds all the lasers, but only one laser is biased on at a particular time. The particular laser biased on is the one coupled to the delay line of choice. The outputs of the different delay lines are coupled into a single high speed detector. In the second approach, shown in Figure 1(b), a single RF modulated beam is split 8 ways and coupled to a detector array through fiber delays. Here a detector array is implemented in a similar

 © Springer-Verlag Berlin, Heidelberg 1990

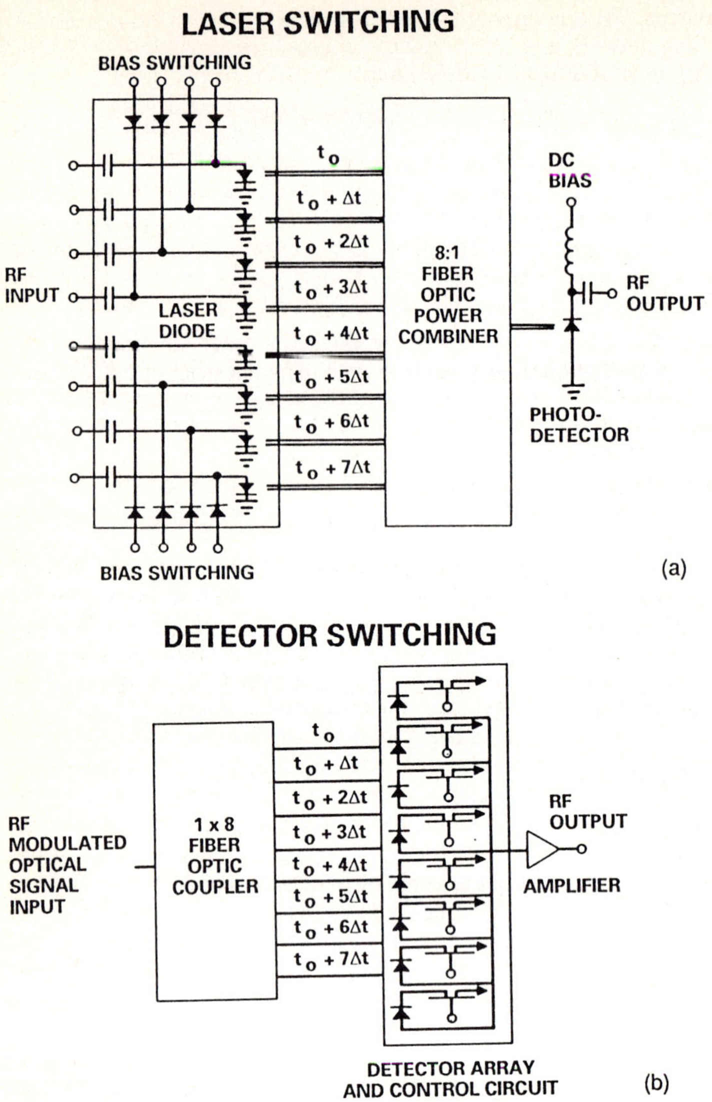

Figure 1. Two approaches to true time delay switching.

way to laser switching, namely only a single detector is biased on at a time, that detector being coupled to the delay line of choice. To handle the wideband signals normally used in radar, the componentry must operate over 11 GHz.

2. Laser Switching Demonstration

Laser switching was implemented in the world's first demonstration of optoelectronic steering of radar transmitter. Three bit time delay modules constructed according to the design in Figure 1(a) were built. GaInAsP X-based lasers were used as the high frequency laser sources. Each laser was interfaced to hybrid circuitry which contained a wideband matching network and TTL logic circuitry for bias switching. The laser response switching was accomplished by simply biasing up the particular laser desired. Unbiasing or low biasing of the unwanted lasers provided excellent RF isolation of the unwanted channels. The lasers were coupled to multimode fiber delays which had been cut to provide the necessary time delay for beam steering. In practice, the incremental delay length of fiber was 1.5 cm. We demonstrated that these delays could be cut and spliced to a fused coupler to 1 mm tolerance. With the laser appropriately biased to dc levels at which the laser resonance frequencies were matched at 7 GHz, excellent phase insertion linearity from 1 to 11 GHz was demonstrated [1]. This experiment proved that the dispersionless delay line switching had been accomplished.

Four identically designed time shifter modules were built for a dual band (L and X) beam steering demonstration. A dual band radar transmitter was built out of four subarrays with a total common aperture size of 52 cm. The arrays consisted of 8 L-band elements and 32 X-band elements. The true time delay steering units were used to provide a common feed network for the wideband steering of both the arrays. At X-band we used phase shifters within the subarray to provide improved beam quality.

The system was fully tested to prove that indeed wideband operation of the transmitter with no beam squinting was achieved using the photonic beam steerers. The transmitter radiation patterns were taken on an indoor range. The radiation patterns for the 2 and 9 GHz were measured with the delays set to a particular angle of transmission. The radiation pattern for two bands directed to -12° are shown in Figure 2. Note that the patterns for the two widely differing frequencies are directed

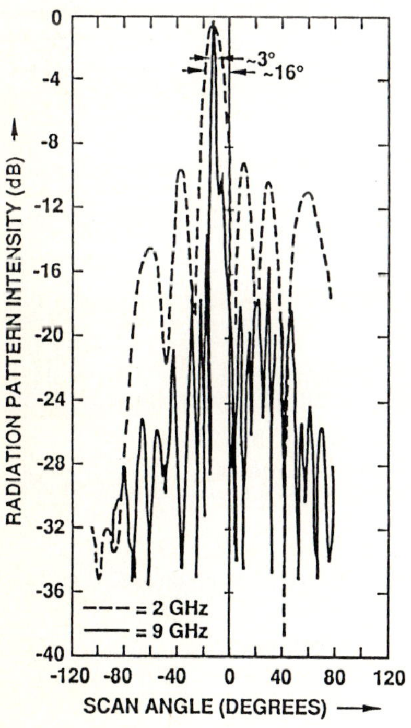

Figure 2. Radiation pattern (for - 12°) of phased array antenna at 2 and 9 GHz.

at the same angle, no steering occurs because of the true time delay steering. We verified that the beam did not steer for frequency variation of 1.8 to 2.5 GHz and 8.5 to 9.5 GHz. We demonstrated that the array could be steered to different angles provided by the 3 bit units. The beams were steered from -28 to +28° in 4° increments, by simply setting the delay lines appropriately. Again we noticed that the programmed angle and the actual measured angle coincided for L- and X- band frequencies.

3. Detector Switching

Bias switching of a detector array can be used for the true time delay steering applications as well. This approach is less developed at this time because of the lack of high speed detector arrays with frequency performance to 11 GHz. We have, however, shown that the diode switching approach can also provide high isolation. An array of MSM diodes has been used to demonstrate >55 dB isolation at 3 GHz [2]. Work is in progress to develop high speed InGaAs PIN diode arrays for this application. In practice we have found that a simple open circuit approach to the bias selection suffices for high isolation and rapid switch settings [3]. The reconfiguration time required is only a microsecond or so.

4. Conclusions

We have described the role of optoelectronic switching for radar applications. In response to the demands for high instantaneous bandwidth, agile frequency of operation and large arrays, true time delay switching provides an optimal steering scheme. Optical delay lines have been shown to have excellent performance for these applications. Using laser switching we have demonstrated for the first time ever true time delay steering of a dual band (L and X) radar transmitter.

ACKNOWLEDGMENTS
We gratefully acknowledge the sponsorship of part of this work by a US Air Force contract with RADC.

REFERENCES

1. W. Ng, A. Walston, G.L. Tangonan, I. Newberg, and J.J. Lee, "Wideband fiber-optic delay network for phased array antenna steering," Electron. Lett. *25*, 1456 (1989).

2. G.L. Tangonan et al. "8x8 optoelectronic crossbar switch," Electron. Lett. *24*, 275 (1988).

3. S. Forrest, G.L. Tangonan, and V. Jones, "A simple 8x8 optoelectronic crossbar switch," J. Lightwave Tech., *7*, 607 (1989).

Index of Contributors